STRING THEORY RESEARCH PROGRESS

STRING THEORY RESEARCH PROGRESS

FERENC N. BALOGH
EDITOR

Nova Science Publishers, Inc.
New York

Copyright © 2008 by Nova Science Publishers, Inc.

All rights reserved. No part of this book may be reproduced, stored in a retrieval system or transmitted in any form or by any means: electronic, electrostatic, magnetic, tape, mechanical photocopying, recording or otherwise without the written permission of the Publisher.

For permission to use material from this book please contact us:
Telephone 631-231-7269; Fax 631-231-8175
Web Site: http://www.novapublishers.com

NOTICE TO THE READER

The Publisher has taken reasonable care in the preparation of this book, but makes no expressed or implied warranty of any kind and assumes no responsibility for any errors or omissions. No liability is assumed for incidental or consequential damages in connection with or arising out of information contained in this book. The Publisher shall not be liable for any special, consequential, or exemplary damages resulting, in whole or in part, from the readers' use of, or reliance upon, this material. Any parts of this book based on government reports are so indicated and copyright is claimed for those parts to the extent applicable to compilations of such works.

Independent verification should be sought for any data, advice or recommendations contained in this book. In addition, no responsibility is assumed by the publisher for any injury and/or damage to persons or property arising from any methods, products, instructions, ideas or otherwise contained in this publication.

This publication is designed to provide accurate and authoritative information with regard to the subject matter covered herein. It is sold with the clear understanding that the Publisher is not engaged in rendering legal or any other professional services. If legal or any other expert assistance is required, the services of a competent person should be sought. FROM A DECLARATION OF PARTICIPANTS JOINTLY ADOPTED BY A COMMITTEE OF THE AMERICAN BAR ASSOCIATION AND A COMMITTEE OF PUBLISHERS.

LIBRARY OF CONGRESS CATALOGING-IN-PUBLICATION DATA

String theory research progress / Ferenc N. Balogh (editor).
    p. cm.
ISBN 978-1-60456-075-6 (hardcover)
1. String models--Research. I. Balogh, Ferenc N.
QC794.6.S85S666   2007
539.7'258--dc22   2007041237

Published by Nova Science Publishers, Inc. ✣ New York

CONTENTS

Preface vii

Chapter 1 String Theory and Its Application to Noncommutative Gauge Theory 1
Hidenori Takahashi

Chapter 2 In Search of the (Minimal Supersymmetric) Standard Model String 81
Gerald B. Cleaver

Chapter 3 Introduction to Non-critical String Theory: Classical and Quantum Aspects 133
Stanislav Klimenko and Igor Nikitin

Chapter 4 String Theory with a Continuous Spin Spectrum 155
J. Mourad

Chapter 5 The Coulomb Gas Realization of $SL(2, \mathbb{R})_k$ Structure Constants and String Interactions in AdS_3 179
Gaston Giribet

Chapter 6 Special Bibliography on String Theory 195
John Harrington

Index 227

PREFACE

String theory is a model of fundamental physics whose building blocks are one-dimensional extended objects called strings, rather than the zero-dimensional point particles that form the basis for the standard model of particle physics. The phrase is often used as shorthand for Superstring theory, as well as related theories such as M-theory. By replacing the point-like particles with strings, an apparently consistent quantum theory of gravity emerges. Moreover, it may be possible to "unify" the known natural forces (gravitational, electromagnetic, weak nuclear and strong nuclear) by describing them with the same set of equations.Studies of string theory have revealed that it predicts higher-dimensional objects called branes. String theory strongly suggests the existence of ten or eleven (in M-theory) spacetime dimensions, as opposed to the usual four (three spatial and one temporal) used in relativity theory; however the theory can describe universes with four effective (observable) spacetime dimensions by a variety of methods.

An important branch of the field is dealing with a conjectured duality between string theory as a theory of gravity and gauge theory. It is hoped that research in this direction will lead to new insights on quantum chromodynamics, the fundamental theory of strong nuclear force.

This new book presents the latest research from around the globe.

Chapter 1 presents a few topics of string theory, including AdS/CFT correspondence, noncommutative space and noncommutative gauge theory. The authors briefly describe the main results of their recent studies as well as the review of the string theory and its application to the gauge theories. In particular, conformal field theory and bosonic string are reviewed. Also discussed the dual supergravity picture of the gauge theories and we see how the confinement should be explained in terms of the AdS/CFT correspondence.

In addition, the authors will survey the noncommutative gauge theory and its application to the quantum mechanics and condensed matter physics. In particular, the noncommutative quantum electrodynamics is discussed. They also review recent calculation about the spectrum of an electron in the noncommutative magnetic fields. It is presented that the noncommutative Hofstadter Butterfly diagram has also a fractal structure even though it shows a new effect of the noncommutativity.

Chapter 2 summarizes several developments in string-derived (Minimal Supersymmetric) Standard Models. Part one reviews the first string model containing solely the three generations of theMinimal Supersymmetric StandardModel and a single pair of Higgs as the matter in the observable sector of the low energy effective field theory. This model was constructed by Cleaver, Faraggi, and Nanopoulos in the $\mathbb{Z}_2 \otimes \mathbb{Z}_2$ free fermionic formulation of weak coupled heterotic strings. Part two examines a representative collection of string/brane-derived MSSMs that followed. These additional models were obtained from various construction methods, including weak coupled $\mathbb{Z}_6$ heterotic orbifolds, strong coupled heterotic on elliptically fibered Calabi-Yau's, Type IIB orientifolds with magnetic charged branes, and Type IIA orientifolds with intersecting branes (duals of the Type IIB). Phenomenology of the models is compared.

As presented in Chapter 3, since the creation of the special theory of relativity, relativistic models have been successfully used to describe complex physical phenomena both in micro- and macro-world. However, the development of such models revealed that all of them, starting from a certain level of complexity, possess a number of common problems, whose investigation continues to the present. These problems are, particularly, the presence of singularities on the classical solutions and anomalies appearing in the quantization.

In Chapter 4, a classical action is proposed which upon quantisation yields massless particles belonging to the continuous spin representation of the Poincaré group. The string generalisation of the action is shown to be given by a tensionless extrinsic curvature action. It shares with the classical string action the two-dimensional local reparametrization and Weyl symmetries as well as the global target space Poincaré invariance. It also introduces a mass scale similar to the string tension. It differs however in two important aspects: it is non-polynomial and it is a higher derivative action. The authors analyse the constraints that arise from the string action and show that they are are bilocal. The continuous spin string action is then quantized in the BRST formalism. The authors show that, in the critical dimension 28, the spectrum is ghost free: the vacuum carries a continuous spin representation of the Poincaré group and the higher level states are all of zero norm. The results prove the consistency of the free theory opening the possibility to obtain continuous spin particles from a string theory. The authors also consider the generalisation to a string action with world-sheet supersymmetry and show that the resulting critical dimension is 12.

In Chapter 5, by considering the integral representation of Liouville five-point function, and considering the four-point functions for $sl^{\hat{}}(2)_k$ admissible representations as the staring point, the authors propose an analytic continuation of these correlators and rederive the generic structure contants of the $SL(2,\mathbb{R})_k$ WZNW theory. The analytic continuation the authors consider here employs standard tricks to adapt the Coulomb gas prescription to the case of non-rational conformal field theory, and it is shown to lead to the exact result. The structure constants they rederive here have direct application in string theory as these describe scattering amplitudes in AdS_3 space and in the 2D black hole background.

Chapter 6 contains a bibliography on string theory.

Chapter 1

STRING THEORY AND ITS APPLICATION TO NONCOMMUTATIVE GAUGE THEORY

Hidenori Takahashi[*]
College of Science and Technology, Nihon University, Japan
and
Department of Dentistry at Matsudo, Nihon University, Japan

Abstract

In this chapter, I present a few topics of string theory, including AdS/CFT correspondence, noncommutative space and noncommutative gauge theory. I briefly describe the main results of our recent studies as well as a review of string theory and its application to the gauge theories. In particular, conformal field theory and bosonic string are reviewed. Also the dual supergravity picture of the gauge theories is discussed and we see how the confinement should be explained in terms of the AdS/CFT correspondence.

In addition, I will survey the noncommutative gauge theory and its application to the quantum mechanics and condensed matter physics. In particular, noncommutative quantum electrodynamics is discussed. I also review recent calculations about the spectrum of an electron in noncommutative magnetic fields. We discuss how the noncommutative Hofstadter Butterfly diagram has a fractal structure even though it shows a new effect of the noncommutativity.

1. Introduction

After the discovery of quantum mechanics in early twentieth century, the quantum physics has been developed in the wide area of nature. These developments of quantum physics have cleared many experimental tests. Nowadays we receive benefit in terms of technologies based on the quantum mechanics in variety of contexts. In the middle of twentieth century, we encountered a difficulty of infinity in the quantum electrodynamics. Since there is an infinite number of freedom in the quantum electrodynamics field theory, it cannot

---

[*]E-mail address: htaka@phys.ge.cst.nihon-u.ac.jp

keep out of the way of infinity for physical observables in the perturbative scheme. Fortunately, it was overcomed in terms of the renormalization theory. Further, the predictions of the renormalization theory in the quantum electrodynamics have been checked in terms of high precision experiment tests. The agreements between theoretical predictions and experimental facts in QED, e.g. in the Lamb shift, amaze us [1].

Regardless of many successes in the quantum mechanics and quantum electrodynamics, there are many mysteries in modern physics. It is believed that there are two types of elementary fermionic particle. One is lepton and another is quark. The leptons are sensitive to weak interaction whereas the quarks are sensitive to the strong interaction as well as the weak interaction. Furthermore, both particles are sensitive to electromagnetic interaction if they have electric charge. Indeed, we can observe isolated leptons such as an electron and a neutrino in laboratory. On the other hand, many hadrons which are classified into meson and baryon are observed in experimentally. Further, there is indication of which both mesons and baryons should have inner structure. Thus, in present time, it is believed that hadrons are composed of elementary particles such as quarks. However, we have not found any isolated quarks until now. This fact should indicate that the quarks are confined into the mesons and baryons. Since the quarks obey the strong interaction, the quarks should be confined by strong interaction. Why do isolated quarks disappear whereas isolated leptons observe?

In recent realization of interactions can be made in terms of gauge theories. In particular, the electromagnetic interaction is described by abelian local gauge theory and it is represented in terms of photon exchange. On the other hand, the strong and weak interactions are described by non-abelian local gauge theory. In this case, gluon and weak boson should be exchanged between the particles. In particular, the gluon field is a SU(3) gauge field and it glue the quarks. Note, further, that gluon field interact itself. Again, any shadows of isolated gluon fields also have not been observed.

It is important to note that even in U(1) gauge field theory (QED), we cannot observe any gauge fields itself since the gauge fields are gauge dependent fields. However, in U(1) case, the electric and magnetic fields are observed due to the gauge invariance of the fields. In this respect, in the gauge theory, the physical observables must be gauge invariant quantities. This fact indicates that we must give care to gauge invariance on the way of the calculation for physical observables of the gauge theories. For example, an energy spectrum of the electron and a Berry phase shift are observables. Indeed, we can observe them experimentally. On the other hand, in QCD, the meson mass and baryon mass are observable but quark mass and gluon field are not since the quarks and gluon fields have color charge. Thus, as long as we take on a gauge principle, we cannot employ any isolated color fields independently in the formulation except we consider the phenomenon such as the symmetry breaking of local gauge symmetry.

The perturbation theory provides one of the powerful frameworks in the investigation of the quantum field theory. Indeed, it makes great successes for understanding of QED [2]. In the QED, the total Hamiltonian H_{QED} should be decomposed into the free Hamiltonian H_{QED}^{free} and the interaction term H_{QED}^{int} as

$$H_{QED} = H_{QED}^{\text{free}} + H_{QED}^{\text{int}},$$

where

$$H_{QED}^{\text{free}} = -\frac{1}{4}F^{\mu\nu}F_{\mu\nu} + \bar{\psi}(i\gamma^\mu\partial_\mu - m)\psi, \qquad H_{QED}^{\text{int}} = ej^\mu A_\mu, \qquad j^\mu = \bar{\psi}\gamma^\mu\psi$$

and $F_{\mu\nu} = \partial_\mu A_\nu - \partial_\nu A_\mu$. It is important to note that H_{QED}^{free} and H_{QED}^{int} are invariant under the U(1) gauge symmetry separately,

$$H_{QED}^{\text{free}\prime} = H_{QED}^{\text{free}}, \qquad H_{QED}^{\text{int}\prime} = H_{QED}^{\text{int}}.$$

Therefore, the perturbative scheme in the QED should be well-defined and the perturbative calculation is reliable.

On the other hand, recent criticism [2, 3] on the perturbation of QCD bare a problem of the gauge invariance in the formulation. In QCD, the total Hamiltonian should be decomposed into the free Hamiltonian H_{QCD}^{free} and the interaction term H_{QCD}^{int} in formal manner as

$$H_{QCD} = H_{QCD}^{\text{free}} + H_{QCD}^{\text{int}},$$

where

$$H_{QCD}^{\text{free}} = -\frac{1}{4}G^{a\mu\nu}G_{a\mu\nu} + \bar{\psi}(i\gamma^\mu\partial_\mu - m)\psi, \qquad H_{QCD}^{\text{int}} = gj_\mu^a A_\mu^a, \qquad j_\mu^a = \bar{\psi}\gamma_\mu T^a\psi$$

and $G_{\mu\nu} = \partial_\mu A_\nu - \partial_\nu A_\mu - gf_{abc}A_\mu^b A_\nu^c$. Here, T^a denotes generator of SU(N) group and f^{abc} denotes a structure constant of generators. It is important to note that, in the QCD case, H_{QCD}^{free} and H_{QCD}^{int} are *not* invariant under the non-abelian gauge symmetry separately,

$$H_{QCD}^{\text{free}\prime} \neq H_{QED}^{\text{free}}, \qquad H_{QCD}^{\text{int}\prime} \neq H_{QED}^{\text{int}}.$$

For example, the quark-quark interaction in the perturbation theory depends on the gauge choices [3]. Accordingly, the perturbative treatments of QCD should not be reliable.

The many efforts toward complete understanding of QCD have been made. However, we have no proper answer of confinement problem yet. This should be partially because perturbative scheme should not provide adequate formulation in QCD. Obviously, it is hard for us to perform the calculation in gauge invariant way for 4 dimensional quantum gauge field theory. But, we must do so since confinement should be nonperturbative phenomenon.

The nonperturbative phenomenon is one of the next challenging issues for physicists. One of the interesting nonperturbative phenomena is spontaneous symmetry breaking [2]. The essential point of the spontaneous symmetry breaking is that the physical vacuum drastically changes after the symmetry breaking. Since the vacuum is quite different from the perturbative vacuum, we cannot employ the perturbation theory in the symmetry breaking. In recent investigation of the chiral symmetry breaking [2, 4, 5, 6], we can find the interesting result of the nonperturbative aspect. Indeed, the exact solutions [5] shows matter in question about the perturbative scheme in the symmetry breaking.

The lattice formulation [7, 8] is believed to provide alternative nonperturbative scheme of the quantum field theory. In particular, it is expected that the lattice formulation may provide the nonperturbative scheme of QCD. Indeed, the lattice calculation provides consistent results with the experimental results such as mass ratios. Further, according to Wilson criteria, the confinement of quarks should occur when the potential of the quark-anti-quark is linear potential. Thus, it is important to establish how the quark-anti-quark behaves. It is interesting that, according to the lattice theory, the confinement should occur even in the case of QED. Remember that the four dimensional QED can only provide the Coulomb type potential [2]. Does really confinement of electrons occur in the strong coupling limit for QED?

In present time, we cannot predict any hadrons masses precisely even by the lattice theory. Furthermore, there is a criticism [9] of which the Wilsonian formulation of the lattice should not work for QCD. For example, the Wilson loop may be meaningless since the isolated quarks are assumed. As it is already commented above, the quark itself is not a gauge invariant object. As long as we embrace the gauge principle and we assume that the local gauge symmetry breaking does not occur dynamically, the picture of static quark as the external source should be chancy object in the physical point of view. Thus, we must keep in mind that such static quark picture in the Wilson loop may not be well-defined.

In addition, the action of the lattice formulation has ambiguity in its definition [2, 9]. There are two definitions of which original QCD action is deducible. One definition of Wilsonian action can predict the area law, but another definition cannot. In this respect, the Wilsonian formulation of lattice should not be well-established.

The quantization of gravity is one of the mysteries of modern physics. The string theory is one of the leading candidates for the quantum gravity theory [10, 11, 12]. Nevertheless there are no evidences of the quantum effect in the gravitational phenomena, many particle physicists have been investigating the quantum gravity toward finding the unified theory. Unfortunately, in present time, the string theory should be far from complete theory as the unified theory. Furthermore, it is far from reliable physics since the energy scale where the string theory and supergravity make sense is very high. Therefore, it is difficult to check that it is indeed the physical theory of nature. In this respect, the superstring theory should be a toy model in physics. Meanwhile, we may find a hint of new physics in terms of these investigations.

Although the string theory is a toy model, it has interesting mathematical structures. In addition, recently, much progress has been made toward the realistic model in the string theory. In particular, the recent progress of the gauge/gravity correspondence attach hopes to us to understand the nonperturbative effect in the gauge theories. For instance, we have learned the quark-anti-quark potential of the supersymmetric gauge theory in terms of the AdS/CFT correspondence. The AdS/CFT conjecture claims that some of the gauge theories should be described in terms of the string on the supergravity solutions as a background metric. For example, four dimensional supersymmetric Yang-Mills theory should be described in terms of the D3-branes gravity solution. However, due to the strong constraint, the model becomes scale free model, and we only find trivial results.

In recent years, many gravity solutions are found. Some of the gravity solutions show the nontrivial behavior. For example, the D(-1)-D3-branes gravity system should be dual to the four dimensional Yang-Mills theory in a constant self-dual field which may be interpreted as the instanton. In this case, the quark-anti-quark potential shows linear potential due to the instanton effect. Thus, the confinement should be occurred in the dual gauge theory of the D(-1)-D3-branes system. Further, we can also show that D1-D5-branes gravity system shows linear potential. Therefore, its dual gauge theory has confinement phase. In this case, the confinement may be caused in terms of the instanton effect similarly.

Recently, original AdS/CFT correspondence is overinterpreted into variety gauge/gravity correspondence. In this conjecture, we could find several mass spectra such as baryon masses and glueball mass. And they are consistent with the lattice calculations. Unfortunately, the lattice calculation does not always guarantee right answers. Further, there is no assurance that general gauge/gravity correspondence is valid. In particular, we must take nontrivial decoupling limit when we have corresponding gauge theory in the D(-1)-D3-branes and D1-D5-branes system. Thus, the explanation of the confinement in terms of the gauge/gravity correspondence conjecture is not so reliable.

There are interesting applications of the string theory. One of them is the physics in the noncommutative space-time. The noncommutative space-time has long history in the theoretical physics [13]. However, any evidences of the noncommutative space have not been observed in nature until now. In addition, there is a problem of which the physical consistency of the noncommutative models is not trivial thing. In particular, when time coordinate is not commutative, we encounter an affair of unitarity of the models. Nevertheless, recently, it is found that the string theory provides the noncommutative space in terms of a constant B field. Thus, the nonommutative models may be established by the string theory.

The quantum mechanics and the condensed matter physics in the noncommutative space are interesting subjects for exploring the new idea in mathematical physics. Although there may be a chance of the noncommutative space in nature, it has been motivated by theoretical interests. There are a lot of interesting studies in this area. For example, we can also find the Landau level spectrum in the noncommutative space except differences in factor. Perhaps, we may find the evidences of the noncommutative space from these investigations.

In this chapter, I will exhibit the nonperturbative aspects of the recent physics from variety perspective. In the next section, we review the conformal field theory and we can find the exact treatments of the field theory for the special benefit of the conformal symmetry. In section 3, we see fundamental aspects of the bosonic string and the recent progress of the AdS/CFT correspondence. Section 4 is devoted to the noncommutative space. In particular, fundamental issues of the noncommutative space and the noncommutative gauge theory will be discussed. Further, we see how the string theory provides the noncommutative space.

In section 5, the Landau level problem and its application to the noncommutative space is discussed. In this section, the star-gauge formulation of the noncommutative electron is given. Further, the recent calculation of the noncommutative Hofstadter butterfly diagram is shown. In section 6, I will summarize what we have learned in this chapter.

2. Conformal Field Theory

Conformal Field Theory (CFT) provides a powerful framework for us to study critical phenomena in the statistical physics [14, 15, 16, 17, 18]. One important aspect of the critical phenomena is that the correlation length becomes infinity. In a massive field theory, mass gap m should be related to the correlation length ξ in a statistical model with $m \sim 1/\xi$. Therefore, the critical phenomena could be described in terms of the massless theories since it becomes massless in the limit $\xi \to \infty$. Indeed, the conformal symmetry has a dilatation invariance which corresponds to the scale invariance. In this respect, the CFT models become scale free models. For example, we can find that each level in the energy spectrum of the free massless Dirac field is placed at even interval above the vacuum. Here, we assume the periodic boundary condition in the field. Thus, since there is no mass gap, the conformal field theory is equivalent to a massless field theory at the thermodynamics limit.

In the conformal field theory, the conformal symmetry plays a central role. Especially, the conformal algebra becomes an infinite dimensional in the two dimensional CFT model. Due to an infinite number of symmetries, it can be solvable in exactly way. Thus, in contrast to conventional field theory, possible operator set is completely determined by the conformal symmetries. This means that, regardless of the concrete form of Lagrangian in the CFT model, we can find CFT data and a classification of the possible models. In particular, there are primary operators and their descendant operators which are called secondary operators. It is known that they construct conformal family $[\phi_n]$ [19]. Note that, to construct the Hilbert space for the physical states of the conformal field theory, we consider the highest weight state with respect to a primary operator. Furthermore, there are descendant states which are related to secondary operators.

Ultimately, the complete set $\{A_i\}$ of local operators in the CFT is composed of the conformal families,

$$\{A_i\} = \sum_n [\phi_n]. \tag{2.1}$$

In addition, these operators obey the operator algebra which is described by the Operator Product Expansion (OPE),

$$A_i(r)A_j(0) = \sum_k C_{ij}^k(r) A_k(0), \tag{2.2}$$

where $C_{ij}^k(r)$ are the structure constants of the algebra and they are single-valued c-number functions. In CFT models, the conformal invariance turn out to give constraints on the structure constants. In particular, we can find operator product expansion of the primary operators.

A central charge and a conformal weight are the important CFT data. In quantum conformal field theory, central charge is introduced as the conformal anomaly. One important point is that the central charge can determines a universality class of the CFT models. For example, the conformal weight is completely determined when $c < 1$. In this cae, the CFT models belong to minimal models. However, when $c = 1$, it becomes complicate since there are an infinite number of primary operators.

The conformal field theory also provides the foundation of the string theory. The bosonic string sweeps out the two-dimensional world sheet. Since the world sheet is an inner space or a tentative space, there should be no scale in the world sheet. Thus, each bosonic string in D-dimensional space should be described in terms of the conformal field theory. In particular, it should be composed of D two-dimensional massless scalar fields which belong to the $c = 1$ CFT. In this subsection, we focus on the topics related to the bosonic string. Thus, we will not consider the supersymmetric CFT. But, there are many reviews or text-books about CFT, for example, [14, 15, 16, 17], so we can find more advanced topics in these reviews.

2.1. Conformal Transformations

Under the coordinate transformation $x'^\mu = x^\mu + \varepsilon^\mu$ ($|\varepsilon^\mu| \ll 1$), the metric $ds^2 = dx^\mu dx_\mu$ becomes

$$ds'^2 = dx_\mu dx^\mu + (\partial^\mu \varepsilon_\nu + \partial_\nu \varepsilon^\mu) dx_\mu dx^\nu. \tag{2.3}$$

In particular, we consider the conformal transformation, $ds'^2 = \Omega(x) ds^2$. In this case, it is satisfied that

$$\partial^\mu \varepsilon_\nu + \partial_\nu \varepsilon^\mu = g^\mu_\nu \partial_\lambda \varepsilon^\lambda. \tag{2.4}$$

From eq.(2.4), we have

$$\partial_0 \varepsilon_0 = -\partial_1 \varepsilon_1, \qquad \partial_0 \varepsilon_1 = -\partial_1 \varepsilon_0. \tag{2.5}$$

It is convenient to use the complex number here because eq(2.5) is just Cauchy-Riemann's relation. First, we take the Euclidean rotation to move into the Euclidean space (τ, σ) as

$$x^0 \to -i\tau, \qquad x^1 \to \sigma. \tag{2.6}$$

Further, we impose the Periodic Boundary Condition (PBC) to σ as $\sigma \equiv \sigma + 2\pi$. Thus, we consider a cylinder coordinate of the τ, σ space. The infinitesimal displacement ε^μ is written in terms of $\varepsilon_\tau = i\varepsilon^0$ and $\varepsilon_\sigma = \varepsilon^1$. Therefore, eq.(2.5) becomes

$$\frac{\partial \varepsilon_\tau}{\partial \tau} = \frac{\partial \varepsilon_\sigma}{\partial \sigma}, \qquad \frac{\partial \varepsilon_\sigma}{\partial \tau} = -\frac{\partial \varepsilon_\tau}{\partial \sigma}. \tag{2.7}$$

Next, we map the cylinder coordinate to the complex plane as

$$w = e^{\tau + i\sigma}, \qquad \bar{w} = e^{\tau - i\sigma}. \tag{2.8}$$

The infinitesimal displacements $\varepsilon_\tau, \varepsilon_\sigma$ are

$$\epsilon = e^{\varepsilon_\tau + i\varepsilon_\sigma} \sim \varepsilon_\tau + i\varepsilon_\sigma, \qquad \bar{\epsilon} \sim \varepsilon_\tau - i\varepsilon_\sigma. \tag{2.9}$$

Finally, eq.(2.5) becomes

$$\frac{\partial \epsilon}{\partial \bar{w}} = 0, \qquad \frac{\partial \bar{\epsilon}}{\partial w} = 0. \tag{2.10}$$

Thus, the conformal transformations are factorized into a holomorphic part and an antiholomorphic part. Here, it is important to note that the infinite past ($\tau \to -\infty$) and the infinite future ($\tau \to \infty$) correspond to $w = 0$ and $w = \infty$ on the plane, respectively. Further, the equal time ($\tau = $ const) lines become the circles of the constant radius on the complex plane.

Hereafter, we consider CFT in the complex plane $(z, \bar{z})$ instead of the Euclidean space (τ, σ). The two coordinates are related to each other with the relations,

$$z = \tau + i\sigma, \qquad \bar{z} = \tau - i\sigma, \tag{2.11}$$

and

$$\partial \equiv \frac{\partial}{\partial z} = \frac{1}{2}\left(\frac{\partial}{\partial \tau} - i\frac{\partial}{\partial \sigma}\right), \qquad \bar{\partial} \equiv \frac{\partial}{\partial \bar{z}} = \frac{1}{2}\left(\frac{\partial}{\partial \tau} + i\frac{\partial}{\partial \sigma}\right). \tag{2.12}$$

Here, z and $\bar{z}$ are independent variables. Note that the complex plane $(w, \bar{w})$ can be transformed into the complex plane $(z, \bar{z})$ with the conformal map $w = e^z$. Now, we consider the conformal transformations in the complex plane ($z, \bar{z}$). They are given by

$$z \longrightarrow f(z), \qquad \bar{z} \longrightarrow \bar{f}(\bar{z}). \tag{2.13}$$

where $f(\bar{f})$ is a holomorphic (anti-holomorphic) function. Since f and $\bar{f}$ are arbitrary functions in the complex plane, there is an infinite number of the symmetries. This means that the conformal transformation in two dimensions is represented in terms of an infinite dimensional algebra.

Infinitesimal transformations of the conformal symmetries can be written by

$$z \to z + \epsilon(z), \qquad \epsilon(z) = \sum_{n \in \mathbf{Z}} \epsilon_n z^{n+1}, \tag{2.14a}$$

$$\bar{z} \to \bar{z} + \bar{\epsilon}(\bar{z}), \qquad \bar{\epsilon}(\bar{z}) = \sum_{n \in \mathbf{Z}} \bar{\epsilon}_n \bar{z}^{n+1}. \tag{2.14b}$$

All of the transformations given by eq.(2.14) are not well-defined whole of the Riemann sphere ($\mathbf{C} \cup \infty$). In order that the conformal transformations are well-defined at $z \to 0$ and $z \to \infty$, n must be greater than -1 as well as it is less than 1. Therefore, the global conformal transformations are given by

$$\epsilon(z) = \epsilon_{-1} + 2\epsilon_0 z + \epsilon_1 z^2, \qquad \epsilon(\bar{z}) = \bar{\epsilon}_{-1} + 2\bar{\epsilon}_0 \bar{z} + \bar{\epsilon}_1 \bar{z}^2, \tag{2.15}$$

where $\epsilon_m, \bar{\epsilon}_m (m = -1, 0, 1)$ are infinitesimal and arbitrary numbers. Note that the global transformation is composed of translation, rotation, dilatation and special conformal transformation. The finite form of the global conformal transformations can be written by

$$z \to \frac{az+b}{cz+d}, \qquad \bar{z} \to \frac{\bar{a}\bar{z}+\bar{b}}{\bar{c}\bar{z}+\bar{d}}, \tag{2.16}$$

where $a, b, c, d, \bar{a}, \bar{b}, \bar{c}, \bar{d} \in \mathbf{C}$ and $ad - bc = 1$, $\bar{a}\bar{d} - \bar{b}\bar{c} = 1$.

2.2. Stress Energy Tensor and Virasoro Algebra

Under the infinitesimal transformation $x^\mu \to x^\mu + \varepsilon^\mu(x)$ ($\mu = 1, 2$) in the Euclidean space, the variation of the action is given by

$$\delta S = -\int \frac{d^2 x}{2\pi} T_{\mu\nu} \partial_\mu \varepsilon_\nu, \tag{2.17}$$

where 2π is introduced by convention. For the translation invariance, the $\varepsilon^\mu(x)$ becomes a constant. In this case, we have the conservation law

$$\partial_\mu T_{\mu\nu} = 0. \tag{2.18}$$

For the rotational invariance, $\varepsilon_\mu = \omega_{\mu\nu} x^\nu$ ($\omega_{\mu\nu} = -\omega_{\nu\mu}$). It implies that

$$T_{\mu\nu} = T_{\nu\mu}. \tag{2.19}$$

For the dilatation invariance, $\varepsilon_\mu = \epsilon x_\mu$ (ϵ is constant). It means $T_{\mu\nu}$ is traceless,

$$\text{Tr}[T_{\mu\nu}] = T_{11} + T_{22} = 0. \tag{2.20}$$

Now, we define the stress energy tensors in the complex plane as

$$T_{zz}(z,\bar{z}) = \frac{1}{4}(T_{11} - T_{22} - 2iT_{12}), \tag{2.21a}$$

$$\bar{T}_{\bar{z}\bar{z}}(z,\bar{z}) = \frac{1}{4}(T_{11} - T_{22} + 2iT_{12}), \tag{2.21b}$$

$$T_{z\bar{z}}(z,\bar{z}) = -\frac{1}{4}(T_{11} + T_{22}), \tag{2.21c}$$

where $z = \tau + i\sigma$ and $\bar{z} = \tau - i\sigma$. By convention, the stress-energy tensors are redefined by

$$T(z,\bar{z}) \equiv -2\pi T_{zz}, \quad \bar{T}(z,\bar{z}) \equiv -2\pi T_{\bar{z}\bar{z}}, \quad \Theta(z,\bar{z}) \equiv -2\pi T_{z\bar{z}}. \tag{2.22}$$

From eq.(2.18), we find

$$\partial_{\bar{z}} T = \partial_z \Theta, \quad \partial_z \bar{T} = \partial_{\bar{z}} \Theta. \tag{2.23}$$

In particular, if there is a dilatation invariance, we can take $\Theta = 0$ since $T_{\mu\nu}$ is traceless. In this case, we have

$$\partial_{\bar{z}} T = 0, \quad \partial_z \bar{T} = 0. \tag{2.24}$$

Thus, since

$$T = T(z), \quad \bar{T} = \bar{T}(\bar{z}), \tag{2.25}$$

the stress energy tensor is separated by two parts, one is a holomorphic function and the other is an anti-holomorphic function. This is general property of the CFT models.

In quantum theory, the stress energy tensor should behave under $z \to w(z)$ as

$$T(z) = \left(\frac{dw}{dz}\right)^2 \tilde{T}(w) + \frac{c}{12}\{w,z\}, \tag{2.26}$$

where c is central charge and $\{w,z\}$ is the Schwartzian derivative which is defined by

$$\{w,z\} \equiv \frac{d^3w}{dz^3}\bigg/\frac{dw}{dz} - \frac{3}{2}\left(\frac{d^2w}{dz^2}\right)^2\bigg/\left(\frac{dw}{dz}\right)^2. \tag{2.27}$$

Note that, for the global transformations, the Schwartzian derivative becomes zero. Further, it can be shown that under the consecutive transformation $z \to w \to u$,

$$\{u,z\} = \{u,w\}\left(\frac{dw}{dz}\right)^2 + \{w,z\}.$$

Thus, the infinitesimal transformation law for $T(z)$ becomes

$$\delta_\epsilon T(z) = \epsilon(z)\partial T(z) + 2\partial\epsilon(z)T(z) + \frac{c}{12}\partial^3\epsilon(z). \tag{2.28}$$

To duplicate the relation of eq.(2.28), we must take the OPE of the stress energy tensor as

$$T(z)T(w) = \frac{c/2}{(z-w)^4} + \frac{2T(w)}{(z-w)^2} + \frac{\partial T(w)}{z-w} + \text{reg.} \tag{2.29}$$

It is convenient to define the mode expansions of $T(z)$ and $\bar{T}(\bar{z})$ by

$$T(z) = \sum_{n \in \mathbf{Z}} z^{-n-2} L_n, \qquad \bar{T}(\bar{z}) = \sum_{n \in \mathbf{Z}} \bar{z}^{-n-2} \bar{L}_n. \tag{2.30}$$

Now, we consider the commutation relations of L_n and $\bar{L}_n$. From eq.(2.30), we can write

$$L_n = \oint \frac{dz}{2\pi i} z^{n+1} T(z), \qquad \oint \frac{dz}{2\pi i} z^{n+2} \partial T(z) = -(n+2) L_n. \tag{2.31}$$

Therefore, the commutation relation of L_n can be evaluated as follow,

$$\begin{aligned}
[L_m, L_n] &= \oint_0 \frac{dz}{2\pi i} z^{m+1} T(z) \oint_0 \frac{dw}{2\pi i} w^{n+1} T(w) \bigg|_{|z|>|w|} - \oint_0 \frac{dw}{2\pi i} w^{n+1} T(w) \oint_0 \frac{dz}{2\pi i} z^{m+1} T(z) \bigg|_{|z|<|w|} \\
&= \oint_0 \frac{dw}{2\pi i} w^{n+1} \oint_w \frac{dz}{2\pi i} z^{m+1} T(z) T(w) \\
&= \oint_0 \frac{dw}{2\pi i} w^{n+1} \oint_w \frac{dz}{2\pi i} z^{m+1} \left[\frac{c/2}{(z-w)^4} + \frac{2T(w)}{(z-w)^2} + \frac{\partial T(w)}{z-w} + \text{reg.} \right] \\
&= \oint_0 \frac{dw}{2\pi i} w^{n+1} \left[\frac{c/2}{3!} \frac{d^3}{dz^3} z^{m+1} \bigg|_{z \to w} + 2 \frac{d}{dz} z^{m+1} \bigg|_{z \to w} T(w) + w^{m+1} \partial T(w) \right] \\
&= \frac{c}{12} (m+1) m (m-1) \delta_{n+m,0} + 2(m+1) \oint_0 \frac{dw}{2\pi i} w^{m+n+1} T(w) + \oint_0 \frac{dw}{2\pi i} w^{m+n+2} \partial T(w) \\
&= \frac{c}{12} m(m^2 - 1) \delta_{n+m,0} + 2(m+1) L_{m+n} - (m+n+2) L_{m+n}.
\end{aligned}$$

Accordingly, the commutation relation of L_n is written by

$$[L_m, L_n] = (m-n) L_{m+n} + \frac{c}{12} m(m^2 - 1) \delta_{m+n,0}. \tag{2.32a}$$

Similarly, we can also find the commutation relation of $\bar{L}_n$ as

$$[\bar{L}_m, \bar{L}_n] = (m-n) \bar{L}_{m+n} + \frac{c}{12} m(m^2 - 1) \delta_{m+n,0}. \tag{2.32b}$$

Further, it can be shown that

$$[L_m, \bar{L}_n] = [\bar{L}_m, L_n] = 0. \tag{2.32c}$$

The relations of eq.(2.32) are called *Virasoro Algebra*, which is a fundamental algebra in the CFT models.

2.3. Primary Operator and Its Operator Product Expansion

Under the global conformal transformations, the quasi-primary operator Φ is transformed according to

$$\Phi(z,\bar{z}) = \left(\frac{dw}{dz}\right)^h \left(\frac{d\bar{w}}{d\bar{z}}\right)^{\bar{h}} \Phi(w,\bar{w}), \tag{2.33}$$

where $\{h, \bar{h}\}$ are called conformal weights. Here, $\bar{h}$ does not mean the complex conjugate of h. Due to the global conformal invariance, the 1,2 and 3-point function of the quasi-primary operators can be determined exactly.

Further, in two dimensions, N-point function can be determined owing to an infinite number of symmetries. In this case, we can consider the primary operator ϕ which is transformed as to

$$\phi(z,\bar{z}) = \left(\frac{dw}{dz}\right)^h \left(\frac{d\bar{w}}{d\bar{z}}\right)^{\bar{h}} \phi(w,\bar{w}), \tag{2.34}$$

under *any* conformal transformations. Note that the primary operator is also quasi-primary operator. For the case of the infinitesimal transformation, eq.(2.34) becomes

$$\delta_\epsilon \phi = \epsilon \partial \phi + h(\partial \epsilon)\phi, \tag{2.35a}$$
$$\delta_{\bar{\epsilon}} \phi = \bar{\epsilon}\bar{\partial}\phi + \bar{h}(\bar{\partial}\bar{\epsilon})\phi. \tag{2.35b}$$

The conformal charges are defined in terms of the stress energy tensors $T(\xi), \bar{T}(\bar{\xi})$ as

$$T_\epsilon \equiv \oint_0 \frac{d\xi}{2\pi i}\epsilon(\xi)T(\xi), \qquad \bar{T}_{\bar{\epsilon}} \equiv \oint_0 \frac{d\bar{\xi}}{2\pi i}\bar{\epsilon}(\bar{\xi})\bar{T}(\bar{\xi}), \tag{2.36}$$

where **0** means that the contour is the circle whose center is at the origin. In general, a variation of the field A under the transformation with respect to the charge Q_ϵ is given by

$$\delta_\epsilon A = [Q_\epsilon, A]. \tag{2.37a}$$

For the anti-holomorphic part, we have

$$\bar{\delta}_{\bar{\epsilon}} \bar{A} = [\bar{Q}_{\bar{\epsilon}}, \bar{A}]. \tag{2.37b}$$

In the complex planes, the brackets [ ] is defined by means of

$$[Q_\epsilon, A(z,\bar{z})] = Q_\epsilon A - A Q_\epsilon$$
$$\equiv \oint_{0,|\xi|>|z|} \frac{d\xi}{2\pi i}\epsilon(\xi)Q(\xi)A(z,\bar{z}) - \oint_{0,|\xi|<|z|} \frac{d\xi}{2\pi i}\epsilon(\xi)A(z,\bar{z})Q(\xi)$$
$$= \oint_z \frac{d\xi}{2\pi i}\epsilon(\xi)R(Q(\xi)A(z,\bar{z})),$$

see also Fig. 1. Here, R denote the radial ordering which is defined by

$$R(A(z)B(w)) = \begin{cases} A(z)B(w) & \text{for } |z| > |w| \\ B(w)A(z) & \text{for } |z| < |w| \end{cases}. \tag{2.38}$$

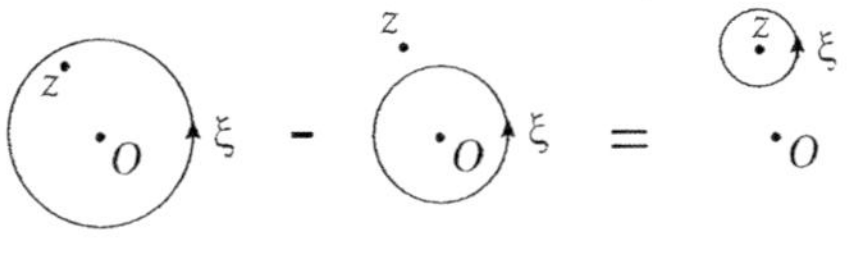

Figure 1. The commutator $[Q_\epsilon, A]$ is equivalent to a contour integral around the point z in the complex plane.

The similar relation holds in the anti-holomorphic part as well. Note that the radial ordering corresponds to the time ordering in the Minkowski space. Hereafter, we omit the R symbol for convenience.

From eq.(2.14) and eq.(2.30), the conformal charges are written by

$$T_\epsilon = \sum_{n \in Z} \epsilon_n L_n, \qquad \bar{T}_{\bar\epsilon} = \sum_{n \in Z} \bar\epsilon_n \bar{L}_n. \tag{2.39}$$

Further, from eq.(2.35) and eq.(2.37), we have

$$[L_n, \phi] = z^{n+1} \partial \phi + (n+1) h z^n \phi. \tag{2.40a}$$
$$[\bar{L}_n, \phi] = \bar{z}^{n+1} \bar{\partial} \phi + (n+1) \bar{h} \bar{z}^n \phi. \tag{2.40b}$$

Alternatively, if we take the OPE between $T(z)$ and the primary operator $\phi(w, \bar{w})$ as

$$T(z)\phi(w, \bar{w}) = \frac{h\, \phi(w, \bar{w})}{(z-w)^2} + \frac{\partial \phi(w, \bar{w})}{z-w} + \text{reg.}, \tag{2.41a}$$

where reg. means non-singular term, we can confirm the relation

$$\delta_\epsilon \phi = \oint_z \frac{d\xi}{2\pi i} \epsilon(\xi) T(\xi) \phi(z, \bar{z}) = \oint_z \frac{d\xi}{2\pi i} \epsilon(\xi) \left[\frac{h\, \phi(z, \bar{z})}{(\xi-z)^2} + \frac{\partial \phi(z, \bar{z})}{\xi - z} \right] = \epsilon \partial \phi + h (\partial \epsilon) \phi.$$

This is nothing but eq.(2.35a). Similarly, we can find eq.(2.35b) from the OPE

$$\bar{T}(\bar{z})\phi(w, \bar{w}) = \frac{\bar{h}\, \phi(w, \bar{w})}{(\bar{z}-\bar{w})^2} + \frac{\bar{\partial} \phi(w, \bar{w})}{\bar{z} - \bar{w}} + \text{reg.} \tag{2.41b}$$

2.4. SL(2,Z) Invariance and Its Representation

From eq.(2.32), it can be easily verified that

$$[L_m, L_n] = (m-n) L_{m+n}, \qquad [\bar{L}_m, \bar{L}_n] = (m-n) \bar{L}_{m+n}, \tag{2.42}$$

where $m, n = 0, \pm 1$. In particular, we have

$$[L_0, L_{-n}] = n L_{-n}, \qquad [L_1, L_{-n}] = (1+n) L_{1-n}, \tag{2.43}$$

where $n > 0$. Therefore, $\{L_{-1}, L_0, L_1\}$ and $\{\bar{L}_{-1}, \bar{L}_0, \bar{L}_1\}$ construct the closed algebra respectively. These algebras are known as the Lie algebra of sl(2,C) which gives rise to the global conformal transformations. For the translation $x_\mu \to x_\mu + \varepsilon_\mu$, which can be expressed by $z \to z + \epsilon_{-1}$ and $\bar{z} \to \bar{z} + \bar{\epsilon}_{-1}$ where $\epsilon_{-1} = \varepsilon_1 + i\varepsilon_2$ and $\bar{\epsilon}_{-1} = \varepsilon_1 - i\varepsilon_2$ in the complex plane. Thus, the variation of the primary operator ϕ is given by

$$\delta^{\text{tra}} \phi(x_\mu) = \epsilon_{-1} \partial \phi(z,\bar{z}) + \bar{\epsilon}_{-1} \bar{\partial} \phi(z,\bar{z}), \tag{2.44}$$

On the other hand, eq.(2.37a) shows

$$\delta^{\text{tra}}_{\epsilon,\bar{\epsilon}} \phi(z,\bar{z}) = \epsilon_{-1} L_{-1} \phi(z,\bar{z}) + \bar{\epsilon}_{-1} \bar{L}_{-1} \phi(z,\bar{z}). \tag{2.45}$$

Accordingly, the generators of the translation in the holomorphic and the anti-holomorphic part are written by

$$L_{-1} = \partial, \qquad \bar{L}_{-1} = \bar{\partial}. \tag{2.46}$$

For the rotation $x_\mu \to x_\mu + \omega_{\mu\nu} x_\nu$ ($\omega_{\mu\nu} = -\omega_{\nu\mu}$), which is equivalent to $z \to z - \epsilon_0 z$ and $\bar{z} \to \bar{z} + \bar{\epsilon}_0 \bar{z}$ where $\epsilon_0 = i\omega_{12}$, the variations of the primary operator are given by

$$\delta^{\text{rot}} \phi(x_\mu) = \epsilon_0 \left(z \partial - \bar{z} \bar{\partial}\right) \phi(z,\bar{z}), \qquad \delta^{\text{rot}}_{\epsilon,\bar{\epsilon}} \phi(z,\bar{z}) = -\epsilon_0 (L_0 - \bar{L}_0) \phi(z,\bar{z}). \tag{2.47}$$

Thus, the generator of the rotation or *spin* operator $\hat{S}$ is written by

$$\hat{S} = L_0 - \bar{L}_0 = -(z \partial - \bar{z} \bar{\partial}). \tag{2.48}$$

In the case of the dilatation $x_\mu \to x_\mu + \varepsilon x_\mu$, which is equivalent to $z \to z + \epsilon_0 z$ and $\bar{z} \to \bar{z} + \bar{\epsilon}_0 \bar{z}$ where $\epsilon_0 = \varepsilon$, we have

$$\delta^{\text{dila}} \phi(x_\mu) = \epsilon_0 \left(z \partial + \bar{z} \bar{\partial}\right) \phi(z,\bar{z}), \qquad \delta^{\text{dila}}_{\epsilon,\bar{\epsilon}} \phi(z,\bar{z}) = \epsilon_0 (L_0 + \bar{L}_0) \phi(z,\bar{z}). \tag{2.49}$$

In this case, the generator of the dilatation $\hat{D}$ is written by

$$\hat{D} = L_0 + \bar{L}_0 = z \partial + \bar{z} \bar{\partial}. \tag{2.50}$$

For the special conformal transformation $x_\mu \to x_\mu + 2(\boldsymbol{x} \cdot \boldsymbol{\varepsilon}) x_\mu + x^2 \varepsilon_\mu$, we can write $z \to z + \epsilon_1 z^2$ and $\bar{z} \to \bar{z} + \bar{\epsilon}_1 \bar{z}^2$ where $\epsilon = \varepsilon_1 - i\varepsilon_1$ and $\bar{\epsilon} = \varepsilon_1 + i\varepsilon_1$, and thereby the variations of the primary operator are

$$\delta^{\text{SCT}} \phi(x_\mu) = (\epsilon_1 z^2 \partial + \bar{\epsilon}_1 \bar{z}^2 \bar{\partial}) \phi(z,\bar{z}), \qquad \delta^{\text{SCT}}_{\epsilon,\bar{\epsilon}} \phi(z,\bar{z}) = (\epsilon_1 L_1 + \bar{\epsilon}_1 \bar{L}_1) \phi(z,\bar{z}). \tag{2.51}$$

Therefore, the generators of the special conformal transformation are written by

$$L_1 = z^2 \partial, \qquad \bar{L}_1 = \bar{z}^2 \bar{\partial}. \tag{2.52}$$

2.5. Vacuum and Its SL(2,Z) Invariance

Before discussing the vacuum of the CFT, we define the adjoint operator of ϕ as

$$\phi^\dagger(z, \bar{z}) = \phi\left(\frac{1}{\bar{z}}, \frac{1}{z}\right) \frac{1}{\bar{z}^{2h}} \frac{1}{z^{2h}}. \tag{2.53}$$

When $\bar{z}$ is the complex conjugate z^* of the variable z, eq.(2.53) defines the hermitian conjugate field of ϕ in Minkowski space. Note that the third term of eq.(2.28) becomes zero under the global conformal transformations. Therefore, the stress energy tensor T has the conformal weight of $(2, 0)$. Similarly, we can find that the conformal weight of $\bar{T}$ is $(0, 2)$. Thus, the adjoint of $T(z)$ is given by

$$T^\dagger(z) = T\left(\frac{1}{\bar{z}}\right) \frac{1}{\bar{z}^4} = \sum \bar{z}^{-n-2} L_n. \tag{2.54}$$

On the other hand, the complex conjugate of T can be written by

$$T^\dagger(z) = \sum z^{*-m-2} L_m^\dagger. \tag{2.55}$$

To impose the hermicity on the stress energy tensor, we have

$$L_n^\dagger = L_{-n} \tag{2.56}$$

since $\bar{z} = z^*$. Obviously, we can also show $\bar{L}_n^\dagger = \bar{L}_{-n}$ in the anti-holomorhic part.

Now, we consider the vacuum $|0\rangle$ of the conformal field theory model. Let the $T(z)$ operate onto the vacuum. We have

$$T(z)|0\rangle = \sum_n z^{-n-2} L_n |0\rangle. \tag{2.57}$$

It can be shown that

$$\langle 0| T^\dagger(z) = \langle 0| \sum_n L_n^\dagger \bar{z}^{-n-2} = \sum_n \langle 0| L_{-n} \bar{z}^{-n-2}. \tag{2.58}$$

By requiring the regularity both of eq.(2.57) and (2.58) at $z = 0$, we have conditions on the L_n and $\bar{L}_n$ as

$$L_n|0\rangle = 0, \quad (n \geq -1) \tag{2.59}$$

and

$$\langle 0|L_m = 0, \quad (m \leq 1). \tag{2.60}$$

Correspondingly, we can have the conditions of the $\bar{L}_n$. Note that the vacuum $|0\rangle$ is annihilated both $L_{0,\pm 1}$ and $\bar{L}_{0,\pm 1}$. Accordingly, the vacuum is SL(2, Z) invariant state.

2.6. Highest Weight States

Let us consider the state
$$|h, \bar{h}\rangle = \lim_{z,\bar{z}\to 0} \phi(z, \bar{z})|0\rangle, \tag{2.61}$$

where $\phi(z, \bar{z})$ is a primary operator of weight $(h, \bar{h})$. It is noted that $z, \bar{z} \to 0$ corresponds to $\tau \to 0$ in the Euclidean space. This should mean that $|h, \bar{h}\rangle$ is *in* state. From eq.(2.40), we can find that, for $n > 0$,

$$L_0|h, \bar{h}\rangle = h|h, \bar{h}\rangle, \qquad L_n|h, \bar{h}\rangle = 0, \tag{2.62a}$$
$$\bar{L}_0|h, \bar{h}\rangle = \bar{h}|h, \bar{h}\rangle, \qquad \bar{L}_n|h, \bar{h}\rangle = 0. \tag{2.62b}$$

The state which satisfies the conditions eq.(2.62) is called a highest state or primary state. Similarly, we have *out* state which satisfies that

$$\langle h, \bar{h}|L_0 = \langle h, \bar{h}|h, \qquad \langle h, \bar{h}|L_m = 0, \tag{2.63a}$$
$$\langle h, \bar{h}|\bar{L}_0 = \langle h, \bar{h}|\bar{h}, \qquad \langle h, \bar{h}|\bar{L}_m = 0, \tag{2.63b}$$

where $m < 0$. In conformal field theory models, the holomorphic and anti-holomorphic parts are separately appeared. Thus, the highest weight state are given by the tensor product $|h, \bar{h}\rangle = |h\rangle \otimes |\bar{h}\rangle$. From eq.(2.62) and (2.63), we have

$$(L_0 + \bar{L}_0)|h, \bar{h}\rangle = (h + \bar{h})|h, \bar{h}\rangle \equiv \Delta|h, \bar{h}\rangle, \tag{2.64a}$$
$$(L_0 - \bar{L}_0)|h, \bar{h}\rangle = (h - \bar{h})|h, \bar{h}\rangle \equiv S|h, \bar{h}\rangle. \tag{2.64b}$$

Since $L_0 - \bar{L}_0$ is the generator of rotation and $L_0 + \bar{L}_0$ is the generator of dilatation (see eq.(2.48) and eq.(2.50)), we can interpret S as *spin* and Δ as scaling dimension, respectively.

Eq.(2.62) remind us of the representation of the angular momentum eigenstate. From eq.(2.43) at $n > 0$, we have

$$L_0(L_{-n}|h\rangle) = (h + n)(L_{-n}|h\rangle). \tag{2.65}$$

Thus, L_{-n} is a *raising operator* by n. Moreover, we can construct the *level $h + N$* state as

$$|h + N\rangle = L_{-k_1}L_{-k_2}\cdots L_{-k_n}|h\rangle, \quad (1 \leq k_1 \leq k_2 \leq \cdots \leq k_n), \tag{2.66}$$

where $N = \sum k_i$. The states $|h + N\rangle$ $(N = 1, 2, \cdots)$ are called *descendant states* or *secondary states* with respect to the highest weight state $|h\rangle$. It is noticed that the state $|h + N\rangle$ is also an eigenstate of L_0,

$$L_0|h + N\rangle = (h + N)|h + N\rangle. \tag{2.67}$$

Since $L_1|h + N\rangle \neq 0$, $|h + N\rangle$ is not a highest weight state. On the other hand, for $n > 0$, we can show

$$L_0(L_n|h + N\rangle) = (h + N - n)(L_n|h + N\rangle) \tag{2.68}$$

thanks to $[L_0, L_n] = -nL_n$. This means that L_n $(n > 0)$ is a *lowering operator* by n.

Here, we summarize the CFT data.

highest weight state : $|h, \bar{h}\rangle$
$L_0|h, \bar{h}\rangle = h|h, \bar{h}\rangle, \bar{L}_0|h, \bar{h}\rangle = \bar{h}|h, \bar{h}\rangle$

descendant states : $|h+N, \bar{h}+\bar{N}\rangle = L_{-k_1} \cdots L_{-k_n} \bar{L}_{-\bar{k}_1} \cdots \bar{L}_{-\bar{k}_n}|h, \bar{h}\rangle$
$(1 \leq k_1, \bar{k}_1 \leq \cdots \leq k_n, \bar{k}_n, N = \sum k_i, \bar{N} = \sum \bar{k}_i)$
$L_0|h+N, \bar{h}+\bar{N}\rangle = (h+N)|h+N, \bar{h}+\bar{N}\rangle$
$\bar{L}_0|h+N, \bar{h}+\bar{N}\rangle = (\bar{h}+\bar{N})|h+N, \bar{h}+\bar{N}\rangle$

raising operators : $L_{-n}, \bar{L}_{-n}$ $(n>0)$
lowering operators : $L_n, \bar{L}_n$ $(n>0)$

$$(2.69)$$

The descendent states upon the highest weight state $|h, \bar{h}\rangle$ can be constructed as

level	weight	state
0	$(h, \bar{h})$	$|h, \bar{h}\rangle$
1	$(h+1, \bar{h}+1)$	$L_{-1}\bar{L}_{-1}|h, \bar{h}\rangle$
2	$(h+2, \bar{h}+2)$	$L_{-2}\bar{L}_{-2}|h\rangle, L_{-1}^2\bar{L}_{-1}^2|h, \bar{h}\rangle$
$\vdots$	$\vdots$	
N	$(h+N, \bar{h}+\bar{N})$	$\prod \bar{L}_{-\bar{k}_i}L_{-k_i}|h, \bar{h}\rangle$
$\vdots$	$\vdots$	

$$(2.70)$$

where $N = \sum k_i$, $\bar{N} = \sum \bar{k}_i$. The number of states at level N is given by $P(N)$, which is the number of partitions of N into positive integer parts. It is known that the generating function of $P(N)$ is given by

$$\prod_{n=1}^{\infty}(1-q^n)^{-1} = \sum_{N=0}^{\infty} P(N)q^N, \qquad (2.71)$$

where $P(0) = 1$. Further, it is known that the generating function of $P(N)$ converges in the region $|q| < 1$.

Consequently, the subset of the Hilbert space of CFT spanned by the highest weight states and its descendants is known as *Verma module* which is denoted by $V(c, h) \otimes \bar{V}(c, \bar{h})$. The full Hilbert space of a CFT model can be composed of the sum of the Verma modules,

$$\text{CFT} = \sum_{h, \bar{h}} V(c, h) \otimes \bar{V}(c, \bar{h}). \qquad (2.72)$$

2.7. Descendant Operator (Secondary Operator)

Obviously, there can be an operator which corresponds to the descendant state. Since the descendant state is associated with the highest weight state, the descendant operator is deduced from the primary operator. First, we expand the $T(z)$ around an arbitrary point w as

$$T(z) = \sum_{n \in \mathbf{Z}} (z-w)^{-n-2} L_n(w). \qquad (2.73)$$

Therefore, the OPE between $T(z)$ and the primary operator $\phi(w, \bar{w})$ is written by

$$T(z)\phi(w) = \sum_{n\in \mathbf{Z}} (z-w)^{-n-2} (L_n\phi)(w). \qquad (2.74)$$

Thus, the descendant operator with respect to the state $L_{-n}|h\rangle$ should be defined by

$$(L_n\phi)(w) = \oint_w \frac{dz}{2\pi i} \frac{T(z)\phi(w)}{(z-w)^{n-1}}. \qquad (2.75)$$

It is covenient to denote $(L_n\phi)(w)$ in $\phi^{(-n)}(w)$. Now, we consider the cases of $n = 0, -1$. From eq.(2.41a), we can find

$$\phi^{(0)}(w) = h\phi(w), \qquad \phi^{(-1)}(w) = \oint_w \frac{dz}{2\pi i} T(z)\phi(w) = \partial_w \phi(w). \qquad (2.76)$$

Thus, we can write the OPE between $\phi^{(-1)}(w)$ and $T(z)$ as

$$T(z)\phi^{(-1)}(w) = \partial_w T(z)\phi(w) = \frac{2h\phi(w)}{(z-w)^3} + \frac{(h+1)\partial\phi(w)}{(z-w)^2} + \frac{\partial^2\phi(w)}{z-w} + \text{reg}. \qquad (2.77)$$

The OPE of eq.(2.77) indicates that $\phi^{(-1)}(w)$ is the level 1 state with respect to the highest weight state $|h\rangle$. On the other hand, from eq.(2.37a), we can show

$$\delta_\epsilon \phi^{(-1)} = \epsilon \partial^2 \phi + (h+1)\partial\phi\,\partial\epsilon + h\phi\partial^2\epsilon = \epsilon\partial\phi^{(-1)} + (h+1)\phi^{(-1)}\partial\epsilon + h\phi\partial^2\epsilon. \qquad (2.78)$$

Further, we have

$$\left[L_n, \phi^{(-1)}(z)\right] = z^{n+1}\partial^2\phi + (h+1)(n+1)z^n \phi^{(-1)} + hn(n+1)z^{n-1}\phi. \qquad (2.79)$$

In particular, for $n = 0$, it means that

$$L_0 \lim_{z\to 0} \phi^{(-1)}(z)|h\rangle = (h+1) \lim_{z\to 0} \phi^{(-1)}(z)|h\rangle. \qquad (2.80)$$

Therefore, $\phi^{(-1)}(z)$ is the level 1 descendant state.

Another example of a descendant operator is the stress energy tensor. It is a level two descendant state of the identity operator $\mathbf{1}$ since

$$(L_{-2}\mathbf{1})(w) = \oint_w \frac{dz}{2\pi i} \frac{T(z)\mathbf{1}}{z-w} = T(w). \qquad (2.81)$$

Accordingly, there are descendant operators with respect to the primary operator ϕ_α. Both of them are usually referred to as *conformal family* $[\phi_\alpha]$. The complete set of the operators $\{A_i\}$ in conformal field theory is composed in terms of the conformal families [19].

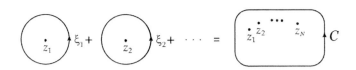

Figure 2. There are all of the singular points $z_1, \cdots, z_N$ in the contour C.

2.8. Correlation Functions and Ward Identity

Here, we consider the N-point correlation function $\langle \phi_1(z_1, \bar{z}_1) \cdots \phi_N(z_N, \bar{z}_N) \rangle = \langle X \rangle$. Using eq.(2.37a), we have

$$
\begin{aligned}
\delta_\epsilon \langle X \rangle &= \sum_i \langle \phi_1(z_1, \bar{z}_1) \cdots \delta_\epsilon \phi_i(z_i, \bar{z}_i) \cdots \phi_N(z_N, \bar{z}_N) \rangle \\
&= \sum_i \langle \phi_1(z_1, \bar{z}_1) \cdots \oint_{z_i} \frac{dw}{2\pi i} \epsilon(w) T(w) \phi_i(z_i, \bar{z}_i) \cdots \phi_N(z_N, \bar{z}_N) \rangle \\
&= \oint_C \frac{dw}{2\pi i} \epsilon(w) \langle T(w) X \rangle,
\end{aligned} \quad (2.82)
$$

where C is a contour which is shown in Fig. 2. On the other hand, from eq.(2.35a), we have

$$
\begin{aligned}
\delta_\epsilon \langle X \rangle &= \sum_i \langle \phi_1(z_1, \bar{z}_1) \cdots \delta_\epsilon \phi_i(z_i, \bar{z}_i) \cdots \phi_N(z_N, \bar{z}_N) \rangle \\
&= \sum_i (h_i \partial \epsilon + \epsilon \partial) \langle \phi_1(z_1, \bar{z}_1) \cdots \phi_i(z_i, \bar{z}_i) \cdots \phi_N(z_N, \bar{z}_N) \rangle \\
&= \oint_C \frac{dw}{2\pi i} \epsilon(w) \sum_i \left[\frac{h_i}{(w-z_i)^2} + \frac{1}{w-z_i} \frac{\partial}{\partial z_i} \right] \langle X \rangle.
\end{aligned} \quad (2.83)
$$

Finally, we find the Ward identity in the CFT model as

$$
\langle T(w) \phi_1(z_1, \bar{z}_1) \cdots \rangle = \sum_i \left[\frac{h_i}{(w-z_i)^2} + \frac{1}{w-z_i} \frac{\partial}{\partial z_i} \right] \langle \phi_1(z_1, \bar{z}_1) \cdots \rangle. \quad (2.84)
$$

In the global symmetry case, we can verify from eq.(2.83) that

$$
\sum_i \partial_i \langle \phi_1(z_1) \cdots \phi_N(z_N) \rangle = 0 \quad (2.85a)
$$

$$
\sum_i (h_i + z_i \partial_i) \langle \phi_1(z_1) \cdots \phi_N(z_N) \rangle = 0 \quad (2.85b)
$$

$$
\sum_i (2h_i z_i + z_i^2 \partial_i) \langle \phi_1(z_1) \cdots \phi_N(z_N) \rangle = 0 \quad (2.85c)
$$

where $\partial_i = \partial/\partial z_i$. Thus, the two and three point correlation functions are exactly determined from eq.(2.85) apart from the normalization factor as

$$
\langle \phi_i(z_1) \phi_j(z_2) \rangle \sim \frac{\delta_{ij}}{(z_1-z_2)^{2h_i}}, \quad \langle \phi_i(z_1) \phi_j(z_2) \phi_k(z_3) \rangle \sim \frac{1}{z_{12}^{\gamma_{ij,k}} z_{23}^{\gamma_{jk,i}} z_{13}^{\gamma_{ik,j}}}, \quad (2.86)
$$

where $z_{mn} \equiv z_m - z_n$ and $\gamma_{ab,c} \equiv h_a + h_b - h_c$.

2.9. Minimal Models

As we have learned in previous section, the representation of the Virasoro Algebra is given by the Verma modules. It is well-known that the representation is irreducible unless the conformal weight h takes special values of h [20, 21]. On the other hand, for the special values of h, there are *null states* whose norm is zero. In this case, the Verma module $V(c, h) \otimes \bar{V}(c, \bar{h})$ becomes reducible. However, removing all of the null states, we obtain the irreducible representation which is called degenerate representation.

In the region of $0 \leq c < 1$, it is known that there are unitary CFT models. In this case, c can be parameterized by m as

$$c = 1 - \frac{6}{m(m+1)}, \qquad (m = 2, 3, 4, \cdots). \tag{2.87}$$

Moreover, the conformal weights $h_{p,q}$ are also written in terms of m, p and q [22]. It is interesting that there is correspondence between the series of the minimal models and the statistical models at their critical points. For instance, $c = 1/2$ ($m = 3$) CFT describes the critical point of the two-dimensional Ising model and $c = 4/5$ ($m = 5$) CFT describes the point of the 3-state (Z_3) Potts model and so on.

2.9.1. Null State

For simplicity, we only consider the holomorhic part here. The norm of the level n state $L_{-n}|h\rangle$ becomes

$$\langle h|L_n L_{-n}|h\rangle = \langle h|\left[L_{-n}L_n + 2L_0 + \frac{c}{12}n(n^2-1)\right]|h\rangle = 2h + \frac{c}{12}n(n^2-1), \tag{2.88}$$

where we define $\langle h | h \rangle = 1$. Note that the norm of the states must be positive for a unitary model. Therefore, in the unitary models, two parameters h, c are suppressed by conditions

$$h \geq 0, \qquad c \geq 0, \tag{2.89}$$

where the former condition is yielded from eq.(2.88) of $n = 1$ and the later is followed from the fact that the norm must be positive at large n. In particular, there are two states, L_{-1}^2 and L_{-2} at the level 2. In this case, we consider the 2×2 matrix $M_2(c, h)$,

$$M_2(c, h) \equiv \begin{bmatrix} \langle h|L_1^2 L_{-1}^2|h\rangle & \langle h|L_1^2 L_{-2}|h\rangle \\ \langle h|L_2 L_{-1}^2|h\rangle & \langle h|L_2 L_{-2}|h\rangle \end{bmatrix} = \begin{bmatrix} 4h(2h+1) & 6h \\ 6h & 4h + c/2 \end{bmatrix}. \tag{2.90}$$

Thus, the determinant of the matrix eq.(2.90) is given by

$$\det M_2(c, h) = 2(16h^3 - 10h^2 + 2h^2 c + hc) = 32h(h - h_{1,2})(h - h_{2,1}), \tag{2.91}$$

where $h_{1,2}, h_{2,1}$ are written by

$$h_{1,2} = \frac{(m-1)^2 - 1}{4m(m+1)}, \quad h_{2,1} = \frac{(m+2)^2 - 1}{4m(m+1)} \tag{2.92}$$

and c is given by eq.(2.87). Now, we consider the state

$$|\chi\rangle \equiv \left(L_{-2} - \frac{3}{2(2h+1)}L_{-1}^2\right)|h\rangle. \tag{2.93}$$

It is easy to prove that the state eq.(2.93) satisfies the following relations

$$L_0|\chi\rangle = (h+2)|\chi\rangle, \qquad L_n|\chi\rangle = 0, \tag{2.94}$$

where $n > 0$. Thus, from eq.(2.62a), we can find that the state eq.(2.93) is the highest weight state of weight $h + 2$. However, we can also find that

$$\langle\chi|\chi\rangle = \langle h|\left(L_2 - \frac{3}{2(2h+1)}L_1^2\right)|\chi\rangle = 0. \tag{2.95}$$

Thus, $|\chi\rangle$ becomes no physical state. $|\chi\rangle$ is called *null state*. Apparently, the null state and its descendant states are orthogonal to all other states. Therefore, the null states and their descendant states are taken to be zero $[|\chi\rangle] = 0$ at the last to remove them. Eliminating all of the null states in the Verma module $V(c, h)$, we can find the irreducible representation $V^{(\text{irr})}(c, h)$ which is referred to as degenerate representation.

We also note that the null field associated with level 2 null state $|\chi\rangle$ can be written by

$$\phi_\chi = L_{-2}\phi - \frac{3}{2(2h+1)}L_{-1}^2\phi, \tag{2.96}$$

where $L_{-1} = \partial$. From this equation, we obtain the differential equation of the correlation function in terms of the property of the null state [19].

2.9.2. Unitary Minimal Models

In the above, we find the null state of the level two. It can be generalized in more higher level. For level N, we must consider the $P(N) \times P(N)$ matrix $M_N(c, h)$ and find zeroes of $M_N(c, h)$ to construct the irreducible representation. Due to Kac [20], the generalized form of eq.(2.91) at the level N is given by

$$\det M_N(c, h) = C_N \prod_{\substack{r,s=1 \\ 1 \leq rs \leq N}} (h - h_{p,q})^{P(N-pq)}, \tag{2.97}$$

where $P(n)$ is the number of partitions of an integer n, which is already introduced at eq.(2.71). Meanwhile C_N is a constant. The determinant eq.(2.97) is referring as to *Kac determinant*. The mathematical proof of (2.97) is given by Feigin and Fuchs [21, 22]. In the Kac determinant $\det M_N(c, h_{p,q})$, the following fact are known,

1. When $\det M_N < 0$, the negative norm states appear. Thus, the representation becomes non-unitary.

2. When $\det M_N \geq 0$,

    - the Kac determinant has no zeroes at any level only if $c > 1, h \geq 0$.

- m is not real and $h_{p,q}$ becomes imaginary or negative only if $1 < c < 25$.
- all the $h_{p,q}$ are negative only if $c \geq 25$.
- the Kac determinant is not negative only if $c = 1$. Therefore, Kac determinant makes no restriction on the unitarity about the representation of the Virasoro algebra. To preserve the unitarity in the $c = 1$ CFT, we take only the conditions $c \geq 1, h \geq 0$. This means that there is an infinite number of the primary fields. However, at the $h_{p,q} = n^2/4$, the determinant vanishes. In this case, we must remove all of the null states.
- the unitary representation is allowed only when (p, q, m) become the specific values [22] in the case of $0 < c < 1$, which will be discussed below.

When $0 < c < 1$, it is convenient to parameterize the central charge c in terms of m as

$$c = 1 - \frac{6}{m(m+1)}, \tag{2.98}$$

where $m = 3, 4, \cdots$. To each value of c, the conformal weights $h_{p,q}$ are given by

$$h_{p,q}(m) = \frac{[(m+1)p - mq]^2 - 1}{4m(m+1)}, \quad (1 \leq p \leq m-1, \ 1 \leq q \leq p). \tag{2.99}$$

Note that there are $m(m-1)/2$ of $h_{p,q}$ at given m. Furthermore, it is easy to show that

$$h_{p,q} = h_{m-p, m+1-q}. \tag{2.100}$$

The symmetry of eq.(2.100) indicates that each $h_{p,q}$ appears twice. In general, when $h = h_{p,q}$, there are null vectors at level pq which satisfy

$$L_0|\chi\rangle = (h_{p,q} + pq)|\chi\rangle, \quad L_n|\chi\rangle = 0, \quad (n \geq 0). \tag{2.101}$$

It is important to note that we can construct the unitary theory for given central charge, conversely. It is known that the representation of the Virasoro algebra is constructed out of the currents of the affine Kac-Moody algebra. For instance, $c < 1$ CFT is constructed via the coset construction [23].

2.10. $C = 1$ Conformal Field Theory

In the previous section, we have learned that there are the discrete set of the unitary minimal models in the region $0 < c < 1$. In this case, the number of the primary operators becomes finite and its conformal weights are restricted to eq.(2.99) due to the unitarity. On the other hand, if c becomes larger than 1, the situation becomes complicate. Apart from $h \geq 0$, there is no restriction on the weights. Therefore, in general, the number of primary operators becomes infinite. However, it is known that the number of primary operators becomes finite at the special value of the conformal weight even though $c \geq 1$. According to [20], there are null states when the conformal dimension satisfies $\Delta = n^2/4$ where n is an integer. For example, the massless boson whose compact radius $R = \sqrt{2}$ is one of the $c = 1$ CFT models and it is represented in terms of the two highest weight states of the $k = 1$ SU(2) current algebra [14].

Alternatively, the string theory is described in terms of the $c = 1$ CFT. In the case of the D-dimensional bosonic string, there is no-ghost theorem [24]. It suggests that the dimension D and conformal weight $(h, \bar{h})$ are restricted to

$$D = 26, \qquad (h, \bar{h}) = (1, 1)$$

In this subsection, we consider the $c = 1$ conformal field theory. In particular, the massless scalar boson is discussed considerately. We will find that the Virasoro algebra of the massless boson is represented in terms of the operators of the U(1) Kac-Moody algebra. Further, we will learn that one of the primary operators in the $c = 1$ CFT models, which are known as the vertex operator, conduce to the highest weight states both of the $c = 1$ Virasoro algebra and U(1) Kac-Moody algebra.

2.10.1. Massless Scalar Field in Two-Dimensions

We consider the free massless scalar field in two-dimensions whose action S is given by

$$S = \frac{g}{2} \int d^2x \left[(\partial_t \Phi)^2 - (\partial_\sigma \Phi)^2 \right]. \tag{2.102}$$

Note that the action S is invariant under the $\Phi \to \Phi + $ constant. When we take a Wick rotation ($t \to -i\tau$), we have a Gaussian model which is well-known model in statistical physics.

The boundary condition of the field Φ is given by the periodic boundary condition as

$$\Phi(t, \sigma + 2\pi) = \Phi(t, \sigma). \tag{2.103}$$

From the variation of the action with respect to the field Φ, we have an equation of motion

$$\left(\frac{\partial^2}{\partial t^2} - \frac{\partial^2}{\partial \sigma^2} \right) \Phi = 0. \tag{2.104}$$

The mode expansion of Φ can be written by

$$\Phi(t, \sigma) = \mathcal{X} + \frac{\mathcal{P}t}{2\pi g} + \frac{i}{\sqrt{4\pi g}} \sum_{n \neq 0} \frac{1}{n} \left[\alpha_n e^{-in(t+\sigma)} + \bar{\alpha}_n e^{-in(t-\sigma)} \right]. \tag{2.105}$$

Now, we take the commutation relations as

$$\left[\Phi(t, \sigma), \Phi(t, \sigma') \right] = 0, \qquad \left[\Phi(t, \sigma), \Pi(t, \sigma') \right] = i\delta(\sigma - \sigma'), \tag{2.106}$$

where $\Pi = \partial_t \Phi$ is the conjugate field of Φ and the delta function $\delta(x)$ is given by

$$\delta(x) = \frac{1}{2\pi} \sum_{n=-\infty}^{\infty} e^{inx}. \tag{2.107}$$

Note that we recover the commutation relations eq.(2.106) if we take

$$[\mathcal{X}, \mathcal{P}] = i, \qquad [\alpha_m, \alpha_n] = m\delta_{m+n, 0}, \qquad [\bar{\alpha}_m, \bar{\alpha}_n] = m\delta_{m+n, 0} \tag{2.108}$$

and the other relations are vanished. This algebra is referred to as $U(1)$ Kac-Moody algebra or $U(1)$ current algebra.

Next, we go to the complex plane in accordance with the conformal map,

$$z = e^{i(t+\sigma)} = e^{\tau + i\sigma}, \qquad \bar{z} = e^{i(t-\sigma)} = e^{\tau - i\sigma}. \tag{2.109}$$

The mode expansion eq.(2.105) become

$$\phi(z) = \frac{1}{\sqrt{\pi g}} \left[\mathcal{Q} - i\alpha_0 \ln z + i \sum_{n \neq 0} \frac{z^{-n}}{n} \alpha_n \right], \tag{2.110a}$$

$$\bar{\phi}(\bar{z}) = \frac{1}{\sqrt{\pi g}} \left[\bar{\mathcal{Q}} - i\bar{\alpha}_0 \ln \bar{z} + i \sum_{n \neq 0} \frac{\bar{z}^{-n}}{n} \bar{\alpha}_n \right], \tag{2.110b}$$

where n is an integer and we define

$$\Phi \equiv \frac{1}{2} \left[\phi(z) + \bar{\phi}(\bar{z}) \right], \qquad \mathcal{X} \equiv \frac{1}{2\sqrt{\pi g}} (\mathcal{Q} + \bar{\mathcal{Q}}), \qquad \alpha_0 = \bar{\alpha}_0 \equiv \frac{\mathcal{P}}{2\sqrt{\pi g}} \tag{2.111}$$

Note that ϕ and $\bar{\phi}$ are not well-defined operators due to the existence of the term $\ln z, \ln \bar{z}$ which are singular at $z, \bar{z} = 0$. In general, it is well-known that there is no massless scalar field in two dimensions [25]. Thereby, we consider $\phi, \bar{\phi}$ as auxiliary fields.

The commutation relation of α_m can convert to the algebra of the conventional harmonic oscillator operators with a map

$$a_{m_+} \equiv \frac{1}{\sqrt{m_+}} \alpha_{m_+}, \qquad a^\dagger_{m_+} \equiv \frac{1}{\sqrt{m_+}} \alpha_{-m_+},$$

where m_+ is a positive integer. In this case, we have famous commutation relation of the harmonic oscillator

$$[a_m, a^\dagger_n] = \delta_{mn}.$$

Thus, $\alpha_{-n}, \bar{\alpha}_{-n}$ $(n > 0)$ are creation operators while $\alpha_n, \bar{\alpha}_n (n > 0)$ are annihilation operators. As long as there is no symmetry breaking, it is natural to define the vacuum $|0\rangle$ as

$$\alpha_n |0\rangle = 0, \qquad \bar{\alpha}_n |0\rangle = 0, \quad (n \geq 0). \tag{2.112}$$

The Hamiltonian of the scalar field is written by

$$H = \int d\sigma \left[\Pi^\mu \dot{X}_\mu - L \right] = \frac{g}{2} \int d\sigma \left[(\partial_t \Phi)^2 + (\partial_\sigma \Phi)^2 \right], \tag{2.113}$$

where L is the Lagrangian of the scalar field. Therefore, the Hamiltonian can be written by

$$H = \frac{\mathcal{P}^2}{4\pi g} + \frac{1}{2} \sum_{m \neq 0} \left(: \alpha_m \alpha_{-m} : + : \bar{\alpha}_m \bar{\alpha}_{-m} : \right), \tag{2.114}$$

where $: \cdots :$ denotes normal ordering of operators. Moreover, it can be rewritten by

$$H = \frac{1}{2} \sum_{m \in \mathbb{Z}} \left(: \alpha_m \alpha_{-m} : + : \bar{\alpha}_m \bar{\alpha}_{-m} : \right) = \sum_{m > 0} \left(\alpha_{-m} \alpha_m + \bar{\alpha}_{-m} \bar{\alpha}_m \right). \tag{2.115}$$

2.10.2. Virasoro Generators of $c = 1$ CFT

Before discussing the spectrum, we will see that the model obey the $c = 1$ Virasoro algebra. First, note that the Hamiltonian eq.(2.114) is reproduced by the operators,

$$T(z) = -\frac{\pi g}{2} :\partial \phi(z) \partial \phi(z): , \qquad \bar{T}(\bar{z}) = -\frac{\pi g}{2} :\bar{\partial} \bar{\phi}(\bar{z}) \bar{\partial} \bar{\phi}(\bar{z}): , \qquad (2.116)$$

where the normal ordering of the operator $\partial \phi$ is defined by

$$:\partial \phi(z) \partial \phi(z): \equiv \lim_{z \to w} \left[\partial \phi(z) \partial \phi(w) - \langle 0 | \partial \phi(z) \partial \phi(w) | 0 \rangle \right]. \qquad (2.117)$$

Here, we take the similar definition in the normal ordering for the anti-holomorphic part $\bar{\partial} \bar{\phi}$. Actually, from eq.(2.31) and eq.(2.124), we reproduce the Hamiltonian eq.(2.114) as

$$H = L_0 + \bar{L}_0 = \oint \frac{dz}{2\pi i} z T(z) + \oint \frac{d\bar{z}}{2\pi i} \bar{z} \bar{T}(\bar{z}), \qquad (2.118)$$

where the formula

$$\oint \frac{dz}{2\pi i} z^{-n-1} = \delta_{n0}$$

is used. Therefore, T and $\bar{T}$ should be the stress-energy tensor of the holomorphic part and the anti-holomorphic part in the 1+1 dimensional massless scalar field, respectively. From eq.(2.31) and eq.(2.124), the mode expansions of $T(z)$ and $\bar{T}(\bar{z})$ are given by

$$T(z) = \sum_{n \in \mathbf{Z}} z^{-n-2} L_n, \qquad L_n = \frac{1}{2} \sum_{m \in \mathbf{Z}} :\alpha_{m+n} \alpha_{-m}: , \qquad (2.119a)$$

$$\bar{T}(\bar{z}) = \sum_{n \in \mathbf{Z}} \bar{z}^{-n-2} \bar{L}_n, \qquad \bar{L}_n = \frac{1}{2} \sum_{m \in \mathbf{Z}} :\bar{\alpha}_{m+n} \bar{\alpha}_{-m}: . \qquad (2.119b)$$

In particular, we have

$$L_0 = \frac{1}{2} \alpha_0^2 + \sum_{n=1}^{\infty} \alpha_{-n} \alpha_n, \qquad \bar{L}_0 = \frac{1}{2} \bar{\alpha}_0^2 + \sum_{n=1}^{\infty} \bar{\alpha}_{-n} \bar{\alpha}_n. \qquad (2.120)$$

It is not so difficult to reproduce the Virasoro algebra eq.(2.32) in terms of eq.(2.108) and eq.(2.119). We can, further, show that

$$L_{0,\pm 1} |0\rangle = 0, \qquad \bar{L}_{0,\pm 1} |0\rangle = 0. \qquad (2.121)$$

Therefore, the vacuum of the U(1) Kac-Moody algebra is also SL(2, Z) invariant. Next, we consider the OPE between $T(z)$ and its own,

$$T(z)T(w) = \frac{(\pi g)^2}{4} \overline{\partial \phi(z) \partial \phi(w)} \; \overline{\partial \phi(z) \partial \phi(w)} 2$$

$$+ \frac{(\pi g)^2}{4} \overline{\partial \phi(z) \partial \phi(w)} :\partial \phi(z) \partial \phi(w): \times 4 + \text{reg.}$$

$$= \frac{1/2}{(z-w)^4} + \frac{\pi g}{(z-w)^2} :\partial \phi(z) \partial \phi(w): + \text{reg.}$$

$$= \frac{1/2}{(z-w)^4} + \frac{2T(w)}{(z-w)^2} + \frac{\partial T(w)}{z-w} + \text{reg.}, \qquad (2.122)$$

where ⌐ denote the contraction and

$$\overline{\partial\phi(z)\partial\phi(w)} \equiv \langle 0|\partial\phi(z)\partial\phi(w)|0\rangle = -\frac{1}{\pi g (z-w)^2} \quad (2.123)$$

The OPE eq.(2.122) is nothing but eq.(2.29) of the $c=1$ CFT models. We have similar OPE in the anti-holomorphic part of the stress energy tensor as well. Finally, the massless free scalar model is one of the $c=1$ CFT models.

2.10.3. Primary Operator (I)

We consider the derivative of ϕ and $\bar{\phi}$,

$$J(z) \equiv i\,\partial\phi(z) = \frac{1}{\sqrt{\pi g}} \sum_{n\in\mathbb{Z}} z^{-n-1}\alpha_n, \qquad \bar{J}(\bar{z}) \equiv i\,\bar{\partial}\bar{\phi}(\bar{z}) = \frac{1}{\sqrt{\pi g}} \sum_{n\in\mathbb{Z}} \bar{z}^{-n-1}\bar{\alpha}_n. \quad (2.124)$$

We can easily evaluate the correlation functions of the operator $J, \bar{J}$ and find that

$$\langle J(z)J(w)\rangle = \frac{1}{\pi g (z-w)^2}, \qquad \langle \bar{J}(\bar{z})\bar{J}(\bar{w})\rangle = \frac{1}{\pi g (\bar{z}-\bar{w})^2}. \quad (2.125)$$

Further, we have

$$\langle J(z)\phi(w)\rangle = -\frac{i}{\pi g}\frac{1}{z-w}, \qquad \langle \phi(z)\phi(w)\rangle = -\frac{1}{\pi g}\ln(z-w) \quad (2.126)$$

and so forth. Eq.(2.125) suggests that J and $\bar{J}$ are the primary operators. To show it, we evaluate the OPE between the operators $J, \bar{J}$ and the stress energy tensors $T, \bar{T}$. In particular, the OPE between $T(z)$ and $J(w)$ is written by

$$T(z)J(w) = \frac{\pi g}{2} :J(z)\overline{J(z)J(w)}: \times 2 + \text{reg.} = \frac{J(w)}{(z-w)^2} + \frac{\partial J(w)}{z-w} + \text{reg.} \quad (2.127)$$

where $\overline{J(z)J(w)} \equiv \langle 0|J(z)J(w)|0\rangle$. eq.(2.41a) and eq.(2.127) indicate that $J(z)$ is a primary operator of weight (1,0). Similarly, we find that $\bar{J}(\bar{z})$ is a primary operator of weight (0,1).

2.10.4. Primary Operator (II): Vertex Operator

There are another primary operators in the $c=1$ CFT models. They are called vertex operator. The vertex operator is given by

$$V_{q,\bar{q}}(z,\bar{z}) = :e^{i\sqrt{\pi g}\,[q\phi(z)+\bar{q}\bar{\phi}(\bar{z})]}:, \quad (2.128)$$

where $\phi(z), \bar{\phi}(\bar{z})$ are given by eq.(2.110) and $q, \bar{q}$ are parameters. Here, the normal ordering has the following meaning

$$V_{q,\bar{q}}(z,\bar{z}) = e^{i(qQ+\bar{q}\bar{Q})}e^{q\alpha_0\ln z + \bar{q}\bar{\alpha}_0\ln\bar{z}}e^{q\varphi_\uparrow(z)}e^{q\varphi_\downarrow(z)}e^{\bar{q}\bar{\varphi}_\uparrow(\bar{z})}e^{\bar{q}\bar{\varphi}_\downarrow(\bar{z})}, \quad (2.129)$$

where

$$\varphi_\uparrow(z) \equiv \sum_{n>0} \frac{z^n}{n}\alpha_{-n}, \quad \varphi_\downarrow(z) \equiv \sum_{n>0} \frac{z^{-n}}{n}\alpha_n. \quad (2.130)$$

The anti-holomorphic parts $\bar{\varphi}_\uparrow(\bar{z})$ and $\bar{\varphi}_\downarrow(\bar{z})$ are also defined in similar fashion. The OPE of the vertex operators is given by

$$V_{q,\bar{q}}(z,\bar{z})V_{q',\bar{q}'}(w,\bar{w}) = (z-w)^{qq'}(\bar{z}-\bar{w})^{\bar{q}\bar{q}'} :e^{i\sqrt{\pi g}\,[q\phi(z)+\bar{q}\bar{\phi}(z)+q'\phi(w)+\bar{q}'\bar{\phi}(\bar{w})]}: . \quad (2.131)$$

In addition, the OPE between the $J(z)$ and the vertex operator $V_q(w,\bar{w})$ can be evaluated in terms of eq.(2.124) and (2.129). Remember that the formula of any operators X, Y,

$$e^X Y e^{-X} = Y + [X,Y] + \frac{1}{2!}\left[X,[X,Y]\right] + \cdots . \quad (2.132)$$

It can be shown that

$$J(z)V_{q,\bar{q}}(w,\bar{w}) = \frac{q}{\sqrt{\pi g}}\frac{1}{z-w}V_{q,\bar{q}}(w,\bar{w}) + \text{reg}. \quad (2.133)$$

Thus, the OPE between the stress energy tensor $T(z)$ and the vertex operator $V_{q,\bar{q}}(w,\bar{w})$ is given by

$$T(z)V_{q,\bar{q}}(w,\bar{w}) = \frac{1}{2}\sum_{m,n\in\mathbf{Z}}z^{-m-n-2}:\alpha_m\alpha_n:V_{q,\bar{q}}(w,\bar{w})$$

$$= \frac{q^2}{2}\frac{1}{(z-w)^2}V_{q,\bar{q}}(w,\bar{w}) + \frac{\partial V_{q,\bar{q}}(w,\bar{w})}{z-w} + \text{reg}. \quad (2.134)$$

Similarly, we have the OPE in the anti-holomorphic part as

$$\bar{J}(\bar{z})V_{q,\bar{q}}(w,\bar{w}) = \frac{\bar{q}}{\sqrt{\pi g}}\frac{1}{\bar{z}-\bar{w}}V_{q,\bar{q}}(w,\bar{w}) + \text{reg}. \quad (2.135a)$$

$$\bar{T}(\bar{z})V_{q,\bar{q}}(w,\bar{w}) = \frac{\bar{q}^2}{2}\frac{1}{(\bar{z}-\bar{w})^2}V_{q,\bar{q}}(w,\bar{w}) + \frac{\bar{\partial} V_{q,\bar{q}}(w,\bar{w})}{\bar{z}-\bar{w}} + \text{reg}. \quad (2.135b)$$

Therefore, $V_{q,\bar{q}}(w,\bar{w})$ is a primary operator of weight

$$\Delta_{q\bar{q}} = \left(\frac{q^2}{2},\frac{\bar{q}^2}{2}\right). \quad (2.136)$$

2.10.5. Highest Weight State

The highest weight state is assigned in tems of the primary operator with eq.(2.61). First, we consider the highest weight states with respect to the primary operators $J, \bar{J}$. The corresponding states are given by

$$|\phi_J\rangle = \lim_{z\to 0} J(z)|0\rangle = \frac{1}{\sqrt{\pi g}}\alpha_{-1}|0\rangle, \quad (2.137a)$$

$$|\bar{\phi}_J\rangle = \lim_{\bar{z}\to 0} \bar{J}(\bar{z})|0\rangle = \frac{1}{\sqrt{\pi g}}\bar{\alpha}_{-1}|0\rangle. \quad (2.137b)$$

We can show that

$$L_0|\phi_J\rangle = |\phi_J\rangle, \qquad L_n|\phi_J\rangle = 0, \quad (2.138a)$$

$$\bar{L}_0|\phi_J\rangle = |\phi_J\rangle, \qquad \bar{L}_n|\phi_J\rangle = 0, \quad (2.138b)$$

where $n > 0$. Therefore, $|\phi_J, \bar{\phi}_{\bar{J}}\rangle \equiv |\phi_J\rangle \otimes |\bar{\phi}_{\bar{J}}\rangle$ is the highest weight state of weight $(1,1)$. We can also show that

$$H|\phi_J, \bar{\phi}_{\bar{J}}\rangle = 2|\phi_J, \bar{\phi}_{\bar{J}}\rangle. \tag{2.139}$$

However, $|\phi_J, \bar{\phi}_{\bar{J}}\rangle$ is highest weight state of the $c = 1$ Virasoro algebra whereas it is not that of the U(1) Kac-Moody algebra since

$$\alpha_0|\phi_J\rangle = 0, \quad \alpha_1|\phi_J\rangle \neq 0, \quad \alpha_n|\phi_J\rangle = 0, \tag{2.140}$$

where $n > 1$. Similar relations hold in the anti-holomorphic part as well.

Next, we consider the highest weight state with respect to the vertex operator $V_{q,\bar{q}}(z,\bar{z})$. The highest weight state of the vertex operator is given by

$$|q, \bar{q}\rangle = \lim_{z,\bar{z} \to 0} V_{q,\bar{q}}(z,\bar{z})|0\rangle = e^{i\sqrt{\pi g}(qQ+\bar{q}\bar{Q})}|0\rangle. \tag{2.141}$$

In this case, we find that

$$L_0|q,\bar{q}\rangle = \frac{q^2}{2}|q,\bar{q}\rangle, \quad L_n|q,\bar{q}\rangle = 0, \tag{2.142a}$$

$$\bar{L}_0|q,\bar{q}\rangle = \frac{\bar{q}^2}{2}|q,\bar{q}\rangle, \quad L_n|q,\bar{q}\rangle = 0, \tag{2.142b}$$

where $n > 0$. Note that the state $|q,\bar{q}\rangle$ satisfies the eigenvalue equation,

$$H|q,\bar{q}\rangle = \left(\frac{q^2}{2} + \frac{\bar{q}^2}{2}\right)|q,\bar{q}\rangle. \tag{2.143}$$

Moreover, in contrast to the state $|\phi_J\rangle$, the highest weight state of the vertex operator $|q,\bar{q}\rangle$ satisfies that

$$\alpha_0|q,\bar{q}\rangle = q|q,\bar{q}\rangle, \quad \alpha_n|q,\bar{q}\rangle = 0, \quad Q|q,\bar{q}\rangle = \frac{q}{\sqrt{\pi g}}|q,\bar{q}\rangle \tag{2.144a}$$

and

$$\bar{\alpha}_0|q,\bar{q}\rangle = \bar{q}|q,\bar{q}\rangle, \quad \bar{\alpha}_n|q,\bar{q}\rangle = 0, \quad \bar{Q}|q,\bar{q}\rangle = \frac{\bar{q}}{\sqrt{\pi g}}|q,\bar{q}\rangle. \tag{2.144b}$$

where $n > 0$. Correspondingly, the state of the vertex operator $|q, \bar{q}\rangle$ is not only the highest weight state of the $c = 1$ Virasoro algebra but also the highest weight state of the $U(1)$ Kac-Moody algebra of the charge $(q/\sqrt{\pi g}, \bar{q}/\sqrt{\pi g})$.

2.10.6. Compactification on a Circle

We restrict the field Φ on a circle of the radius R. In this case, the boundary condition can be written by

$$\Phi(t, \sigma + 2\pi) = \Phi(t, \sigma) + 2\pi R N, \tag{2.145}$$

where N is an arbitrary integer and N is known as a winding number. In this case, the mode expansion of Φ can be written by

$$\Phi(t, \sigma) = \mathcal{X} + \frac{\mathcal{P}t}{2\pi g} + RN\sigma + \frac{i}{\sqrt{4\pi g}} \sum_{n \neq 0} \frac{1}{n} \left[\alpha_n e^{-in(t+\sigma)} + \bar{\alpha}_n e^{-in(t-\sigma)}\right]. \tag{2.146}$$

Therefore, the zero modes become

$$\alpha_0 \equiv \frac{\mathcal{P}}{2\sqrt{\pi g}} + \sqrt{\pi g}RN, \qquad \bar{\alpha}_0 \equiv \frac{\mathcal{P}}{2\sqrt{\pi g}} - \sqrt{\pi g}RN. \tag{2.147}$$

According to the modular invariance property of the $c = 1$ partition function [26], the highest weight states of the vertex operator $|q, \bar{q}\rangle$ can be parameterized in terms of the integer M, N. In this respect, the eigenstate of α_0 and $\bar{\alpha}_0$ can be written by

$$\alpha_0|M,N\rangle = q_{M,N}|M,N\rangle, \qquad \bar{\alpha}_0|M,N\rangle = \bar{q}_{M,N}|M,N\rangle, \tag{2.148}$$

where

$$q_{M,N} = \frac{M}{2R'} + R'N, \qquad \bar{q}_{M,N} = \frac{M}{2R'} - R'N. \tag{2.149}$$

Here, we define $R' = \sqrt{\pi g}R$. Obviously, eq.(2.142a) and eq.(2.142b) become

$$L_0|M,N\rangle = h_{M,N}|M,N\rangle, \qquad \bar{L}_0|M,N\rangle = h_{M,-N}|M,N\rangle, \tag{2.150}$$

where

$$h_{M,N} = \frac{1}{2}\left(\frac{M}{2R'} + R'N\right)^2. \tag{2.151}$$

Further, we can show that

$$L_n|M,N\rangle = 0, \qquad \bar{L}_n|M,N\rangle = 0, \tag{2.152}$$

where $n > 0$. Accordingly, the $|M, N\rangle$ is the highest weight state both of the $U(1)$ Kac-Moody algebra, and the $c = 1$ Virasoro algebra and its weight is given by $(h_{M,N}, h_{M,-N})$. Note that the conformal weight eq.(2.151) depends on the continuous parameter R.

We also have the *spin* and scaling dimension of the highest weight state with respect to the vertex operator $V_{M,N}(z, \bar{z})$ as

$$\Delta = h_{M,N} + h_{M,-N} = \left(\frac{M}{2R'}\right)^2 + (R'N)^2, \qquad s = h_{M,N} - h_{M,-N} = MN. \tag{2.153}$$

See eq.(2.64a) and (2.64b).

3. String Theory and AdS/CFT Correspondence

3.1. Bosonic String

The string $X^\mu(\tau, \sigma)$ ($\mu = 0, 1, \cdots D-1$) is a one-dimensional object which make a movement in the two-dimensional surface. The two dimensional surface is known as a world sheet. The string propagates in D-dimensional Minkowski space together with the string sweeps out the world sheet. The world sheet is described in terms of two parameters (τ, σ). τ is a time-like evolution parameter and σ is a space-like position parameter. It is important to note that τ and σ are both dimensionless parameters and have no any physical meanings. Nevertheless, we should consider the dynamics of the string in the world sheet as if τ is time in formal manner.

There are two types of string, one is a open string and another is a closed string. For the open string, the two ends of the string are conventionally fixed by $\sigma = 0, 2\pi$. For the closed string, there is no end or the two ends of the open string is tied as $X^\mu(\tau, 0) = X^\mu(\tau, 2\pi)$. Thus, the spatial direction of the string is described by σ with $0 \leq \sigma \leq 2\pi$.

The string should be a 1+1 dimensional scalar field as the dynamical object in the world sheet. On the other hand, the string should also behave as a vector under D dimensional Poincaré transformation. Here, we only consider a bosonic string. The bosonic string $X^\mu(\tau, \sigma)$ ($\mu = 0, 1, \cdots D - 1$) can be described by the Nambu-Goto action

$$S_{\text{NG}} = -\frac{1}{4\pi\alpha'} \int d\tau d\sigma \sqrt{-\eta_{\mu\nu} \det\left(\frac{\partial X^\mu}{\partial \xi_\alpha} \frac{\partial X^\nu}{\partial \xi_\beta}\right)}, \qquad (\alpha, \beta = 0, 1), \qquad (3.1)$$

where $(\xi_0, \xi_1) = (\tau, \sigma)$ are two parameters of the two dimensional world-sheet and α' is given by

$$\alpha' = \ell_s^2, \qquad (3.2)$$

where ℓ_s is called string length or fundamental length. The determinant is taken in 2×2 matrix whose element is labeled by α, β. Note that a factor $1/4\pi\alpha'$ is known of the string tension. The Nambu-Goto action should mean the area of a sheet embedded in D dimensional Minkowski space.

The Nambu-Goto action is classically equivalent to a Polyakov action

$$S_{\text{P}} = -\frac{1}{4\pi\alpha'} \int d^2\xi \sqrt{-g} g^{\alpha\beta} \partial_\alpha X^\mu \partial_\beta X_\mu, \qquad (3.3)$$

where $g^{\alpha\beta}$ is two-dimensional metric of the string world sheet and $g = \det(g_{\alpha\beta})$. The Euler-Lagrange equation with respect to $g^{\alpha\beta}$ shows that

$$T_{\alpha\beta} = \partial_\alpha X_\mu \partial_\beta X^\mu - \frac{1}{2} g_{\alpha\beta} g^{\alpha'\beta'} \partial_{\alpha'} X^\mu \partial_{\beta'} X_\mu = 0, \qquad (3.4)$$

where the relation $\delta g = -g_{\alpha\beta}(\delta g^{\alpha\beta})g$ is used. Note that $T_{\alpha\beta}$ is traceless, $g^{\alpha\beta} T_{\alpha\beta} = 0$. The relation $T_{\alpha\beta} = 0$ indicates that

$$\det\left(\partial_\alpha X_\mu \partial_\beta X^\mu\right) = \frac{g}{4} \left(g^{\alpha'\beta'} \partial_{\alpha'} X^\mu \partial_{\beta'} X_\mu\right)^2.$$

Substituting the relation into S_p, we recover the Nambu-Goto action. On the other hand, varying the Polyakov action with respect to X^μ, we have

$$\frac{1}{\sqrt{-g}} \partial_\alpha \left(\sqrt{-g} g^{\alpha\beta} \partial_\beta X^\mu\right) = 0. \qquad (3.5)$$

This equation gives suggestion that the scalar field X^μ is coupled to the two-dimensional gravity.

Next, we consider symmetries of the Polyakov action. The Polyakov action is invariant under the reparametrization

$$\tau' = f(\tau, \sigma), \qquad \sigma' = g(\tau, \sigma) \qquad (3.6\text{a})$$

and the Weyl rescaling (conformal transformation)

$$g'_{\alpha\beta} = e^{2h(\tau,\sigma)} g_{\alpha\beta}, \tag{3.6b}$$

where f, g, h are arbitrary functions on the worldsheet. The transformations of reparametrization are equivalent to

$$\delta X^\mu = \epsilon^\alpha \partial_\alpha X^\mu, \tag{3.7a}$$

$$\delta g_{\alpha\beta} = \epsilon^{\alpha'} \partial_{\alpha'} g_{\alpha\beta} + \partial_\alpha \epsilon^{\alpha'} g_{\alpha'\beta} + \partial_\beta \epsilon^{\beta'} g_{\alpha\beta'}, \tag{3.7b}$$

$$\delta \sqrt{-g} = \partial_{\alpha'} \left(\epsilon^{\alpha'} \sqrt{-g} \right) \tag{3.7c}$$

and the transformation of the Weyl rescaling is to be

$$\delta g_{\alpha\beta} = 2\Lambda g_{\alpha\beta}, \tag{3.7d}$$

$$\delta X^\mu = 0. \tag{3.7e}$$

Here, ϵ^α and Λ are arbitrary infinitesimal functions of the world sheet. In addition, the action is also invariant under the Poincaré transformation,

$$\delta X^\mu = a^\mu{}_\nu X^\nu + b^\mu \tag{3.8}$$

with preserving $g_{\alpha\beta}$ unchanged. Here, $a_{\mu\nu}$ is anti-symmetric tensor in the D dimensional Minkowski space and $a^\mu{}_\nu, b^\mu$ are infinitesimal constant parameters.

To proceed to quantize the string, we fix the freedom of the metric $g_{\alpha\beta}$, called *gauge fixing*. Thanks to the reparametrization invariance as well as the conformal invariance in the action, we can take the conformal gauge

$$g_{\alpha\beta} = \eta_{\alpha\beta}, \tag{3.9}$$

where $\eta_{\alpha\beta}$ is a metric of the two-dimensional Minkowski space and $\eta_{\alpha\beta} = \mathrm{diag}(-1, 1)$. In this case, the Polyakov action becomes

$$S_\mathrm{P} = -\frac{1}{4\pi\alpha'} \int d^2\xi\, \eta^{\alpha\beta} \partial_\alpha X^\mu \partial_\beta X_\mu = \frac{1}{4\pi\alpha'} \int d^2\xi \left[\dot{X}^\mu \dot{X}_\mu - X^{\mu\prime} X'_\mu \right], \tag{3.10}$$

where

$$\dot{X}_\mu \equiv \frac{\partial X_\mu}{\partial \tau}, \qquad X'_\mu \equiv \frac{\partial X_\mu}{\partial \sigma}.$$

From the Euler-Lagrange equation with respect to the scalar field X^μ,

$$\frac{\partial}{\partial \tau}\left(\frac{\partial L}{\partial \dot{X}^\mu}\right) + \frac{\partial}{\partial \sigma}\left(\frac{\partial L}{\partial X^{\mu\prime}}\right) = \frac{\partial L}{\partial X^\mu},$$

the equation of motion of the free massless scalar field becomes

$$\left(\frac{\partial^2}{\partial \tau^2} - \frac{\partial^2}{\partial \sigma^2} \right) X_\mu = 0, \tag{3.11}$$

where L is defined by $S = \int d^2\xi \, L$. In standard formulation, the Euler-Lagrange equation insures the minimum of action $S(\phi, \partial_\mu \phi)$ since the surface terms of the variation of the action δS becomes zero due to zero of ϕ at the boundary or periodicity of ϕ. For the case of the closed string, the string satisfies

$$X_\mu(\tau, \sigma + 2\pi) = X_\mu(\tau, \sigma). \tag{3.12}$$

Therefore, there is no surface term. On the other hand, in the case of open string, the surface term

$$\delta S_{\text{surface}} = -\frac{1}{2\pi\alpha'} \int d\tau \left[X'_\mu \delta X^\mu \right]_{\sigma=0}^{\sigma=2\pi} \tag{3.13}$$

has nontrivial value. To vanish the surface term, we impose a Neumann boundary condition on the open string X^μ as

$$\left. \frac{\partial X_\mu(\tau, \sigma)}{\partial \sigma} \right|_{\sigma=0, 2\pi} = 0 \tag{3.14}$$

or a Dirichlet boundary condition on the open string X^μ as

$$\left. \delta X_\mu \right|_{\sigma=0, 2\pi} = 0. \tag{3.15a}$$

Note that the condition eq.(3.15a) is equivalent to

$$\left. \frac{\partial X_\mu(\tau, \sigma)}{\partial \tau} \right|_{\sigma=0, 2\pi} = 0. \tag{3.15b}$$

Strictly speaking, two-dimensional massless scalar field is not well defined in quantum level because of the infrared singularity of the field [25]. This fact makes us concern about a definition of the string. As it is already known in the conformal field theory, the boson field Φ itself cannot be written by a well-defined function while the derivative of field $\partial \Phi$ can be expressed in terms of a well-defined function. However, it is important to note that there is no proof that $\partial \Phi$ describes one physical degree of freedom something like particle field. In this respect, the two-dimensional massless scalar field may be introduced as a provisional field into the model. Here, we does not go any further into this point and we would think that X_μ should be an auxiliary field.

To quantize the string, we take the equal *time*(τ) commutation relations of the D scalar fields $X^\mu(\tau, \sigma)$ in formal manner as

$$\left[X^\mu(\tau, \sigma), X^\nu(\tau, \sigma') \right] = 0, \quad \left[\Pi^\mu(\tau, \sigma), \Pi^\nu(\tau, \sigma') \right] = 0, \tag{3.16a}$$
$$\left[X^\mu(\tau, \sigma), \Pi^\nu(\tau, \sigma') \right] = i\eta^{\mu\nu} \delta(\sigma - \sigma'), \tag{3.16b}$$

where the canonical conjugate field of X_μ is defined by

$$\Pi^\mu = \frac{\partial L}{\partial \dot{X}_\mu} = \frac{1}{2\pi\alpha'} \dot{X}^\mu.$$

Further, the Hamiltonian of the bosonic string is defined by

$$H = \int_0^{2\pi} d\sigma \left[\Pi^\mu \dot{X}_\mu - L \right] = \frac{1}{4\pi\alpha'} \int_0^{2\pi} d\sigma \left[\dot{X}^\mu \dot{X}_\mu + X^{\mu'} X'_\mu \right]. \tag{3.17}$$

3.2. Open and Closed String

3.2.1. Mode Expansion

Classically, the solution of the wave equation eq.(3.11) is generally given by the d'Alembert solution,

$$X(\tau, \sigma) = f(\tau - \sigma) + g(\tau + \sigma), \qquad (3.18)$$

where f, g are arbitrary functions of τ, σ. Note that f, g are determined by the initial and boundary condition. For the case of the periodic boundary condition eq.(3.12), there should be no reflected wave and the right and left moving wave are propagating independently. On the other hand, we must consider the reflected wave in the case of the open boundary. For the Neumann boundary, the end of the string move freely, and then the end of the string becomes loop. In this case, we can find that

$$g(\tau + \sigma) = f(\tau + \sigma). \qquad (3.19)$$

On the other hand, for the Dirichlet boundary, the end of string is fixed. In this case, the end of the string becomes node and we can find that

$$g(\tau + \sigma) = -f(\tau + \sigma). \qquad (3.20)$$

Therefore, the Fourier decomposition of the closed string should be written by

$$X^\mu(\tau, \sigma) = \sum_n \left(a_n^\mu e^{-in(\tau - \sigma)} + b_n^\mu e^{-in(\tau + \sigma)} \right). \qquad (3.21a)$$

At the same time, the Fourier decomposition of the open string with the Neumann boundary should be written by

$$X^\mu(\tau, \sigma) = \sum_n \left(a_n^\mu e^{-in(\tau - \sigma)} + a_n^\mu e^{-in(\tau + \sigma)} \right) = \sum_n a_n^\mu e^{-in\tau} \cos(n\sigma). \qquad (3.21b)$$

Meanwhile, the Fourier decomposition of the open string with the Dirichlet boundary should be written by

$$X^\mu(\tau, \sigma) = \sum_n \left(a_n^\mu e^{-in(\tau - \sigma)} - a_n^\mu e^{-in(\tau + \sigma)} \right) = \sum_n a_n^\mu e^{-in\tau} \sin(n\sigma). \qquad (3.21c)$$

3.2.2. Closed String

The closed string can be written in terms of the creation operators $\alpha^\mu_{-m}, \beta^\mu_{-m}$ $(m > 0)$ and annihilation operators $\alpha^\mu_m, \beta^\mu_m$ $(m > 0)$ as

$$X^\mu(\tau, \sigma) = x^\mu + \alpha' p^\mu \tau + i\sqrt{\frac{\alpha'}{2}} \sum_{m \neq 0} \frac{1}{m} \left[\alpha_m^\mu e^{-im(\tau + \sigma)} + \beta_m^\mu e^{-im(\tau - \sigma)} \right], \qquad (3.22)$$

where

$$[x^\mu, p^\nu] = i\eta^{\mu\nu} \qquad (3.23a)$$

and

$$[\alpha_m^\mu, \alpha_n^\nu] = m\delta_{m+n,0}\eta^{\mu\nu}, \qquad [\beta_m^\mu, \beta_n^\nu] = m\delta_{m+n,0}\eta^{\mu\nu}, \qquad [\alpha_m^\mu, \beta_n^\nu] = 0. \qquad (3.23b)$$

As already discussed in the subsection 2.10.1., these commutation relations are equivalent to the commutation relations of the harmonic oscillator

$$[a_{m_+}^\mu, a_{n_+}^{\nu\dagger}] = \eta^{\mu\nu}\delta_{m_+, n_+}, \qquad [b_{m_+}^\mu, b_{n_+}^{\nu\dagger}] = \eta^{\mu\nu}\delta_{m_+, n_+},$$

where m_+, n_+ are positive integers. Therefore, we can construct the Fock space build up by applying the raising operator $\alpha_{-m_+}^\mu, \beta_{-m_+}^\mu$ to the vacuum $|0\rangle$. The vacuum satifies that

$$\alpha_m^\mu |0\rangle = 0, \qquad \beta_n^\nu |0\rangle = 0, \qquad (m > 0). \qquad (3.24)$$

Note, however, that the norm of the time-like component of the string becomes negative since $\eta^{00} = -1$.

It is convenient to decompose $X^\mu(\tau, \sigma)$ into a right mover $X_R(\tau - \sigma)$ and a left mover $X_L(\tau + \sigma)$ as

$$X^\mu(\tau, \sigma) = X_L^\mu(\tau + \sigma) + X_R^\mu(\tau - \sigma), \qquad (3.25a)$$

where

$$X_L^\mu(\tau + \sigma) = x_L^\mu + \frac{\alpha' p^\mu}{2}(\tau + \sigma) + i\sqrt{\frac{\alpha'}{2}} \sum_{m \neq 0} \frac{1}{m} \alpha_m^\mu e^{-im(\tau+\sigma)}, \qquad (3.25b)$$

$$X_R^\mu(\tau - \sigma) = x_R^\mu + \frac{\alpha' p^\mu}{2}(\tau - \sigma) + i\sqrt{\frac{\alpha'}{2}} \sum_{m \neq 0} \frac{1}{m} \beta_m^\mu e^{-im(\tau-\sigma)}, \qquad (3.25c)$$

where $x^\mu = x_L^\mu + x_R^\mu$. The Hamiltonian of the closed string is written by

$$H = \frac{\alpha'}{2} p^\mu p_\mu + \frac{1}{2} \sum_{m \neq 0} \left(: \alpha_m^\mu \alpha_{-m\,\mu} : + : \beta_m^\mu \beta_{-m\,\mu} : \right). \qquad (3.26)$$

As seen before, the Virasoro generators are constructed in terms of the mode operators α_m, β_m of D free massless scalar fields. Indeed, if we take the Virasoro generators as

$$L_n = \frac{1}{2} \sum_{m \in \mathbb{Z}} : \alpha_{m+n}^\mu \alpha_{-m\,\mu} :, \qquad \bar{L}_n = \frac{1}{2} \sum_{m \in \mathbb{Z}} : \beta_{m+n}^\mu \beta_{-m\,\mu} :, \qquad (3.27)$$

we find the Virasoro algebra

$$[L_m, L_n] = (m-n)L_{m+n} + \frac{D}{12}m(m^2-1)\delta_{m+n,0}, \qquad (3.28a)$$

$$[\bar{L}_m, \bar{L}_n] = (m-n)\bar{L}_{m+n} + \frac{D}{12}m(m^2-1)\delta_{m+n,0}. \qquad (3.28b)$$

Note that the Hamiltonian is rewritten in terms of the Virasoro generator $L_0, \bar{L}_0$ as

$$H = L_0 + \bar{L}_0. \qquad (3.29)$$

Here, we define

$$\alpha_0^\mu = \beta_0^\mu \equiv \sqrt{\frac{\alpha'}{2}} p^\mu \quad (3.30)$$

and

$$L_0 = \frac{1}{2}\alpha_0 \cdot \alpha_0 + \sum_{n=1}^{\infty} \alpha_{-n} \cdot \alpha_n, \quad \bar{L}_0 = \frac{1}{2}\beta_0 \cdot \beta_0 + \sum_{n=1}^{\infty} \beta_{-n} \cdot \beta_n, \quad (3.31)$$

where "$\cdot$" denotes the inner product of D-dimensional vector space. Eventually, the closed string is composed of the holomorphic part and the anti-holomorphic part of the Virasoro algebra.

3.2.3. Open String

The open string in the Neumann boundary condition can be written in terms of the creation operators $\tilde{\alpha}_{-m}^\mu$, ($m > 0$) and annihilation operators $\tilde{\alpha}_m^\mu$ ($m > 0$) as

$$X_N^\mu(\tau, \sigma) = x^\mu + 2\alpha' p^\mu \tau + i\sqrt{2\alpha'} \sum_{m \neq 0} \frac{1}{m} \tilde{\alpha}_m^\mu e^{-im\tau} \cos m\sigma, \quad (3.32)$$

where

$$[x^\mu, p^\nu] = i\eta^{\mu\nu}, \quad [\tilde{\alpha}_m^\mu, \tilde{\alpha}_n^\nu] = m\delta_{m+n,0}\eta^{\mu\nu}, \quad \delta(\sigma - \sigma') = \frac{1}{\pi} \sum_m \cos m\sigma \cos m\sigma'. \quad (3.33)$$

The Hamiltonian of the open string is written by

$$H = 2\alpha' p^\mu p_\mu + \sum_{m \neq 0} : \tilde{\alpha}_m^\mu \tilde{\alpha}_{-m\,\mu} : . \quad (3.34)$$

In the open string, the Virasoro generators are constructed in terms of the mode operators $\tilde{\alpha}_m$. In this case, the Virasoro generators are written by

$$L_n = \frac{1}{2} \sum_{m \in \mathbf{Z}} : \tilde{\alpha}_{m+n}^\mu \tilde{\alpha}_{-m\,\mu} :, \quad (3.35)$$

where n is an integer. Indeed, we also find the Virasoro algebra eq.(3.28a) from eq.(3.35). In particular, the Hamiltonian is rewritten in terms of the Virasoro generator L_0 as

$$H = 2L_0, \quad (3.36)$$

where

$$L_0 = \frac{1}{2}\tilde{\alpha}_0 \cdot \tilde{\alpha}_0 + \sum_{n=1}^{\infty} \tilde{\alpha}_{-n} \cdot \tilde{\alpha}_n. \quad (3.37)$$

Here, we define

$$\tilde{\alpha}_0^\mu \equiv \sqrt{2\alpha'} p^\mu. \quad (3.38)$$

In contrast to the closed string, there is only the holomorphic part of the Virasoro algebra.

Similarly, we can quantize a open string in the Dirichlet boundary condition. The mode expansion is written by

$$X_D^\mu(\tau,\sigma) = c_0^\mu + 2\alpha' d_0^\mu \sigma + \sqrt{2\alpha'} \sum_{m \neq 0} \frac{1}{m} \tilde{\beta}_m^\mu e^{-im\tau} \sin m\sigma, \qquad (3.39)$$

where c_0^μ and d_0^μ are constants. This string is fixed c_0^μ at $\sigma = 0$ and $c_0^\mu + 4\pi\alpha' d_0^\mu$ at $\sigma = 2\pi$. In this case, the commutation relations are given by

$$[\tilde{\beta}_m^\mu, \tilde{\beta}_n^\nu] = m\delta_{m+n,0}\eta^{\mu\nu}, \qquad \delta(\sigma - \sigma') = \frac{1}{\pi} \sum_{m \neq 0} \sin m\sigma \sin m\sigma'. \qquad (3.40)$$

The Hamiltonian is written by

$$H = 2\alpha' d_0^\mu d_{0\mu} + \sum_{m \neq 0} : \tilde{\beta}_m^\mu \tilde{\beta}_{-m\,\mu} :. \qquad (3.41)$$

In this case, the Hamiltonian becomes

$$H = 2L_0, \qquad (3.42)$$

where

$$L_0 = \frac{1}{2}\tilde{\beta}_0 \cdot \tilde{\beta}_0 + \sum_{n=1}^{\infty} \tilde{\beta}_{-n} \cdot \tilde{\beta}_n, \qquad \tilde{\beta}_0^\mu \equiv \sqrt{2\alpha'} d_0^\mu. \qquad (3.43)$$

Here, zero mode $\tilde{\beta}_0$ is nothing to do with the center-of-mass momentum of the string. Further, the string in the Dirichlet boundary breaks the translation invariance due to the existence of the constant c_0^μ. In this respect, the Dirichlet boundary configuration had been given up as a solution of the string until recent progress of the D-brane dynamics [12].

3.3. Spectrum of Bosonic String

Now, we consider the spectrum of the bosonic string. Previously, we have learned the Hilbert space of the CFT models is composed of the height weight states and its descendants states. In $c = 1$ CFT models, there are at least two species of the primary operators. One is a derivative of the scalar field and another is a vertex operator. In particular, for the open string, we consider the highest weight states of the holomorphic vertex operator. The highest weight state of weight $(q, 0)$ is written by

$$|q\rangle = V_q(0)|0\rangle = e^{iq_\mu x^\mu}|0\rangle, \qquad (3.44a)$$

where the corresponding vertex operator V_q are given by

$$V_q =: e^{2iq_\mu X_N^\mu} : \qquad (3.44b)$$

In this case, the highest weight state $|q\rangle$ obeys eq.(2.142a). Thus, we have

$$L_0|q\rangle = \frac{q^\mu q_\mu}{2}|q\rangle, \qquad L_n|q\rangle = 0, \tag{3.45}$$

where $n > 0$. Further, since the highest state satisfies eq.(2.144a), the state obeys the relations

$$\widetilde{\alpha}_0^\mu|q\rangle = q^\mu|q\rangle, \qquad \widetilde{\alpha}_n^\mu|q\rangle = 0, \qquad (n > 0). \tag{3.46}$$

On the other hand, for the closed string, the vertex operator is given by

$$V_{q\bar{q}} = V_q \otimes V_{\bar{q}} =: e^{2iq_\mu X_L^\mu + 2i\bar{q}_\mu X_R^\mu} :. \tag{3.47}$$

In this case, the highest weight state is written by

$$|q, \bar{q}\rangle = |q\rangle \otimes |\bar{q}\rangle = e^{iq_\mu x^\mu + i\bar{q}_\mu x^\mu}|0\rangle, \tag{3.48}$$

Similarly, the highest weight state of the closed string satisfies

$$L_0|q, \bar{q}\rangle = \frac{q \cdot q}{2}|q, \bar{q}\rangle, \quad \bar{L}_0|q, \bar{q}\rangle = \frac{\bar{q} \cdot \bar{q}}{2}|q, \bar{q}\rangle, \quad L_n|q, \bar{q}\rangle = 0, \quad \bar{L}_n|q, \bar{q}\rangle = 0 \tag{3.49}$$

as well as

$$\alpha_0^\mu|q, \bar{q}\rangle = q^\mu|q, \bar{q}\rangle, \qquad \bar{\beta}_m^\mu|q, \bar{q}\rangle = \bar{q}^\mu|q, \bar{q}\rangle, \qquad \alpha_n^\mu|q\rangle = 0, \qquad \beta_m^\mu|q, \bar{q}\rangle = 0, \tag{3.50}$$

where $n > 0$. Obviously, the anti-holomorphic part shows almost same results with holomorphic part. Thus, it is sufficient to consider the holomorphic part selectively.

Until now, we does not treat the string X^μ as a dynamical object in D-dimensional space while X^μ should be a dynamical object in the two-dimensional world sheet. Strictly speaking, X^μ is not dynamical object even in two-dimensional world sheet since τ is not the time coordinate but the evolution parameter in the world sheet. To consider the dynamics of the string in D-dimensional space-time, we must assign X^0 to the time coordinate. Furthermore, let the string move freely in the D-dimensional space. In this case, p_μ should be interpreted as the momentum of the center-of-mass of the string.

When a relativistic particle of mass m move freely in the space with the energy $E(\equiv k^0)$ and the momentum $\boldsymbol{k}$, it is satisfied the dispersion relation of

$$E^2 = \boldsymbol{k}^2 + m^2. \tag{3.51}$$

Accordingly, the mass of the particle is given by

$$m^2 = -k^\mu k_\mu, \tag{3.52}$$

where $k^\mu = (E, \boldsymbol{k})$.

Now, we consider the mass $\mathcal{M}$ of the bosonic string. In present case, the mass can be written as $\mathcal{M}^2 = -p^\mu p_\mu$. In the string theory, there is a no-ghost theorem. The no-ghost theorem indicates that the dimension of the string D becomes 26 and the conformal weight of the highest weight state is given by one. These conditions are also derived from the fact that the string obeys the Poincaré algebra. Ultimately, for the open string, we can set

$$L_0|\text{phys}\rangle = |\text{phys}\rangle, \tag{3.53}$$

where $|\text{phys}\rangle$ is a physical state. For the highest weight state, we have

$$\sum_\mu \frac{q^\mu q_\mu}{2} = 1, \qquad (\mu = 0, 1, 2, \cdots, D-1). \tag{3.54}$$

Finally, there are 24 degrees of freedom $(q_1, q_2, \cdots, q_{24})$ in the momentum space of the 26-dimensional bosonic string since there are two constraint, eq.(3.52) and eq.(3.54).

For the open string, we consider the $\widetilde{L_0}$ of eq.(3.37). Thus, L_0 can be written in terms of the mass $\widetilde{\mathcal{M}}$ of the open string as

$$L_0 = -\alpha'\widetilde{\mathcal{M}}^2 + \sum_{n=1}^{\infty} \widetilde{\alpha}_{-n} \cdot \widetilde{\alpha}_n. \tag{3.55}$$

Therefore, for the highest state, eq.(3.53) lead to

$$-\alpha'\widetilde{\mathcal{M}}^2 = 1. \tag{3.56}$$

Unfortunately, the mass becomes negative, which is known as a tachyon. For this reason, the bosonic string is unstable physically and we must abandon the bosonic string in physical point of view. Though, it gets better in the superstring. For the first excited state (level 1 descendant state) $|q;1\rangle = \alpha^i_{-1}|q\rangle$ ($i = 1, 2, \cdots 24$), eq.(3.53) leads to $-\alpha'\widetilde{\mathcal{M}}^2 + 1 = 1$. Thus,

$$-\alpha'\widetilde{\mathcal{M}}^2 = 0. \tag{3.57}$$

Accordingly, the first excited state shows the massless vector field in the 26 dimensions and invokes a photon. In general, for the level N descendant state, eq.(3.53) leads to

$$-\alpha'\widetilde{\mathcal{M}}^2 + N = 1. \tag{3.58}$$

For the closed string, we consider both L_0 and $\bar{L}_0$ which are given by eq.(3.31). eq.(3.31) can be rewritten in terms of the mass $\mathcal{M}$ of the closed string as

$$L_0 = -\frac{\alpha'}{4}\mathcal{M}^2 + \sum_{n=1}^{\infty} \alpha_{-n} \cdot \alpha_n, \qquad \bar{L}_0 = -\frac{\alpha'}{4}\mathcal{M}^2 + \sum_{n=1}^{\infty} \beta_{-n} \cdot \beta_n. \tag{3.59}$$

The no-ghost theorem indicates that

$$L_0|\text{phys}\rangle = |\text{phys}\rangle, \qquad \bar{L}_0|\text{phys}\rangle = |\text{phys}\rangle. \tag{3.60}$$

Similar to the open string, there are also 24 degrees of freedom. In this case, for the highest weight state, eq.(3.60) leads to

$$-\frac{\alpha'}{4}\widetilde{\mathcal{M}}^2 = 1. \tag{3.61}$$

Again, we have tachyon state. For first excited state (level 1 descendant state) $|q, \bar{q}; 1\rangle = \alpha^i_{-1}\beta^j_{-1}|q, \bar{q}\rangle$ ($i, j = 1, 2, \cdots 24$), eq.(3.60) leads to $-\alpha'\mathcal{M}^2/4 + 1 = 1$. Thus,

$$-\alpha'\frac{\mathcal{M}^2}{4} = 0. \tag{3.62}$$

Similarly, the massless state also emerges. Note that the massless state is composed of a rank two symmetric tensor S, the rank two anti-symmetric tensor S and a scalar 1 as

$$24 \otimes 24 = S \oplus A \oplus 1. \tag{3.63}$$

They should be interpreted as the spin 2 graviton field $g_{\mu\nu}$, the spin 2 Neveu-Schwartz B field $B_{\mu\nu}$ and the spin 0 dilaton field ϕ, respectively. The extension to the higher excited states is straightforward. In general, for the level N descendant state, eq.(3.60) leads to

$$-\frac{\alpha'}{4}\mathcal{M}^2 + N = 1. \tag{3.64}$$

3.3.1. Strings in Curved Space-Time and Nonlinear σ Model

Here, we consider a case of which a string move in the D-dimensional curved space-time. In this case, the string can be described in terms of a nonlinear sigma model [27] whose action is given by

$$S_P = -\frac{1}{4\pi\alpha'} \int d^2\xi \sqrt{-g} g^{\alpha\beta} G_{\mu\nu}(X) \partial_\alpha X^\mu \partial_\beta X^\nu, \tag{3.65}$$

where $g^{\alpha\beta}$ is two-dimensional metric of the string world sheet and $g = \det(g_{\alpha\beta})$. Further, $G_{\mu\nu}(X)$ is a general metric of the curved space-time. In the closed string, there are two other massless states, one is represented by the anti-symmetric tensor field $B_{\mu\nu}$ and another is represented by the dilaton field Φ. Therefore, the action of which there are graviton $G_{\mu\nu}(X)$, anti-symmetric tensor field $B_{\mu\nu}(X)$ and dilaton $\Phi(X)$ as the background fields should be written by

$$S_P = -\frac{1}{4\pi\alpha'} \int d^2\xi \sqrt{-g} \left[\left(g^{\alpha\beta} G_{\mu\nu} + i\epsilon^{\alpha\beta} B_{\mu\nu} \right) \partial_\alpha X^\mu \partial_\beta X^\nu + \alpha' R^{(2)} \Phi \right], \tag{3.66}$$

where $\epsilon^{\alpha\beta}$ is a two rank anti-symmetric tensor with $\epsilon^{12} = 1$ and $R^{(2)}$ is a Ricci scalar in the world sheet. Due to the conformal invariance as the two-dimensional model, we require the vanishing of the β functions for G, B, Φ:

$$\beta^G_{\mu\nu}/\alpha' = \mathcal{R}_{\mu\nu} + 2\nabla_\mu \nabla_\nu \Phi - \frac{1}{4} H_{\alpha\beta\nu} H_\nu^{\alpha\beta} + O(\alpha') = 0, \tag{3.67a}$$

$$\beta^B_{\mu\nu}/\alpha' = \nabla^\lambda \left(e^{-2\Phi} H_{\mu\nu\lambda} \right) + O(\alpha') = 0, \tag{3.67b}$$

$$\beta^\Phi/\alpha' = -4\nabla^2\Phi + 4\nabla_\omega \Phi \nabla^\omega \Phi + \mathcal{R} - \frac{1}{12} H_{\alpha\beta\nu} H^{\alpha\beta\nu} + O(\alpha') = 0, \tag{3.67c}$$

where ∇_μ is a covariant derivative in the metric $G_{\mu\nu}$ and

$$H_{\lambda\mu\nu} = \partial_\lambda B_{\mu\nu} + \partial_\mu B_{\nu\lambda} + \partial_\nu B_{\lambda\mu}. \tag{3.68}$$

Note that eq.(3.67) should be derived out of the action in the string frame,

$$S = \frac{1}{2\kappa_0^2} \int d^{26}x \sqrt{-G} e^{-2\Phi} \left[\mathcal{R} - \frac{1}{12} H_{\mu\nu\lambda} H^{\mu\nu\lambda} + 4 \partial_\mu \Phi \partial^\mu \Phi \right] + O(\alpha'), \tag{3.69}$$

where κ_0 is a parameter. When we take the Einstein frame, the metric in the Einstein frame $\widetilde{G}_{\mu\nu}$ is written in terms of that of the string frame $G_{\mu\nu}$ as

$$\widetilde{G}_{\mu\nu} = e^{-(\Phi-\Phi_0)/6} G_{\mu\nu}. \tag{3.70}$$

Thus, the action in the Einstein frame becomes

$$S = \frac{1}{2\kappa^2} \int d^{26}x \sqrt{-\widetilde{G}} \left[\widetilde{\mathcal{R}} - \frac{1}{12} e^{-\Phi/3} H_{\mu\nu\lambda} H^{\mu\nu\lambda} - \frac{1}{6} \partial_\mu \Phi \partial^\mu \Phi \right] + O(\alpha'), \tag{3.71}$$

where $\kappa = \kappa_0 e^{\Phi_0}$ which is related to the gravitational coupling constant G_N as $\kappa = (8\pi G_N)^{1/2}$. Since eq.(3.67) as well as eq.(3.69) are valid at the low energy $\alpha' \ll 1$, the superstring should have the supergravity as a low energy effective theory. Note, further, that we could also find the effective action of the D-branes which play a important role in the AdS/CFT conjecture [32].

3.4. Toroidal Compactification

3.4.1. Compactification of Closed String

The consistency of the bosonic string theory demands 26 dimensions. Although it is reduced to 10 dimensions due to the supersymmetry, it is still redundant in comparison with real world. One simple method of the dimension reduction is given by toroidal compactification. For the closed string, it can be given by

$$X(\tau, \sigma + 2\pi) = X(\tau, \sigma) + 2\pi R N, \tag{3.72}$$

where N is the winding number. As we have already discussed the compactification of the $c = 1$ CFT in subsection 2.10.6., the center-of-mass momentum of the string is quantized as

$$P = P_L + P_R = \frac{M}{R}, \tag{3.73}$$

where M is an integer and

$$P_L = \sqrt{\frac{\alpha'}{2} \frac{M}{R}} + \frac{RN}{\sqrt{\alpha'}}, \qquad P_R = \sqrt{\frac{\alpha'}{2} \frac{M}{R}} - \frac{RN}{\sqrt{\alpha'}}. \tag{3.74}$$

It is easily verified that the CFT data are invariant under the duality

$$R \longrightarrow \frac{\alpha'}{R}, \qquad M \longleftrightarrow N \tag{3.75}$$

except P_R since it is transformed in the way of

$$P_R \longrightarrow -P_R \tag{3.76}$$

In the string theory, the symmetry of eq.(3.75) is called T-duality [29]. Moreover, it is apparently extended to an incomplete parity transformation of the string $X = X_L + X_R$ as

$$X_L \longrightarrow X_L, \qquad X_R \longrightarrow -X_R. \tag{3.77}$$

However, eq.(3.77) seems to be a little artificial symmetry.

There are several consistent models in the superstring theory. It is interesting that T-duality can map the type IIA string theory to the type IIB string theory, and vice versa. Furthermore, the T-duality relates the five superstring theories in the ten-dimensions to each other together with a S-duality which inverts the string coupling constant $g_s \leftrightarrow 1/g_s$. At the final stage, all of the string theories may be unified in one theory in terms of the dualities. The unified theory is called M theory. Although it is suggested the existence of the M theory in eleven dimensions, the detail have been veiled until now.

3.4.2. Compactification of Open String and D-brane

The open string with the Neumann boundary should not be compactified on the circle because the free moving at the end of the string is inconsistent with the compactification. Nevertheless, since we would like to consider the four dimensional physics, there should be compactification in the open string. Furthermore, the T-duality should be a fundamental duality in the superstring theory. To overcome the fear of the unification of the superstring, we consider the alternative open string configuration what the Dirichlet boundary condition is imposed. Its configuration had been disregard until the D-brane configuration was discovered. It should be partially because the string configuration of the Dirichlet boundary violates the translation invariance. Note, however, that it is not forbidden configuration in mathematical point of view. Recent progress shows that the D-brane configuration should be a fundamental object in the string theory [12, 30].

Here, it should be important to comment on the T-duality of the open string. There is an assertion of which the T-duality is given by eq.(3.77) even though X_R and X_L are not independent variables in the open string. Indeed, we have the D-brane configuration in terms of the formal replacement of eq.(3.77) from the open string of the Neumann boundary. Nevertheless, eq.(3.77) for the open string should be accidentally equivalent to the T-duality transformation. It should be too artificial to presume what eq.(3.77) is a true symmetry in the open string. In this respect, there may be no physical meaning in the symmetry of eq.(3.77) for the case of the open string.

The compactification of the open string should demand the condition

$$X_D(\tau, \sigma + 2\pi) = X_D(\tau, \sigma) + 2\pi RN, \tag{3.78}$$

where R is a compact radius. Thus, the mode expansion is given by

$$X_D(\tau, \sigma) = c_0 + 2\alpha' d_0 \sigma + 2\pi R + \sqrt{2\alpha'} \sum_{m \neq 0} \frac{1}{m} \tilde{\beta}_m e^{-im\tau} \sin m\sigma, \tag{3.79}$$

where c_0 is a constant and

$$d_0 = \frac{RN}{2\alpha'}. \tag{3.80}$$

That is, the open string can be compactified in terms of the fixed boundary. The object what the end of the open string is attached is called D-brane. In general, the configuration of the D-dimensional string $X^\mu(\tau, \sigma)$ which is represented by

$$X^\mu(\tau, \sigma) = \begin{cases} X_N^\mu(\tau, \sigma) & \text{for } (\mu = 0, 1, 2, \cdots, p), \\ X_D^\mu(\tau, \sigma) & \text{for } (\mu = p+1, p+2, \cdots, D-1) \end{cases} \tag{3.81}$$

requestes the Dp-brane configuration which is spreading in the p-dimensional space-time. Furthermore, the D-brane may be a solitonic object with RR charge [30].

3.5. AdS/CFT Correspondence

There is an interesting idea in the superstring theory, which is known as AdS/CFT correspondence [31, 32, 33]. Since the open string have vectorial massless spectrum, the open string should describe the gauge theory by means of Chan-Paton factor. However, it is not obvious that the gauge theory given by the string theory is indeed conventional gauge theory since the string scale is far from real world. Nevertheless, one may hope that the string theory is a dual theory of QCD by means of 't Hooft limit.

SU(N) Yang-Mills theory should be described in terms of the perturbation method with respect to $1/N$ while keeping $\lambda \equiv g_{YM}^2 N$. Therefore, the large N limit is corresponding to the weak coupling limit $g_{YM}^2 \ll 1$. The expansion term with respect to the diagram whose genus of the closed surface is given by g is proportional to

$$N^{2-2g} \lambda^n$$

where n is an integer. Thus, the non-planar diagrams are suppressed by powers of $1/N^2$ in the large N limit, and then the QCD model should simplify at the large N. Note that the large N expansion should be well established in small λ region in the QCD.

On the contrary, at the large λ limit, the diagram of $1/N$ expansion of the gauge theory seems to be the perturbative diagram of the string [34]. In this region, stringy picture may be adequate. In particular, the D-branes configuration should play a fundamental role in the duality of the gauge/gravity correpondence. The open string attaching the D p-brane could describe the (p+1)-dimensional supersymmetric Yang-Mills theory. Moreover, the open string attaching the N almost coincident Dp-branes may be dual to the U(N) supersymmetric Yang-Mills theory. If we consider the string at the low energy limit where the string scale becomes small ($\ell_s \ll 1$), the effective action of the string should be written by

$$S = S_{\text{bulk}} + S_{\text{brane}} + S_{\text{int}}, \quad (3.82)$$

where S_{bulk} is action of the ten-dimensional supergravity and S_{brane} describes the brane which contains the super Yang-Mills theory. On the other hand, S_{int} is the action which describes the interaction between the bulk gravity and branes. It is known that, except the gauge theory on the branes, any terms, e.g. the S_{int} and any higher derivative terms, vanish in the decoupling limit,

$$g_{\text{YM}} = \text{fix}, \quad \ell_s \to 0, \quad (3.83)$$

where g_{YM} is a Yang-Mills coupling. Note that when we consider the ten-dimensional type IIB superstring theory with the D3-branes, the Newton constant κ in the ten dimensions and g_{YM} in the four dimensions are given by

$$\kappa^2 \sim \ell_s^8 g_s^2, \quad g_{\text{YM}}^2 \sim g_s. \quad (3.84)$$

In addition, the supergravity theory in the bulk could become free. Therefore, there are two theories which are decoupled each other in the system, one is the supergravity in the bulk and another is the supersymmetric Yang-Mills theory at the "boundary" in the low energy limit. On the other hand, the tension T_p of the Dp-brane becomes infinity,

$$T_p \sim \frac{1}{\ell_s^{p+1}} \to \infty, \tag{3.85}$$

at the limit $\ell_s \to 0$. This means that the fluctuation of the brane is suppressed due to the large tension. Therefore, the Dp-brane at the decoupling limit becomes classical supergravity object.

In the supergravity, the Dp-brane should be interpreted as a solitonic object. The type II supergravity solution describing N coincident extremal Dp-branes is given by

$$ds^2 = f_p^{-1/2}(-dt^2 + dx_1^2 + \cdots + dx_p^2) + f_p^{1/2}(dx_{p+1}^2 + \cdots + dx_9^2), \tag{3.86a}$$

$$e^{-2(\phi-\phi_\infty)} = f_p^{(p-3)/2}, \qquad A_{0\ldots p} = -\frac{1}{2}(f_p^{-1} - 1), \qquad r^2 \equiv x^{p+1} + \cdots + x^9, \tag{3.86b}$$

where f_p is a harmonic function of the transverse coordinate $x_{p+1}, \cdots x_9$ which is given by

$$f_p = 1 + \frac{d_p g_{YM}^2 N}{\alpha'^2 U^{7-p}}, \qquad U \equiv \frac{r}{\alpha'} \tag{3.87}$$

where d_p and g_{YM} are given by

$$d_p = 2^{7-2p} \pi^{\frac{9-3p}{2}} \Gamma\left(\frac{7-p}{2}\right), \qquad g_{YM} = (2\pi)^{(p-2)/2} \alpha'^{(p-3)/4} \sqrt{g_s}. \tag{3.88}$$

Further, the string coupling constant g_s is given by $g_s = e^{\phi_\infty}$. It is known that $p = 0, 2, 4, 6$ exist in the type IIA theory and $p = -1, 1, 3, 5, 7$ exist in the type IIB theory. Now, we assume that the decoupling limit eq.(3.83) corresponds to the near horizon limit,

$$\alpha' = 0, \qquad U = \text{fixed}. \tag{3.89}$$

In this limit, the metric eq.(3.86a) becomes

$$ds^2/\alpha' = \frac{U^{(7-p)/2}}{\sqrt{d_p g_{YM}^2 N}}(-dt^2 + dx_1^2 + \cdots + dx_p^2) + \frac{\sqrt{d_p g_{YM}^2 N}}{U^{(7-p)/2}} dU^2 + \frac{\sqrt{d_p g_{YM}^2 N}}{U^{(3-p)/2}} d\Omega_{8-p}^2. \tag{3.90}$$

Therefore, the metric becomes $AdS_p \times S^{8-p}$. Consequently, we are led to the conjecture of the AdS/CFT duality which suggests that supersymmetric Yang-Mills theory should be described in terms of the superstring theory on the D-brane. Note, however, that the corresponding Yang-Mills theory is a model of the conformal field theory. For example, $\mathcal{N}=4$ supersymmetric Yang-Mills theory in (3+1)-dimensions could be described by the type IIB string theory on $AdS_5 \times S^5$ with the appropriate limit. Note that the isometry of $AdS_5 \times S^5$ is equivalent to the superconformal group in (3+1)-dimensions. Furthermore, the N coincident extremal Dp-branes system may describe the $(p+1)$-dimensional $U(N)$ supersymmetric Yang-Mills theory with sixteen supercharges [33].

3.5.1. AdS$_5$/CFT$_4$ Correspondence and Wilson Loop

One advantage of the AdS/CFT correspondence is that we could consider QCD in terms of the string technique. For example, we could estimate the Wilson loop of the QCD by virtue of the correspondence [35, 36]. The Wilson loop is defined by

$$\hat{W}(C) = \exp\left(ie \oint_C A^\mu dx_\mu\right), \tag{3.91}$$

where A^μ is a gauge field. According to the Wilson criterion, the Wilson loop $W(T, L)$ in the plaqutte whose lengths are L, T (T is a Euclidean time) may show the potential between two static quark by means of

$$-\ln \lim_{T\to\infty} W(T, R) = E(L) - E_0, \tag{3.92}$$

where $E(L)$ is the energy of the system. Here, we assume that a static quark and a static anti-quark are placed at the distance L. On the other hand, E_0 is vacuum energy of which the quark and the anti-quark are absent. Here, the static quarks mean that there is a assumption of which quarks are not to be dynamical objects. Thus, in this formulation, the quarks should be included as the external source. Note, however, that there are intrinsic problems and criticisms on the Wilson loop formulation [9], which will be discussed below.

There is a claim of which the expectation value of the Wilson loop W becomes

$$W \sim \exp(-c_0 LT) \tag{3.93a}$$

in the strong coupling limit whereas it becomes

$$W \sim \exp(-c_1(L+T)) \tag{3.93b}$$

in the opposite limit. Here, c_0, c_1 are constant parameters. It may be valid that the quark-anti-quark potential $V(L)$ in the large loop is related to the Wilson loop as

$$e^{-V(L)T} \sim W. \tag{3.94}$$

Hence, in the case of eq.(3.93a), the potential becomes

$$V(L) = c_0 L \tag{3.95a}$$

and then the linear potential emerge. In this case, we obtain the area law and the quarks should be confining. On the other hand, in the case of eq.(3.93b), the potential becomes

$$V(L) = c_1. \tag{3.95b}$$

In this case, we obtain the perimeter law and the quark may be deconfining.

Now, we consider the Wilson loop of the $\mathcal{N}=4$ supersymmetric Yang-Mills theory in terms of the dual picture [35, 36]. In the large gN, the expectation value of the Wilson loop should be

$$\left\langle W(C) \right\rangle \sim e^{-S}, \tag{3.96}$$

where S is the proper area of a fundamental string world sheet which is given by the Nambu-Goto action,

$$S = \frac{1}{2\pi\alpha'} \int d\tau d\sigma \sqrt{\det G_{MN} \partial_\alpha X^M \partial_\beta X^N}. \tag{3.97}$$

The supergravity solution carrying D3 brane charge is written by

$$ds^2 = f^{-1/2}\left(-dt^2 + dx_1^2 + dx_2^2 + dx_3^2\right) + f^{1/2}\left(dr^2 + r^2 d\Omega_5^2\right), \tag{3.98}$$

where

$$f = 1 + \frac{4\pi g N}{\alpha'^2}\left(\frac{\alpha'}{r}\right)^4 = 1 + \frac{4\pi g N}{\alpha'^2}\frac{1}{U^4}, \qquad (U \equiv r/\alpha'). \tag{3.99}$$

Since, in the near horizon limit eq.(3.89),

$$f \to \frac{4\pi g N}{\alpha'^2}\frac{1}{U^4},$$

the metric becomes

$$ds^2 = \alpha'\left[\frac{u^2}{R^2}\left(-dt^2 + dx_1^2 + dx_2^2 + dx_3^2\right) + R^2 \frac{du^2}{u^2} + R^2 d\Omega_5^2\right], \tag{3.100}$$

where $R = (4\pi g N)^{1/4}$ and we obtain the AdS$_5 \times S^5$. Note that R is a radius both of AdS$_5$ space and S^5 sphere. Further, R over string length ℓ_s is given by

$$R^4/\ell_s^4 \sim gN \sim g_{\rm YM}^2 N.$$

Therefore, the 't Hooft limit ($g_{\rm YM}^2 N \gg 1$) of the Yang-Mills theory corresponds to $R \gg \ell_s$ region. Since the supergravity description should be valid in the $R \gg \ell_s$ region, the strong coupling limit of the Yang-Mills theory should be described in terms of the low energy limit of the superstring theory.

Now, we would evaluate the Wilson loop. First, we take the static configuration,

$$X_0 = \tau, \qquad X_1 = \sigma, \qquad U = U(\sigma). \tag{3.101}$$

In this case, since the metric $h_{\mu\nu} \equiv G_{MN} \partial_\mu X^M \partial_\nu X^N$ becomes

$$h_{\tau\tau} = -\alpha' \frac{U^2}{R^2}(1 - v^2), \qquad h_{\sigma\sigma} = \frac{\alpha'}{R^2 U^2}(R^4 (\partial_\sigma U)^2 + U^4), \tag{3.102}$$

we can write

$$\det h_{\mu\nu} = -\alpha'\left[(\partial_\sigma U)^2 + (U/R)^4\right]. \tag{3.103}$$

Hence, the action becomes

$$S = \frac{1}{2\pi}\int_0^T d\tau \int d\sigma \sqrt{(\partial_\sigma U)^2 + \left(\frac{U}{R}\right)^4}. \tag{3.104}$$

Therefore, the Hamiltonian should be written by

$$\mathcal{H} = \partial_\sigma U \frac{\partial \mathcal{L}}{\partial(\partial_\sigma U)} - \mathcal{L} = -\frac{T}{4\pi}\frac{(U/R)^4}{\sqrt{(\partial_\sigma U)^2 + (U/R)^4}}. \tag{3.105}$$

Since the action do not depend on σ explicitly, we can take $\partial_\sigma \mathcal{H} = 0$. Thus, we have

$$\frac{\partial}{\partial \sigma}\left(\frac{(U/R)^4}{\sqrt{(\partial_\sigma U)^2 + (U/R)^4}} \right) = 0. \tag{3.106}$$

Here, we define U_0 to be the point where $\partial_\sigma U = 0$. In this case, we find

$$\partial_\sigma U = \frac{U^2}{R^2 U_0^2} \sqrt{U^4 - U_0^4}. \tag{3.107}$$

Note that U is symmetric under $\sigma \to -\sigma$. This means that the minimum of U becomes U_0 at $\sigma = 0$. Therefore, σ is determined by

$$\sigma = \int_0^\sigma d\sigma = \frac{R^2}{U_0^2} \int_{U_0}^U \frac{dU}{y^2 \sqrt{y^4 - 1}} = \frac{R^2}{U_0} \int_1^{U/U_0} \frac{dy}{y^2 \sqrt{y^4 - 1}}, \tag{3.108}$$

where y is defined in terms of the relation $U = U_0 y$. Further, U_0 is determined by

$$\frac{L}{2} = \int_0^{L/2(U=\infty)} d\sigma = \frac{R^2}{U_0} \frac{\sqrt{2}\pi^{3/2}}{\Gamma(1/4)^2}.$$

Thus, we find

$$U_0 = \frac{2\sqrt{2}\pi^{3/2}}{\Gamma(1/4)^2} \frac{R^2}{L}, \tag{3.109}$$

where L should be the quark-anti-quark distance. Finally, the action becomes

$$S = \frac{T}{2\pi} \int_{-L/2}^{L/2} d\sigma \sqrt{(\partial_\sigma U)^2 + \left(\frac{U}{R}\right)^4} = \frac{2T}{2\pi R^2} \int_{0(U \to U_0)}^{L/2(U \to \infty)} \frac{dU}{\partial_\sigma U} \frac{U^4}{U_0^2} = \frac{T}{\pi} \int_{U_0}^\infty dU \frac{U^2}{\sqrt{U^4 - U_0^4}}.$$

Now, we evaluate the energy

$$E = \frac{S}{T} = \frac{1}{\pi} \int_{U_0}^\infty dU \frac{U^2}{\sqrt{U^4 - U_0^4}} = \frac{U_0}{\pi} \int_1^\infty dy \frac{y^2}{\sqrt{y^4 - 1}}. \tag{3.110}$$

Note that the energy at $U_0 = 0$ becomes

$$E_0 \equiv E(U_0 \to 0) = \frac{1}{\pi} \int_0^\infty dU \to \infty, \tag{3.111}$$

which may be interpreted the energy of two separated quarks [32]. Thus, the total energy of the configuration should be regularized in terms of subtraction of the infinity E_0 as

$$E_{\text{reg}} = E(U_0) - E_0 = E(U_0) - \frac{1}{\pi}\left(\int_1^\infty + \int_0^1 \right) dU = \frac{U_0}{\pi}\left[\int_1^\infty dy \left(\frac{y^2}{\sqrt{y^4 - 1}} - 1 \right) - 1 \right].$$

According to the relation

$$f(\lambda) = \int_1^\infty dy \left(\frac{y^2}{\sqrt{y^4 - 1}} - 1 \right) y^\lambda = \frac{1}{4}\left[B\left(\frac{1}{2}, -\frac{\lambda + 1}{4}\right) - B\left(1, -\frac{\lambda + 1}{4}\right) \right], \tag{3.112}$$

we have
$$E_{\text{reg}} = \frac{U_0}{\pi} f(0) = -\frac{4\pi^2 R^2}{\Gamma(1/4)^4} \frac{1}{L}, \quad (3.113)$$

where $B(a, b)$ is the Beta function. Finally, we have the Coulomb type potential in the $\mathcal{N}=4$ supersymmetric Yang-Mills theory.

Next, we take the alternative configuration [37],
$$X^0 = \tau, \quad X^2 = \sigma, \quad X^3 = v\tau, \quad U = U(\sigma). \quad (3.114)$$

In this case, the action becomes
$$S = \frac{T\sqrt{1-v^2}}{2\pi} \int d\sigma \sqrt{(\partial_\sigma U)^2 + \left(\frac{U}{R}\right)^4}. \quad (3.115)$$

The extremum condition of the action ($\partial_\sigma \mathcal{H} = 0$) suggests that
$$(Ru)^4 \sqrt{\frac{1-v^2}{(\partial_\sigma u)^2 + u^4}} = \text{const.} = R^4 u_0^2 \sqrt{1-v^2}, \quad (3.116)$$

where $\lambda = R^4, U = \sqrt{\lambda} u$ and $\partial_\sigma u(u_0) = 0$. Thus, σ is given by
$$\sigma = \frac{1}{u_0} \int_1^{u/u_0} \frac{dy}{y^2 \sqrt{y^4 - 1}}, \quad (3.117)$$

where $y = u/u_0$. Further, u_0 is written in terms of the quark-anti-quark distance L as
$$u_0 = \frac{2\sqrt{2}\pi^{3/2}}{\Gamma(1/4)^2} \frac{1}{L}. \quad (3.118)$$

Finally, the action becomes
$$S = \frac{TR^2 u_0 \beta}{\pi} \int_1^\infty dy \frac{y^2}{\sqrt{y^4 - 1}}, \quad (3.119)$$

where $\beta = \sqrt{1-v^2}$. In this configuration, we also have the Coulomb potential,
$$E_{\text{reg}} = E(u_0) - E(u_0 \to 0) = -\frac{4\pi^2 R^2 \beta}{\Gamma(1/4)^4} \frac{1}{L}. \quad (3.120)$$

Accordingly, in terms of the AdS/CFT correspondence, we can find the quark-anti-quark potential in the $\mathcal{N} = 4$ supersymmetric Yang-Mills theory. However, eq.(3.120) is trivial result since there must be $1/L$ dependence in the CFT models due to the dimensional analysis. In next subsection, we consider alternative supergravity solutions which can show non-Coulomb type potential.

3.5.2. D1-D5-branes System

In type IIB string theory, there is a black hole solution with non zero horizon area, which is known as D1-D5-branes system [38, 39]. In the context of the AdS/CFT correspondence, the D1-D5-branes system whose near horizon geometry is $AdS_3 \times S^3 \times M^4$ could describe the (1+1)-dimensional conformal field theory which has eight supersymmetries [31]. The CFT which corresponds to AdS_3 may be related to the moduli space of the instantons [38]. Thus, small fluctuations of the instanton moduli may be described by the (1+1)-dimensional sigma model. In this respect, the D1-D5-branes system should have the dual string description of the (5+1)-dimensional gauge theory with the solitonic object. In this case, D1-branes may be interpreted as the instantons.

The metric of the D1-D5-branes system is given by

$$ds^2 = f_1^{-1/2} f_5^{-1/2}(-dt^2 + dx_1^2) + f_1^{1/2} f_5^{-1/2}(dx_2^2 + \cdots + dx_5^2) + f_1^{1/2} f_5^{1/2}(dr^2 + r^2 d\Omega_3^2), \tag{3.121}$$

where

$$f_1 = 1 + \frac{g_6 Q_1}{\sqrt{V} \alpha'} \left(\frac{\alpha'}{r}\right)^2, \qquad f_5 = 1 + \frac{g_6 Q_5 \sqrt{V}}{\alpha'} \left(\frac{\alpha'}{r}\right)^2. \tag{3.122}$$

In this case, there are many ways of taking the decoupling limit. For example, we could take the decoupling limit [31] as

$$\alpha' \to 0, \qquad \frac{r}{\alpha'} = \text{fixed}, \qquad V = \frac{V_4}{(2\pi)^4 \alpha'^2} = \text{fixed}, \qquad g_6 = \frac{g}{\sqrt{V}} = \text{fixed}, \tag{3.123}$$

where V_4 is the volume of M^4. In this case,

$$f_1 \to \frac{g_6 Q_1}{\sqrt{V} \alpha'} \frac{1}{U^2}, \qquad f_5 \to \frac{g_6 Q_5 \sqrt{V}}{\alpha'} \frac{1}{U^2}, \tag{3.124}$$

where $U = r/\alpha'$. Therefore, the metric becomes

$$ds^2 = \alpha' \frac{U^2}{\widetilde{R}^2}(-dt^2 + dx_1^2) + \sqrt{\frac{Q_1}{V Q_5}}(dx_2^2 + \cdots + dx_5^2) + \alpha' \widetilde{R}^2 \left(\frac{dU^2}{U^2} + d\Omega_3^2\right), \tag{3.125}$$

where $\widetilde{R}^4 = g_6^2 Q_1 Q_5$. Note that the second part of the metric is not proportional to α'. This metric may describe 1+1 dimensional supersymmetric CFT model.

On the other hand, we could take alternative limiting procedure [40]. In this case, we could impose the condition

$$\widetilde{V} = \frac{V_4}{(2\pi)^4} = \alpha'^2 V = \text{fixed} \tag{3.126}$$

instead of V = fixed. It is convenient to write

$$f_1 = 1 + \frac{g_6 Q_1}{\sqrt{\widetilde{V}} U^2} = 1 + \frac{R_1^2}{U^2}, \qquad f_5 = 1 + \frac{g_6 Q_5 \sqrt{\widetilde{V}}}{\alpha'^2 U^2} = 1 + \frac{R_5^2}{\alpha'^2 U^2}, \tag{3.127}$$

where we define that

$$R_1^2 \equiv \frac{gQ_1}{\alpha'V} = \frac{\alpha'gQ_1}{\widetilde{V}}, \qquad R_5^2 \equiv \alpha'gQ_5, \qquad \frac{R_1^2}{R_5^2} = \frac{Q_1}{\widetilde{V}Q_5}, \qquad \widetilde{R}^4 = R_1^2 R_5^2. \qquad (3.128)$$

Now, we consider the case $R_1 \gg 1$. In the new decoupling limit, we have

$$f_1 \to \frac{g_6 Q_1}{\sqrt{\widetilde{V}} U^2} = \frac{R_1^2}{U^2}, \qquad f_5 \to \frac{g_6 Q_5 \sqrt{\widetilde{V}}}{\alpha'^2 U^2} = \frac{R_5^2}{\alpha'^2 U^2}, \qquad (3.129)$$

Therefore, the metric becomes

$$ds^2 = \alpha' \left[\frac{U^2}{\widetilde{R}^2} (-dt^2 + dx_1^2) + \frac{R_1}{R_5}(dx_2^2 + \cdots + dx_5^2) + \widetilde{R}^2 \left(\frac{dU^2}{U^2} + d\Omega_3^2 \right) \right]. \qquad (3.130)$$

First, we consider the configuration,

$$X^0 = \tau, \qquad X^1 = \sigma, \qquad U = U(\sigma). \qquad (3.131)$$

In this case, the action becomes

$$S = \frac{T}{2\pi} \int d\sigma \sqrt{(\partial_\sigma U)^2 + \left(\frac{U}{R}\right)^4}, \qquad (3.132)$$

Therefore, the Wilson loop of the D1-D5-branes system in the configuration eq.(3.131) is same as that of D3-brane system if

$$4\pi g N = g_6^2 Q_1 Q_5. \qquad (3.133)$$

Next, we consider the configuration,

$$X^0 = \tau, \qquad X^2 = \sigma, \qquad X^3 = v\tau, \qquad U = U(\sigma). \qquad (3.134)$$

In this case, it is convenient to use a parameterization such as

$$\sqrt{\lambda} = \widetilde{R}^2 = R_1 R_5, \qquad \widetilde{h} = \frac{g_6 Q_1}{\sqrt{\widetilde{V}}} = R_1^2, \qquad \frac{\widetilde{h}}{\sqrt{\lambda}} = \sqrt{\frac{Q_1}{\widetilde{V}Q_5}} = \frac{R_1}{R_5}, \qquad \widetilde{h} = h U^2. \qquad (3.135)$$

The action becomes

$$S = \frac{T\sqrt{\lambda}}{2\pi} \int d\sigma \sqrt{(1 - v^2 h)((\partial_\sigma u)^2 + hu^4)}, \qquad (3.136)$$

where $U = \widetilde{R}^2 u$. This action is similar to that of D3-branes system in the noncommutative space [37]. The extremum condition of the action is given by

$$hu^4 \sqrt{\frac{1 - v^2 h}{(\partial_\sigma u)^2 + hu^4}} = \text{const.} = u_0^2 \sqrt{h_0} \sqrt{1 - v^2 h_0}, \qquad (3.137)$$

where u_0 is a point of $\partial_\sigma u = 0$ and $h_0 = h(u_0)$. Hence,

$$u = u_0 \sec \frac{u_0 \sqrt{h_0}}{b} \sigma = u_0 \sec \frac{\sqrt{\tilde{h}}}{b\sqrt{\lambda}} \sigma. \tag{3.138}$$

where $b = \sqrt{1 - v^2 h_0}$. Finally, the total energy of the configuration is

$$E = \frac{S}{T} = \frac{\tilde{R}^2}{2\pi} \left[u_0 \tan \frac{u_0 \sqrt{h_0}}{b} \sigma - \frac{v^2 h_0^{3/2} u_0^2}{b} \sigma \right]. \tag{3.139}$$

On the other hand, the quark-anti-quark distance L should be given by

$$\frac{L}{2} = \frac{b}{\sqrt{h_0}} \int_{u_0}^{\infty} \frac{du}{u\sqrt{u^2 - u_0^2}} = \frac{\pi b}{2 u_0 \sqrt{h_0}} = \frac{\pi b R_5}{2}. \tag{3.140}$$

Note that this is consistent with the result in [41] at $p = 5$ for $v = 0$ case. In this case, the regularized energy is given by

$$E_{\text{reg}} = E(u_0) - E(u_0 = 0) = -\frac{v^2 R_1}{2} \sqrt{h_0}. \tag{3.141}$$

Finally, we can write

$$E_{\text{reg}} = -\frac{vR_1}{2} \sqrt{1 - \frac{L^2}{\pi^2 \alpha' g Q_5}}. \tag{3.142}$$

Here, note that the relation between a coupling constant of the Yang-Mills theory and the Dp-branes system should be written by [33]

$$g_{YM}^{2\,(p)} = (2\pi)^{p-2} g \alpha'^{(p-3)/2}. \tag{3.143}$$

In particular, for D5-branes system, we have

$$g_{YM}^{2\,(5)} = (2\pi)^3 g\alpha'. \tag{3.144}$$

Therefore, the quark-anti-quark energy should be given by

$$E_{\text{reg}} = -\frac{vR_1}{2} \sqrt{1 - \frac{8\pi}{g_{YM}^{2\,(5)} Q_5} L^2}. \tag{3.145}$$

Thus, the distance L is suppressed by

$$L < \sqrt{\frac{g_{YM}^{2\,(5)} Q_5}{8\pi}}. \tag{3.146}$$

The previous calculation, we consider the region of $R_1 \gg 1$. Next, we consider the contrary region of R_1 ($R_1 \ll 1$). In this case, we can write

$$f_1 = 1 + \frac{g_6 Q_1}{\sqrt{\tilde{V}} U^2} = 1 + \frac{R_1^2}{U^2} = 1 + h, \qquad f_5 = \frac{R_5^2}{\alpha'^2 U^2}. \tag{3.147}$$

Thus, the action becomes

$$S = \frac{T\sqrt{\lambda}}{2\pi} \int d\sigma \; \sqrt{(1-v^2 f_1)\left((\partial_\sigma u)^2 + hu^4\right)}. \tag{3.148}$$

In this case, it can be shown that u is written by

$$u = u_0 \sec\left[\frac{u_0\sqrt{h_0}}{\bar{b}}\sigma\right] = u_0 \sec\left[\frac{\sqrt{\tilde{h}}}{\bar{b}\sqrt{\lambda}}\sigma\right], \tag{3.149}$$

where

$$\bar{b} = \sqrt{1 - \frac{v^2}{1-v^2} h_0}. \tag{3.150}$$

Finally, the energy is given by

$$E = \frac{S}{T} = \frac{R^2}{2\pi}\left[u_0\sqrt{1-v^2}\tan\frac{u_0\sqrt{h_0}}{\bar{b}}\sigma - \frac{v^2 h_0^{3/2} u_0^2}{\sqrt{1-v^2 f_1(h_0)}}\sigma\right]. \tag{3.151}$$

On the other hand, in this case, the quark and anti-quark distance should be given by

$$\frac{L}{2} = \frac{\bar{b}}{\sqrt{h_0}}\int_{u_0}^\infty \frac{du}{u\sqrt{u^2-u_0^2}} = \frac{\pi\bar{b}\sqrt{\alpha'gQ_5}}{2} = \frac{\pi\bar{b}R_5}{2}. \tag{3.152}$$

Thus, the regularized energy should be given by

$$E_{\text{reg}} = -\frac{vR_1}{2}\sqrt{1 - \frac{L^2}{\pi^2 R_5^2}} = -\frac{vR_1}{2}\sqrt{1 - \frac{L^2}{\pi^2 \alpha' gQ_5}}. \tag{3.153}$$

At the last, we consider the configuration of

$$X^0 - \tau, \qquad X^2 - \sigma, \qquad X^1 - v\tau, \qquad U = U(\sigma). \tag{3.154}$$

In this case, the action becomes

$$S = \frac{T\sqrt{\lambda}}{2\pi}\sqrt{1-v^2}\int du \; \frac{u}{\sqrt{u^2-u_0^2}} = \frac{T\sqrt{\lambda}}{2\pi}\sqrt{(1-v^2)(u^2-u_0^2)}, \tag{3.155}$$

where

$$\sqrt{\lambda} = g_6\sqrt{Q_1Q_5}, \quad \tilde{h}(\equiv R_1^2) = \frac{g_6 Q_1}{\sqrt{\tilde{V}}}\left(\frac{\tilde{h}}{\sqrt{\lambda}} = \sqrt{\frac{Q_1}{\tilde{V}Q_5}}\right), \quad \tilde{h} = hu^2 \tag{3.156}$$

and u is wriiten by

$$u = u_0 \sec u_0\sqrt{h_0}\sigma. \tag{3.157}$$

Since the quark-anti-quark distance should be given by

$$\frac{L}{2} = \frac{1}{\sqrt{h_0}}\int_{w_0}^\infty \frac{dw}{w\sqrt{w^2-w_0^2}} = \frac{\pi}{2w_0\sqrt{h_0}} = \frac{\pi\sqrt{\alpha'gQ_5}}{2} = \frac{\pi R_5}{2}. \tag{3.158}$$

the regularized energy is given by

$$E_{\text{reg.}} = E(u_0) - E(u_0 = 0) = 0. \tag{3.159}$$

3.6. Linear Potential and Confinement by Dual String Picture

3.6.1. Spatial Wilson Loop of D1-D5-branes System

In this subsection, we consider spatial Wilson loop of the D1-D5-branes system [40]. In this case, we could take

$$X^4 = \tau, \qquad X^2 = \sigma, \qquad X^3 = v\tau, \qquad U = U(\sigma). \tag{3.160}$$

In this particular case, we have

$$h_{\tau\tau} = f_1^{1/2} f_5^{-1/2}(1+v^2), \qquad h_{\sigma\sigma} = f_1^{1/2} f_5^{1/2}\left(f_5^{-1} + \alpha'^2(\partial_\sigma U)^2\right), \tag{3.161}$$

where

$$f_1 = 1 + \frac{R_1^2}{U^2}, \qquad f_5 = \frac{R_5^2}{\alpha'^2 U^2}, \tag{3.162}$$

Therefore, the Nambu-Goto action becomes

$$S = \frac{T}{2\pi}\sqrt{1+v^2}\int d\sigma\, f_1^{1/2}\sqrt{(\partial_\sigma U)^2 + \frac{U^2}{R_5^2}}. \tag{3.163}$$

The extremum condition of the action ($\partial_\sigma \mathcal{H} = 0$) indicates that

$$\frac{U^2}{R_5^2}\frac{f_1^{1/2}}{\sqrt{(\partial_\sigma U)^2 + (U/R_5)^2}} = \frac{U_0}{R_5}\sqrt{f_1^0}, \tag{3.164}$$

where $\partial_\sigma U = 0$ at $U = U_0$ and $f_1^0 = f_1(U = U_0)$. Introducing new variable u in terms of $U = R_1 u$, we have

$$\partial_\sigma u = \frac{|u|}{R_5^2(1+u_0^2)}\sqrt{u^2 - u_0^2}. \tag{3.165}$$

In this case, the quark-anti-quark distance L should be given by

$$\frac{L}{2} = \frac{R_5\sqrt{1+u_0^2}}{u_0}\int_1^\infty dy\,\frac{1}{y\sqrt{y^2-1}} = \frac{\pi R_5\sqrt{1+u_0^2}}{2u_0}. \tag{3.166}$$

Thus, we have

$$\frac{1}{u_0} = \sqrt{\frac{L^2}{\pi^2 R_5^2} - 1}. \tag{3.167}$$

On the other hands, the total energy of the configuration is given by

$$E = \frac{S}{T} = 2\frac{\sqrt{1+v^2}}{2\pi}R_1\int_{u_0}^\infty du\,\frac{u^2+1}{u\sqrt{u^2-u_0^2}}. \tag{3.168}$$

Thus, the regularized energy becomes

$$E_{\text{reg}} = \frac{\sqrt{1+v^2}}{\pi u_0}R_1\int_1^\infty dy\,\frac{1}{y\sqrt{y^2-1}} = \frac{\sqrt{1+v^2}}{2u_0}R_1. \tag{3.169}$$

Finally, the quark-anti-quark potential energy should be written in terms of distance L as

$$E = \frac{R_1\sqrt{1+v^2}}{2}\sqrt{\frac{L^2}{\pi^2 R_5^2} - 1} \sim \frac{R_1\sqrt{1+v^2}}{2\pi R_5}L. \qquad (3.170)$$

Therefore, we have the linear potential in (5+1)-dimensional Yang-Mills theory in terms of the dual gravity picture of the D1-D5-branes system. Thus, according to the Wilson criterion, the confinement should occur in this system. Further, since the energy is proportinal to the R_1, the confinement may occur due to the instanton effect. Fortunately, this fact should be confirmed in terms of alternative gravity solution, which is given by D(-1)-D3-branes system [42]. It is known that the D1-D5-branes system is T-dual to the D(-1)-D3-branes system (D-instantons + D3-branes system). It is believed that general gauge/gravity corresponding should be valid [32] even though the correponding gauge theories are not conformal theory. In [42], it is suggested that the D(-1)/D3-branes system is dual to the (nonconformal) gauge theory and the quark-anti-quark potential becomes linear potential in L due to the existence of the D-instantons.

3.6.2. D(-1)-D3-branes System

The D(-1)-D3-branes system should describe the effective action of $\mathcal{N} = 4$ supersymmetric Yang-Mills theory in a constant self-dual gauge field background. It is known that the D(-1)-D3-branes system is T-dual to the D1-D5-branes system as well as the D0-D4-branes system. In the D(-1)-D3-branes system, D-instanton is smeared over the D3-brane. The metric in the string frame of the D3-D(-1)-branes system [42] is given by

$$ds^2 = H_{-1}^{1/2}\left(H_3^{-1/2}dx_m dx_m + H_3^{1/2}dx_s dx_s\right), \qquad (3.171)$$

where $m = 0 \sim 3$, $s = 4 \sim 9$ and

$$H_3 = 1 + \frac{Q_3}{r^4} = 1 + \frac{\lambda}{\alpha'^2 U^4}, \qquad H_{-1} = 1 + \frac{Q^{(4)}_{-1}}{r^4} = 1 + \frac{q\lambda}{U^4}, \qquad (3.172)$$

where

$$r^2 = x_s x_s, \qquad U = \frac{r}{\alpha'}, \qquad Q_3 = 4\pi N g_s \alpha'^2, \qquad Q^{(4)}_{-1} = \frac{K}{N}\frac{(2\pi)^4 \alpha'^2}{V_4}Q_3, \qquad (3.173a)$$

and

$$\lambda \equiv 4\pi g_s N, \qquad q \equiv \frac{K}{N}\frac{(2\pi)^4 \alpha'^2}{V_4}. \qquad (3.173b)$$

Here, N and K are integer values, which represent the number of D3-branes and D-instantons respectively. In the decoupling limit ($\alpha' \to 0$ with keeping U, λ, q=fixed), we have

$$H_3 \to \frac{q\lambda}{U^4}.$$

Thus, the metric becomes

$$ds^2 = \alpha'\sqrt{\lambda}\sqrt{H_{-1}}\left[\lambda^{-1}U^2 dx_m dx_m + \left(\frac{dU^2}{U^2} + d\Omega_5^2\right)\right]. \qquad (3.174)$$

We can show that the Nambu-Goto action in the D(-1)-D3-branes supergravity background at the decoupling limit is written by

$$S = \frac{\sqrt{q}}{2\pi} TL, \qquad (3.175)$$

where L should be the quark-anti-quark distance. Therefore, the quark-anti-quark potential is given by

$$V = \frac{\sqrt{q}}{2\pi} L. \qquad (3.176)$$

Note that the linear potential emerges and the coefficient of the potential depends only on the instanton parameter q. Accordingly, the confinement may be caused by the instanton effect [42].

3.6.3. Does We Understand the Confinement ?

Thanks to the conformal symmetry, we can hope that the superstring theory in the supergravity background should be able to describe the corresponding supersymmetric Yang-Mills theory. In the CFT models, there is no scale and the exponents such as central charge and conformal weight are only important parameters. Remember that the exponents are determined without detail structure of the CFT models. In this respect, the AdS/CFT correspondence should not be surprising aspect.

The energy spectrum of the CFT model is meaningless since there is no measure of the energy scale. Nonetheless, it should be interesting to estimate the finite size correction of the CFT models [15, 16, 18] since we can find a central charge in the correction. In the CFT models, the finite size correction is written in terms of the $1/L$ correction, where L is a system size. Thus, if we introduce a length scale such as the quark-anti-quark distance, the energy must be inversely proportional to the distance. In this respect, it is not so surprising fact of which we find the Coulomb type potential in the (3+1)-dimensional $\mathcal{N} = 4$ supersymmetric Yang-Mills theory.

In contrast to the trivial fact of the AdS_5/CFT_4 correspondence, the D1-D5-branes system as well as D(-1)-D3-branes system show the linear potential. Why? This is partially because we take nontrivial decoupling limit in this case. In the new decoupling limit, we fix the volume V_4 instead of the dimensionless parameter V_4/α'^2. Indeed, the quark-anti-quark potential in the two system depends on the dimensionful parameter V_4. Note, however, that it is not so obvious that the decoupling limit makes sense. Thus, in present time, we only conjecture that the D-brane picture may also describe the corresponding gauge theory at the new decoupling limit.

In the very optimistic point of view, the present results show a realization on the way of complete understanding of the confinement mechanism. Do we really stand on the way to the truth? In the pessimistic point of view, we should not. Note that the external quarks was introduced here regardless of what the isolated quarks should be nonsense physically since the quarks have color charges. As long as we believe the QCD which is formulated in terms of the SU(3) gauge theory, the isolated quarks are not observable due to non gauge invariance. Furthermore, there are intrinsic problems in the Wilson loop formulation [9]. Thus, we may be still on the way far from the complete understanding of the confinement.

3.7. More on Gauge/Gravity Correspondences and Dual Descriptions of Gauge Theories

Nowadays the AdS/CFT correspondence conjecture is extended to genral conjecture of gravity/non-conformal gauge theory correspondence. Its general correspondence is questionable, nevetheless there are many suggestions about the consistency with the lattice calculation of QCD. Note, however, that the consistency with the lattice calculation does not always suggest that it provides a truth.

As long as we believe the general gravity/gauge theory correspondence, the noncommutative gauge theory may have corresponding gravity solution in the string theory [41, 43, 44, 45, 46, 47, 48, 49]. In this case, the background metric of the string theory becomes AdS $\times S \times X$, where X should be noncommutative manifold [48]. Similar to the AdS/CFT correspondence in the commutative space, the large N limit of the noncommutative gauge theories are described in terms of the Dp-branes system with a constant B field in the near horizon region. It is well-known that the noncommutative Yang-Mills theory is equivalent to ordinary Yang-Mills theory with the perturbative higher dimension operator [45]. At the long distance (IR regime), the dual gravity description of the noncommutative supersymmetric Yang-Mills theories become the AdS $\times S$. Thus, the noncommutative gauge theories correspond to the commutative ones at IR regime. On the other hand, closing to the boundary (UV regime) with taking B field to infinity, the noncommutative effect becomes strong. In this region, the dual gravity solutions behave differently from the commutative ones due to the noncommutative effect. For example, the Wilson loop of the D3-branes with a constant B field shows different behavior from that of the commutative D3-branes solution [37]. For the large loop, which corresponds to the IR limit, the Wilson loop provides a Coulomb law in the quark-anti-quark potential since there is small noncommutativity effect [41, 44]. On the other hand, for the small loop, which corresponds to the UV limit, we find the area law and the string tension is controlled by the noncommutativity parameter [37].

The D1-D5-branes system in the constant B field may correspond to the noncommutative gauge theory [47, 48]. In this case [40], the Wilson loop provides the quark-anti-quark potential V as

$$V = \frac{1}{2\pi a R_5} L, \qquad (3.177a)$$

for UV limit, where a is a noncommutative parameter. On the other hand, at the IR limit, the quark-anti-quark potential V becomes

$$V = \frac{R_1}{2\pi R_5 \sqrt{1 + a^2 R_1^2}} L. \qquad (3.177b)$$

Thus, in the noncommutative D1-D5-branes system, the correponding gauge theory shows the linear potential, which one should be caused by the noncommutative effect and another should be caused in terms of the instanton effect.

Alternative advantage of the AdS/CFT correspondence is that we can hope it should be possible to estimate the particle spectrums such as a meson and a baryon in the QCD. For example, we may find the glueball spectrum, which, to be exact, we can only find the ratio of the spectrum. The glueball is believed to be composed of the gluon fields even though the glueball has not been observed yet. The lattice calculation predicts a glueball spectrum in terms of the Monte Calro method or other techniques [8]. According to the calculation of the glueball spectrum in terms of the conjecture of the AdS/CFT correspondence, it shows consistent results with the lattice calculations [50, 51]. On the other hand, the glueball spectrum in the four dimensional noncommutative gauge theory is also calculated by similar strategy of the commutative gauge theory [52]. According to the results of [52], There is slightly difference between the commutative space and the noncomutative space in the glueball spectrum. Again, it is important to note that the results of the string and the lattice do not promise the existence of the glueball although we could expect its existence.

4. Noncommutative Space

The idea of space-time noncommutativity or the endeavor of the quantization of the space-time was investigated by [13] in early pioneering stage of quantum field theory. There were known the essential difficulties in the quantum field theory. The point interaction between fields and particles should give rise to trouble us with the divergences of the physical observables. It is, therefore, natural to consider a finite resolution a of space-time like as lattice theory. One interpretation of the finite resolution is that it may be caused in terms of the space-time fluctuation. In this idea, the space-time becomes dynamical object and the fluctuation may be quantized in terms of a space-space or space-time noncommutativity such as quantum mechanics.

However, we cannot take advantage of its formulation since there is no evidence of minimum length in space-time until now. This fact indicates that the lattice field theory is just for one of the regularization schemes of the continuous model, and we must take zero for lattice spacing a in final stage except the condensed matter physics. Further, there are excellent agreement between quantum electrodynamics in the renomarization theory and experiments, e.g. Lamb shift. Therefore, there is no reason to consider an unconventional space except for mathematical interest.

There are a lot of works making progress on the noncommutative space theory, recently [45, 53, 54, 55]. In particular, it has been developed in terms of the string theory. The noncommutative space is naturally formulated in terms of D-branes as well as B fields in the string theory. As long as we believe the string theory, the noncommutative space should be well established.

Alternatively, the noncommutative space can be formulated in terms of the star-product. In this viewpoint, a model in the noncommutative space is straightforwardly given by a model in the commutative space in terms of the replacement of the multiplication into the star product. Thus, the noncommutative models are described by the commutative language in accordance with a map between the commutative space and the noncommutative space. Moreover, it is well-known that gauge theories in the noncommutative space should be described by gauge theories in the commutative space with the Seiberg-Witten map [45]. The Seiberg-Witten map is proved in terms of the perturbation series with respect to a noncommutative parameter θ [56]. Here, though, we only consider up to first order in θ for simplicity.

There is alternative map between a commutative gauge field and a noncommutative gauge field [57, 58]. This map is called by Bopp shift. The Bopp shift provides the simple way for performing the calculation of the noncommutative problems. In particular, we will discuss a noncommutative Landau problem in terms of the Bopp shift in the next section.

In this section, first, we will discuss the mathematical interpretation of the noncommutative space. Next, we see how the string theory provides the noncommutative space. At the end, we will discuss the gauge theory in the noncommutative space.

4.1. Hermitian Operator for Space-Time Coordinates and Lorentz Invariance in Noncommutative Space

Relativistic theory must be invariant under Lorentz transformation. According to the principle of the special relativity, the interval between two events O and P with coordinates separation

$$s^2 = -c^2 t^2 + x^2 + y^2 + z^2 \tag{4.1}$$

is invariant under transformation from one inertial frame to another, where c is velocity of light. Indeed, we can easily show that eq.(4.1) is invariant under the Lorentz transformation,

$$ct' = \frac{t - \beta x}{\sqrt{1 - \beta^2}}, \quad x' = \frac{x - \beta t}{\sqrt{1 - \beta^2}}, \quad \left(\beta \equiv \frac{v}{c}\right), \tag{4.2}$$

where v is relative velocity of the inertial frame $\mathcal{S}'(t', x')$ with respect to the inertial frame $\mathcal{S}(t, x)$ and $\mathcal{S}'$ move parallel to the x axis of $\mathcal{S}$. Further, eq.(4.1) is also invariant under rotation about the z axis through an angle θ,

$$x' = x \cos\theta + y \sin\theta, \quad y' = -x \sin\theta + y \cos\theta,. \tag{4.3}$$

4.1.1. Space-time Noncommutativity

First, we consider the case of which the commutation relations of space-time becomes nonzero,

$$[t, x] \neq 0.$$

For noncommutative space, it is nontrivial to provide the invariance of eq.(4.1) since the space and time become nontrivial c-numbers. Here, we consider the case of which time coordinate and space coordinate of x axis are only noncommutative each other. Now, we assume [13] that t and x are Hermitian operators on de Sitter space $\mathbf{dS}_2(R_0; \xi_0, \xi_1, \xi_2)$ which is described by

$$R_0^2 = -\xi_0^2 + \xi_1^2 + \xi_2^2, \tag{4.4}$$

where R_0 is a constant and ξ_0, ξ_1, ξ_2 are real variables. The two-dimensional de Sitter space $\mathbf{dS}_2$ has the SO(2,1) symmetry whose generators are written by

$$J_{ab} = -J_{ba} = \xi_a \eta_{bc} \frac{\partial}{\partial \xi_c} - \xi_b \eta_{ac} \frac{\partial}{\partial \xi_c}, \tag{4.5}$$

where $a, b, c, d = 0, 1, 2$ and $\eta_{ab} = \text{diag}(-1, 1, 1)$ is a metric tensor of Minkowski space. The generators satisfy the commutation relation

$$[J_{ab}, J_{cd}] = \eta_{bc} J_{ad} - \eta_{ac} J_{bd} - \eta_{bd} J_{ac} + \eta_{ad} J_{bc}. \tag{4.6}$$

Further, we define

$$\mathcal{X}_0 \equiv J_{20} = -\xi_2\partial_0 - \xi_0\partial_2, \quad \mathcal{X}_1 \equiv J_{21} = \xi_2\partial_1 - \xi_1\partial_2, \quad \mathcal{N} \equiv J_{01} = \xi_0\partial_1 - \xi_1\partial_0. \quad (4.7)$$

In this case, we have

$$[\mathcal{X}_\mu, \mathcal{X}_\nu] = -J_{\mu\nu}, \tag{4.8a}$$

$$[\mathcal{X}_\mu, J_{\nu\rho}] = \eta_{\mu\nu}\mathcal{X}_\rho - \eta_{\mu\rho}\mathcal{X}_\nu, \tag{4.8b}$$

where $\mu, \nu, \rho = 0, 1$. This algebra is reminiscent of Poincaré algebra. It is actually known that de Sitter space and anti-de Sitter space have symmetry of the Poincaré group in suitable limit. In **dS**$_2$ space, there are three generators $\{\mathcal{X}_0, \mathcal{X}_1, \mathcal{N}\}$ and they construct the closed algebra,

$$[\mathcal{X}_0, \mathcal{X}_1] = -\mathcal{N}, \quad [\mathcal{X}_0, \mathcal{N}] = -\mathcal{X}_1, \quad [\mathcal{X}_1, \mathcal{N}] = -\mathcal{X}_0. \tag{4.9}$$

Instead of eq.(4.9), we can rewrite the algebra as

$$[\mathcal{X}_\pm, \mathcal{N}] = \mp \mathcal{X}_\pm, \quad [\mathcal{X}_+, \mathcal{X}_-] = 2\mathcal{N}, \quad [\mathcal{X}^2, \mathcal{N}] = 0, \tag{4.10}$$

where

$$\mathcal{X}_\pm = \mathcal{X}_0 \pm \mathcal{X}_1, \quad \mathcal{X}^2 = \mathcal{X}_0^2 - \mathcal{X}_1^2. \tag{4.11}$$

Therefore, the spectrum of the algebra is specified in terms of the eigenvalues of $\mathcal{X}^2$ and $\mathcal{N}$. In addition, $\mathcal{X}_+, \mathcal{X}_-$ are raising and lowering operator of the eigenvalue $\mathcal{N}$ by factor 1, respectively. Now, we take

$$ct = ia\mathcal{X}_0, \quad x = ia\mathcal{X}_1, \tag{4.12}$$

where a is a constant of length dimension. In this case, we have

$$[ct, x] = -a^2\mathcal{N}, \quad [ct, \mathcal{N}] = -x, \quad [x, \mathcal{N}] = -ct \tag{4.13}$$

and

$$[x_\pm, \mathcal{N}] = \mp ix_\pm, \quad [x_+, x_-] = 2a^2\mathcal{N}, \tag{4.14}$$

where $x_\pm = ct \pm x$. Further,

$$[s^2, \mathcal{N}] = 0, \tag{4.15}$$

where $s^2 = -(ct)^2 + x^2 = a^2\mathcal{X}^2$. Note that the Lorentz invariant distance s^2 is simultaneously determined by an eigenvalue of $\mathcal{N}$. This means that the spectrum of the noncommutative space is composed of discrete values with preserving the Lorentz invariance. Note, further, that we recover the Lorentz transformation eq.(4.2) with

$$\xi_0' = \frac{\xi_0 - \beta\xi_1}{\sqrt{1-\beta^2}}, \quad \xi_1' = \frac{\xi_1 - \beta\xi_0}{\sqrt{1-\beta^2}}. \tag{4.16}$$

On the other hand, we have uncertainty relations of space-time due to the commutation relation eq.(4.13). When we define

$$c\delta t = ct - c\langle t \rangle, \quad \delta x = x - \langle x \rangle, \tag{4.17}$$

it can be shown that
$$\langle (c\delta t)^2 \rangle \langle (\delta x)^2 \rangle \geq \frac{1}{4} |\langle [ct, x] \rangle|^2, \qquad (4.18)$$

where $\langle \hat{O} \rangle$ is an expectation value of operator $\hat{O}$. Therefore, the space-time uncertainty relation is written by
$$(c\Delta t) \cdot \Delta x \geq \frac{a}{2} N_0, \qquad (4.19)$$

where
$$N_0 = \langle \mathcal{N} \rangle, \quad c\Delta t = \sqrt{\langle (c\delta t)^2 \rangle}, \quad \Delta x = \sqrt{\langle (\delta x)^2 \rangle}. \qquad (4.20)$$

4.1.2. Space-Space Noncommutativity

Next, we consider the case of which the commutation relations of space-space becomes nonzero,
$$[x, y] \neq 0,$$
with preserving the rotational invariance. Here, we consider the case of which x coordinate and y coordinate are only noncommutative each other for simplicity. We again assume that x and y are Hermitian operators on de Sitter space eq.(4.4). In the space-space noncommutative case, we could take
$$\mathcal{Y}_m \equiv J_{0m} = \xi_0 \partial_m + \xi_m \partial_0, \quad \mathcal{M} \equiv J_{12} = \xi_1 \partial_2 - \xi_2 \partial_1, \qquad (4.21)$$

where $m = 1, 2$. In this case, we have
$$[\mathcal{Y}_m, \mathcal{Y}_n] = J_{mn}, \qquad (4.22a)$$
$$[\mathcal{Y}_\ell, J_{mn}] = \eta_{\ell m} \mathcal{Y}_n - \eta_{\ell n} \mathcal{Y}_m, \qquad (4.22b)$$

where $\ell, m, n = 1, 2$. In this case, the closed algebra is composed of three generators $\{\mathcal{Y}_1, \mathcal{Y}_2, \mathcal{M}\}$ as
$$[\mathcal{Y}_1, \mathcal{Y}_2] = \mathcal{M}, \quad [\mathcal{Y}_1, \mathcal{M}] = \mathcal{Y}_2, \quad [\mathcal{Y}_2, \mathcal{M}] = -\mathcal{Y}_1. \qquad (4.23)$$

We can rewrite the algebra eq.(4.23) as
$$[\mathcal{Y}_\pm, \mathcal{M}] = \mp i \mathcal{Y}_\pm, \quad [\mathcal{Y}_+, \mathcal{Y}_-] = -2i\mathcal{M}, \quad [\mathcal{Y}^2, \mathcal{M}] = 0, \qquad (4.24)$$

where
$$\mathcal{Y}_\pm = \mathcal{Y}_1 \pm i \mathcal{Y}_2, \quad \mathcal{Y}^2 = \mathcal{Y}_1^2 + \mathcal{Y}_2^2. \qquad (4.25)$$

Therefore, the spectrum is specified in terms of the eigenvalues of $\mathcal{Y}^2$ and $\mathcal{M}$. In addition, $\mathcal{Y}_+, \mathcal{Y}_-$ are raising and lowering operator of the eigenvalue $\mathcal{M}$ by factor i, respectively. Here, we define
$$x = ia\mathcal{Y}_1, \quad y = ia\mathcal{Y}_2, \qquad (4.26)$$

where a is a constant of length dimension. Therefore, the commutation relations of the coordinates (x, y) become
$$[x, y] = -a^2 \mathcal{M}, \quad [x, \mathcal{M}] = y, \quad [y, \mathcal{M}] = -x \qquad (4.27)$$

and
$$[u_\pm, \mathcal{M}] = \mp i u_\pm, \quad [u_+, u_-] = 2ia^2 \mathcal{M}, \tag{4.28}$$

where $u_\pm = x \pm y$. Additionally,
$$[s^2, \mathcal{M}] = 0, \tag{4.29}$$

where $s^2 = x^2 + y^2 = -a^2 \mathcal{Y}^2$. Therefore, the spectrum of the rotational invariant noncommutative space is composed of discrete values. Note, further, that we recover the rotational transformation eq.(4.3) by rotation in the **dS**$_2$ space

$$\xi_1' = \xi_1 \cos\theta + \xi_2 \sin\theta, \quad \xi_2' = -\xi_1 \sin\theta + \xi_2 \cos\theta. \tag{4.30}$$

Again, due to the commutation relation eq.(4.27), the uncertainty relation between x coordinate and y coordinate is given by

$$\Delta x \cdot \Delta y \geq \frac{a}{2} M_0, \tag{4.31}$$

where

$$\Delta x = \sqrt{\langle (\delta x)^2 \rangle} = \sqrt{\langle (x - \langle x \rangle)^2 \rangle}, \quad \Delta y = \sqrt{\langle (\delta y)^2 \rangle} = \sqrt{\langle (y - \langle y \rangle)^2 \rangle}, \quad M_0 = \langle \mathcal{M} \rangle.$$

4.2. Noncommutative Space and String Theory

As shown in the previous subsection, we derive the noncommutative space in terms of the internal de Sitter space with preserving the Lorentz invariance in formal fashion. Obviously, it does not mean that the noncommutative theory make sense physically. In addition, there should be troubles in space-time noncommutative theory since there is no assurance of the unitarity [59, 60]. Therefore, it only shows the consistency for the noncommutativity of the space in the Lorentz invariant theory for the mathematical point of view.

The noncommutativity of space-time should show the uncertainty relation of the space-time due to the nonzero commutation relation between spaces. Intuitively, the relation seems to intend a fact of the fluctuating space-time. However, we should not take such a picture since the spaces are not dynamical object in aforesaid formulation. Moreover, it has intrinsic problem with respect to a indistinctness of the meaning of the de Sitter space in the formulation.

In this subsection, we discuss alternative formulation of the noncommutative space, which is based on the string theory. The noncommutative space is reproduced in terms of the background field $B_{\mu\nu}$ in the string theory. In this respect, it should be straightforwardly to show the consistency of the noncommutative theory if we admit the string theory.

Now, we consider the 10 dimensional string moving in the flat space with a constant B field which is on the D3-brane [46]. For simplicity, the B field $B_{\mu\nu}$ has only nonzero component B_{23}. In the presence of the D3-brane, the string should be written by

$$X^\mu(\tau, \sigma) = \begin{cases} X_N^\mu(\tau, \sigma) & \text{for } (\mu = 0, 1, 2, 3) \\ X_D^\mu(\tau, \sigma) & \text{for } (\mu = 4, 5, \cdots, 9) \end{cases} \tag{4.32}$$

Furthermore, due to the existence of the B field, the string should obey the action of eq.(3.66) without a dilaton as

$$S_P = -\frac{1}{4\pi\alpha'} \int d^2\xi \left(\eta^{\alpha\beta} G_{\mu\nu} - 2\pi\alpha' i \epsilon^{\alpha\beta} B_{\mu\nu} \right) \partial_\alpha X^\mu \partial_\beta X^\nu, \quad (4.33)$$

where we take the conformal gauge $g^{\alpha\beta} = \eta^{\alpha\beta}$ and we redefine the $B_{\mu\nu}$ as $-2\pi\alpha' B_{\mu\nu}$. In this case, the equation of motion of the string is written by

$$\left(\partial_\tau^2 - \partial_\sigma^2 \right) X^\mu = 0.$$

On the other hand, the boundary condition is deformed as

$$g_{mn}\partial_\sigma X^n + 2\pi i \alpha' B_{mn} \partial_\tau X^n \Big|_{\sigma=0,2\pi} = 0, \quad (4.34a)$$

$$\tilde{g}_{ab}\partial_\tau X^b \Big|_{\sigma=0,2\pi} = 0, \quad (4.34b)$$

where $G_{\mu\nu} = (G_{mn}, \tilde{G}_{ab})$ and $m, n = 0, 1, 2, 3, a, b = 4, \cdots, 9$. Further, we take $G_{22} = G_{33} = (G^{22})^{-1} = (G^{33})^{-1} \equiv \eta$. Therefore, the mode expansions are given by

$$X^M(\tau,\sigma) = x^M + 2\alpha' p^M \tau + i\sqrt{2\alpha'} \sum_{k\neq 0} \frac{1}{k} \tilde{\alpha}_k^M e^{-ik\tau} \cos k\sigma, \quad (M=0,1) \quad (4.35a)$$

$$X^r(\tau,\sigma) = x^r + 2\alpha' \left(p^r \tau - \frac{2\pi i \alpha' B_{23} \epsilon^{rs} p_s}{\eta} \sigma \right)$$
$$+ i\sqrt{2\alpha'} \sum_{k\neq 0} \frac{e^{-ik\tau}}{k} \left[\tilde{\alpha}_k^r \cos k\sigma - \frac{2\pi\alpha' B_{23}}{\eta} \epsilon^{rs} \tilde{\alpha}_{ks} \sin k\sigma \right], \quad (r,s=2,3)$$
$$(4.35b)$$

$$X^a(\tau,\sigma) = c_0^a + 2\alpha' d_0^a \sigma + \sqrt{2\alpha'} \sum_{k\neq 0} \frac{1}{k} \tilde{\beta}_k^a e^{-ik\tau} \sin k\sigma, \quad (4.35c)$$

where $\epsilon^{23} = -\epsilon^{32} = 1$. The quantization should be implemented out of the commutation relation,

$$[X^\mu(\tau,\sigma), \Pi^\nu(\tau,\sigma')] = i\delta(\sigma - \sigma'), \quad (4.36)$$

where the canonical momentum Π^μ is given by

$$\Pi^\mu(\tau,\sigma) = \frac{1}{2\pi\alpha'} (\partial_\tau X^\mu + 2\pi i \alpha' B^\mu_\nu \partial_\sigma X^\nu). \quad (4.37)$$

Thus, for $r = 2, 3$, we have

$$\Pi^r(\tau,\sigma) = \pi^{-1} \zeta \left(p^r + \frac{1}{\sqrt{2\alpha'}} \sum_{k\neq 0} e^{-ik\tau} \tilde{\alpha}_k^r \cos k\sigma \right), \quad (4.38)$$

where ζ is defined by

$$\zeta = 1 - \frac{(2\pi\alpha' B_{23})^2}{\eta^2}. \quad (4.39)$$

Finally, we can find the commutation relations

$$[x^m, p^n] = i\mathcal{G}^{mn}, \qquad [\widetilde{\alpha}_k^m, \widetilde{\alpha}_{k'}^n] = k\delta_{k+k',0}\mathcal{G}^{mn}, \qquad (4.40)$$

where the metric $\mathcal{G}^{mn}$ is given by $\mathcal{G}^{mn} = \mathrm{diag}(-1, 1, \zeta^{-1}, \zeta^{-1})$. Furthermore, another commutation relation

$$[X^\mu(\tau, \sigma), X^\nu(\tau, \sigma')] = 0 \qquad (4.41)$$

should also hold. However, in this case, we can find a nontrivial term in the commutation relations of $\mu, \nu = 2, 3$ as

$$[X^2(\tau, \sigma), X^3(\tau, \sigma')] = [x^2, x^3] - \frac{8\pi\alpha'^2 B_{23}}{\eta^2}\frac{\sigma + \sigma'}{2} - \frac{8\pi\alpha'^2 B_{23}}{\eta^2}\sum_{K\neq 0}\frac{1}{K}\sin\frac{\sigma + \sigma'}{2}K.$$

From the well-known series expansion,

$$\sum_{n\neq 0}^\infty \frac{\sin nx}{n} = \pi - x, \qquad (0 < x < 2\pi),$$

we obtain

$$[X^2(\tau, \sigma), X^3(\tau, \sigma')] = [x^2, x^3] - \frac{8\pi^2\alpha'^2 B_{23}}{\eta^2}, \qquad (0 < \sigma + \sigma' < 4\pi), \qquad (4.42a)$$

$$[X^2(\tau, \sigma), X^3(\tau, \sigma')] = [x^2, x^3], \qquad (\sigma = \sigma' = 0), \qquad (4.42b)$$

$$[X^2(\tau, \sigma), X^3(\tau, \sigma')] = [x^2, x^3] - \frac{16\pi^2\alpha'^2 B_{23}}{\eta^2}, \qquad (\sigma = \sigma' = 2\pi). \qquad (4.42c)$$

Therefore, we should take the commutation relation as

$$[x^2, x^3] = i\theta, \qquad (4.43)$$

where $B_{23} \equiv iB$ and

$$\theta \equiv \frac{8\pi^2\alpha'^2 B}{\eta^2}. \qquad (4.44)$$

However, the ends of the string do not commute

$$[X^\mu(\tau, 0), X^\nu(\tau, 0)] = i\theta, \qquad [X^\mu(\tau, 2\pi), X^\nu(\tau, 2\pi)] = -i\theta. \qquad (4.45)$$

Consequently, the noncommutativity of the space-time should emerge on the D-brane. According to the conjecture of the string/gauge theory correspondence, the noncommutative gauge theory may be given rise to the open string with the B field on D-brane.

4.3. Moyal Brackets and Star Product

As seen in previous subsection, there could be a noncommutative space in theoretical point of view. Further, we should expect the existence of the noncommutative gauge theory. The space-time coordinates of the noncommutative space ($\hat{x}_\mu$) obey the commutation relations

$$[\hat{x}_\mu, \hat{x}_\nu] = i\theta_{\mu\nu}, \qquad \theta_{\mu\nu} = -\theta_{\nu\mu} \qquad (4.46)$$

with a parameter $\theta_{\mu\nu}$ which is a two-rank anti-symmetric tensor. The noncommutative models are given in terms of the replacement of multiplication such as

$$f(\hat{x}) \star g(\hat{x}) = \exp\left[\frac{i}{2}\theta^{\mu\nu}\frac{\partial}{\partial \xi^{\mu}}\frac{\partial}{\partial \zeta^{\nu}}\right] f(x+\xi)g(x+\zeta)\bigg|_{\xi=\zeta=0}$$
$$= f(x)g(x) + \frac{i}{2}\theta^{\mu\nu}\partial_{\mu}f\partial_{\nu}g + O(\theta^2), \qquad (4.47)$$

where f, g are any functions in the noncommutative space and x is a coordinate of the commutative space. The new multiplication of eq.(4.47) is called star($\star$)-product. In general, we can define the star product in the higher order of θ. However, for simplicity, we will dismiss the higher order term which is bigger than the second order of $\theta_{\mu\nu}$. Obviously, we can find a noncommutativity of the coordinates in terms of the star-product,

$$[\hat{x}^{\mu}, \hat{x}^{\nu}]_{\star} = \hat{x}^{\mu} \star \hat{x}^{\nu} - \hat{x}^{\nu} \star \hat{x}^{\mu} = i\theta^{\mu\nu},$$

where $[\ ]_{\star}$ is called by Moyal brackets.

Alternatively, the Moyal bracket between the positions of a quantum particle in the noncommutative spaces is also induced by the replacement (Bopp shifts) [58]

$$\hat{x}^{\mu} = x^{\mu} - \frac{\theta^{\mu\nu}}{2}p_{\nu} \qquad (4.48)$$

in a formal manner, where x^{μ} and p^{μ} are assumed to be the position coordinate and the momentum of the particle in the commutative space, respectively. Further, they satisfy the commutation relations

$$[x^{\mu}, p^{\nu}] = i\delta^{\mu\nu}, \qquad [x^{\mu}, x^{\nu}] = 0. \qquad (4.49)$$

Therefore, the physics of the noncommutative space should be represent in terms of language of the commutative space.

Below, we summarize some properties of the $\star$-product. Here, the noncommutative coordinate is denoted with $\hat{\ }$ and $f(\hat{x}), g(\hat{x}), h(\hat{x})$ are the functions in the noncommutative space. Further, $\hat{A}, \hat{B}, \hat{C}$ are operators in the noncommutative space.

1. Moyal brackets:

$$[f, g]_{\star} = f \star g - g \star f = i\theta_{\mu\nu}\partial_{\mu}f\partial_{\nu}g. \qquad (4.50)$$

2. Combination property:

$$\left(f(\hat{x}) \star g(\hat{x})\right) \star h(\hat{x}) = f(\hat{x}) \star \left(g(\hat{x}) \star h(\hat{x})\right). \qquad (4.51)$$

3. Conjugation:

$$\left(\hat{A} \star \hat{B}\right)^{\dagger} = \hat{B}^{\dagger} \star \hat{A}^{\dagger}. \qquad (4.52)$$

4. Differentiation and ordering of a differential operator and a function:

$$\hat{\partial}_\mu \star f = \partial_\mu f + \frac{i}{2}\theta_{ab}\partial_a\partial_\mu\partial_b f = \partial_\mu f. \tag{4.53a}$$

$$(\hat{\partial}_\mu \star \hat{\partial}_\nu) \star f = \hat{\partial}_\mu \star (\hat{\partial}_\nu \star f) = \partial_\mu\partial_\nu f. \tag{4.53b}$$

$$\hat{\partial}_\mu \star (f \star g) = (\partial_\mu f) \star g + f \star (\partial_\mu g). \tag{4.53c}$$

On the other hand,

$$\hat{\partial}_\mu \star (f \star g) \neq (\partial_\mu f) \star g + (f \star \hat{\partial}_\mu) \star g. \tag{4.53d}$$

Eq.(4.53d) is straightforwardly verified in terms of the relation,

$$(f \star \hat{\partial}_\mu) \star g = f \star (\hat{\partial}_\mu \star g) - \frac{i}{2}[f, \partial_\mu g], \tag{4.53e}$$

where

$$(f \star \hat{\partial}_\mu) \star g = f\partial_\mu g + i\theta_{ab}\partial_a f\partial_b\partial_\mu g, \quad f \star (\hat{\partial}_\mu \star g) = f\partial_\mu g + \frac{i}{2}\theta_{ab}\partial_a f\partial_b\partial_\mu g.$$

5. Diamond operation ($\diamond$-operation)

It is convenient to define a diamond operation between $\Phi = \hat{A} \star \hat{B}$ and $\hat{C}$

$$\Phi \diamond \hat{C} \equiv \left(\hat{A} \star (\hat{B} \star \hat{C})\right). \tag{4.54}$$

There are several definitions about the diamond operation. Here, it is defined by means of eq.(4.54). It is interesting to note that when we employ

$$[\partial_x, x]_\star \diamond f \equiv (\partial_x \star x) \diamond f - (x \star \partial_x) \diamond f = \partial_x \star (x \star f) - x \star (\partial_x \star f), \tag{4.55}$$

we can find that

$$[\partial_x, x]_\star = -[x, \partial_x]_\star = 1 \tag{4.56}$$

by means of

$$[\partial_x, x]_\star \diamond f = f, \qquad [x, \partial_x]_\star \diamond f = -f. \tag{4.57}$$

In the conventional quantum mechanics, the commutation relation is given by

$$[x, p]\psi = x(p\psi) - p(x\psi) = x(-i\partial\psi) - (-i\partial)(x\psi) = i\psi, \tag{4.58}$$

where x, p are position and momentum of a quantum particle respectively and ψ is a wave function. Thus, the commutation relation eq.(4.56) resembles a conventional commutation relation between the position and momentum in the quantum mechanics.

On the other hand, when we adopt the star product on behalf of the diamond operation as

$$\left([\partial_x, x]_\star\right) \star f \equiv \partial_x \star (x \star f) - (x \star \partial_x) \star f, \tag{4.59}$$

the commutation relation becomes

$$([\partial_x, x]_\star) \star f = f - \frac{i}{2}\theta_{xb}\partial_x\partial_b f. \tag{4.60}$$

In this case, there is redundant term. In this respect, it may be natural to use the diamond operation in the commutation relations for the noncommutative models.

Although, in mathematical point of view, we can consider the space/time noncommutativity, there is no promise of the physical consistency in the space/time noncommutative models [59, 60]. Therefore, we only consider the D-dimensional space/space noncommutative models, below. Moreover, we assume $\theta_{\mu\nu}(\mu, \nu = 1, 2, \cdots D-1)$ are constant scale parameters, which are given by

$$\theta_{\mu\nu} = \theta\epsilon_{\mu\nu}, \tag{4.61}$$

where $\epsilon_{\mu\nu}$ satisfies $\epsilon_{\mu\nu} = 1$ for $\mu < \nu$.

4.4. Noncommutative Gauge Theory

4.4.1. Gauge Transformation and U(N) Gauge Theory in Noncommutative Space

The U(N) gauge transformation in the noncommutative space should be natural to define in terms of the star-product as

$$\hat{\psi}' = \hat{U} \star \psi = e^{i\lambda^a(\hat{x})T_a} \star \hat{\psi}, \tag{4.62}$$

where $\hat{\psi}$ is a fermion field in the noncommutative language and T_a ($a = 1, 2, \cdots, N^2$) are generators of U(N) group which satisfy

$$[T_a, T_b] = if_{abc}T_c, \qquad \mathrm{tr}(T_aT_b) = \frac{1}{2}. \tag{4.63}$$

where f_{abc} is structure constant. Here, $\hat{U}$ and its conjugation $\hat{U}^\dagger$ are defined by

$$\hat{U} = e_\star^{i\lambda^a(\hat{x})T_a} \equiv 1 + i\lambda^a T_a + \frac{i^2}{2!}\lambda^a \star \lambda^b T_a T_b + \frac{i^3}{3!}\lambda^a \star \lambda^b \star \lambda^c T_a T_b T_c + \cdots, \tag{4.64a}$$

$$\hat{U}^\dagger = e_\star^{-i\lambda^a(\hat{x})T_a} \equiv 1 - i\lambda^a T_a + \frac{i^2}{2!}\lambda^a \star \lambda^b T_a T_b - \frac{i^3}{3!}\lambda^a \star \lambda^b \star \lambda^c T_a T_b T_c + \cdots, \tag{4.64b}$$

respectively. We can show that $\hat{U}$ has (star-)unitarity,

$$\hat{U} \star \hat{U}^\dagger = \hat{U}^\dagger \star \hat{U} = 1. \tag{4.65}$$

On the other hand, the U(N) star-gauge transformation of a gauge field $\hat{A}_\mu$ is defined by

$$\hat{A}_\mu \to \hat{U} \star \hat{A}_\mu \star \hat{U}^\dagger + i\hat{U} \star \partial_\mu \hat{U}^\dagger \tag{4.66}$$

where $\hat{A}_\mu = \hat{A}_\mu^a T_a$. The action of U($N$) Yang-Mills fields in the noncommutative space (NC) should be given by

$$S = -\frac{1}{2g^2}\int d^4x\, \mathrm{tr}\left(\hat{\mathcal{F}}_{\mu\nu} \star \hat{\mathcal{F}}^{\mu\nu}\right) = -\frac{1}{4g^2}\int d^4x\, \hat{\mathcal{F}}_{\mu\nu}^a \star \hat{\mathcal{F}}^{a\mu\nu}, \tag{4.67}$$

where the field strength $\hat{\mathcal{F}}_{\mu\nu}$ in the NC space should be given by

$$\hat{\mathcal{F}}_{\mu\nu} = \hat{\partial}_\mu \hat{A}_\nu - \hat{\partial}_\nu \hat{A}_\mu - i \left[\hat{A}_\mu, \hat{A}_\nu\right]_\star. \tag{4.68}$$

Here, the Moyal brackets between the gauge fields becomes

$$\left[\hat{A}_\mu, \hat{A}_\nu\right]_\star \equiv \hat{A}_\mu \star \hat{A}_\nu - \hat{A}_\nu \star \hat{A}_\mu = [A_\mu, A_\nu] + i\theta_{ij}\partial_i A_\mu \partial_j A_\nu. \tag{4.69}$$

Therefore, the NC field strength is written in terms of the commutative language as

$$\hat{\mathcal{F}}_{\mu\nu} = \partial_\mu A_\nu - \partial_\nu A_\mu - i[A_\mu, A_\nu] + \theta_{ij}\partial_i A_\mu \partial_j A_\nu + O(\theta^2). \tag{4.70}$$

It is important to note that the transformation of the NC field strength $\hat{\mathcal{F}}_{\mu\nu}$ under the star gauge transformation becomes

$$\hat{\mathcal{F}}'_{\mu\nu} = \hat{U} \star \hat{\mathcal{F}}_{\mu\nu} \star \hat{U}^\dagger. \tag{4.71}$$

Thus, in contrast to the conventional gauge theory, the field strength is not (star-)gauge invariant even when the gauge group is U(1).

On the other hand, the covariant derivative in the NC space should be given by

$$\hat{D}_\mu = \hat{\partial}_\mu - i\hat{A}^a_\mu T_a. \tag{4.72}$$

Therefore, the Lagrangian density of the Dirac fields in the NC space should be given by

$$\mathcal{L} = \hat{\bar{\psi}} \star \left[(i\gamma^\mu \hat{D}_\mu - m) \star \hat{\psi}\right], \tag{4.73}$$

where $\hat{\bar{\psi}} = \hat{\psi}^\dagger \gamma^0$.

4.4.2. Noncommutative U(1) Gauge Theory

In U(1) star-gauge theory, the infinitesimal star-gauge transformations are given by

$$\psi' = e^{i\lambda}\left(1 - \frac{1}{2}\theta_{ab}\partial_a\lambda\partial_b\right)\psi + O(\lambda^2, \theta^2), \quad A'_\mu = A_\mu + \partial_\mu\lambda - \theta_{ab}\partial_a\lambda\partial_b A_\mu + O(\lambda^2, \theta^2). \tag{4.74}$$

Further, we can find

$$\hat{D}_\mu \star \hat{\psi} = D_\mu \psi - \frac{i}{2}[A_\mu, \psi]_\star. \tag{4.75}$$

Under the infinitesimal star-gauge transformations, we can show that

$$\hat{D}'_\mu \star \hat{\psi}' = e^{i\lambda} \star (\hat{D}_\mu \star \hat{\psi}), \tag{4.76}$$

where $\hat{D}_\mu = \hat{\partial}_\mu - i\hat{A}_\mu$. On the other hand, since the NC field strength is transformed under the star-gauge transformation as

$$\hat{\mathcal{F}}'_{\mu\nu} = \hat{U} \star \hat{\mathcal{F}}_{\mu\nu} \star \hat{U}^\dagger \simeq \mathcal{F}_{\mu\nu} + \frac{i}{2}[\lambda, \mathcal{F}_{\mu\nu}] = \mathcal{F}_{\mu\nu} - \frac{\theta_{ab}}{2}\partial_a\lambda\partial_b \mathcal{F}_{\mu\nu},$$

we find
$$\hat{\mathcal{F}}'_{\mu\nu} \star \hat{\mathcal{F}}^{\mu\nu\prime} = \hat{\mathcal{F}}_{\mu\nu} \star \hat{\mathcal{F}}^{\mu\nu} + \frac{i}{2}[\lambda, \mathcal{F}_{\mu\nu}\mathcal{F}^{\mu\nu}].$$

Finally, the gauge invariant Lagrangian density of the noncommutative U(1) gauge theory is given by

$$\mathcal{L} = -\frac{1}{4g^2}\hat{\mathcal{F}}_{\mu\nu} \star \hat{\mathcal{F}}^{\mu\nu} + \hat{\bar{\psi}} \star \left[(i\gamma^\mu \hat{D}_\mu - m) \star \hat{\psi}\right], \qquad (4.77)$$

In the gauge theory, there is a gauge freedom. It means that we must consider the gauge invariant quantities as a physical observable. For instance, in conventional electromagnetic field theory, the gauge field (vector potential) A_μ is not a physical observable whereas electric fields and magnetic fields can be observables due to the gauge invariance of the field strength $F^{\mu\nu}$ since electric magnetic fields E and magnetic fields B are defined in terms of the field strength as

$$gE_j = F_{j0} = \partial_j A_0 - \partial_0 A_j, \qquad gB_j = -\frac{1}{2}\epsilon_{jkl}F_{kl} = -\frac{1}{2}\epsilon_{jkl}(\partial_k A_l - \partial_l A_k). \qquad (4.78)$$

In particular, the symmetric gauge

$$A_\mu = \left(0, -\frac{gB}{2}y, \frac{gB}{2}x, 0\right) \qquad (4.79a)$$

and the Landau gauge

$$A_\mu = (0, 0, gBx, 0) \qquad (4.79b)$$

show the same configuration of which there is a constant magnetic fields along z axis, $B = (0, 0, B)$. As long as we evaluate a gauge invariant quantity, the final answer is independent on the gauge choice. However, it is notice that the final result is *not* independent on the gauge choice when we perform a calculation in loosing way of the gauge invariance. For example, contrary to the case of QED, the perturbative calculation of QCD violates the gauge invariance since the perturbative term is not gauge invariant [3]. In this respect, there should be no assurance of consistency of the gauge choice in QCD. This should be why it is difficult to understand QCD physics.

Similarly, the electric fields and the magnetic fields in the noncommutative space are defined by

$$g\hat{E}_j = \hat{\mathcal{F}}_{j0}, \qquad g\hat{B}_j = -\frac{1}{2}\epsilon_{jkl}\hat{\mathcal{F}}_{kl}. \qquad (4.80)$$

In contrast to the ordinary U(1) gauge theory, the NC electric fields and the NC magnetic fields are not gauge invariant due to eq.(4.71). This fact indicates that the NC electric and magnetic fields depend on the gauge choice, and then they cannot be observables in the NC space. Therefore, we must care about whether we violate gauge invariant property on the way to the evaluation of the physical observables. Thus, as well as considering what the physical observable is in the model, we must find gauge invariant quantities in the gauge theory model.

It is interesting that we can show that the relation

$$\left[\hat{D}_\mu, \hat{D}_\nu\right]_\star \diamond \hat{\psi} = c_0 \hat{\psi} \qquad (4.81)$$

is gauge invariant, where c_0 is a constant number. Here, the diamond product between the commutation relation of the covariant derivative and the Dirac field $\hat{\psi}$ in the NC space is given by

$$\left[\hat{D}_\mu, \hat{D}_\nu\right]_\star \diamond \hat{\psi} \equiv \hat{D}_\mu \star (\hat{D}_\nu \star \hat{\psi}) - \hat{D}_\nu \star (\hat{D}_\mu \star \hat{\psi}). \quad (4.82)$$

It can be straightforwardly proved out of the following star-gauge covariance,

$$\hat{D}'_\nu \star (\hat{D}'_\mu \star \hat{\psi}') = e^{i\lambda} \star \left(\hat{D}_\nu \star (\hat{D}_\mu \star \hat{\psi})\right), \quad (4.83)$$

since

$$e^{i\lambda} \star \left(\left[\hat{D}_\mu, \hat{D}_\nu\right]_\star \diamond \hat{\psi}\right) = e^{i\lambda} \star (c_0 \hat{\psi}).$$

Note that eq.(4.82) can be explicitly written by

$$\left[\hat{D}_\mu, \hat{D}_\nu\right]_\star \diamond \hat{\psi} = -i\hat{\mathcal{F}}_{\mu\nu} \star \hat{\psi} + \frac{1}{2}\theta_{ab}\partial_a\left[(\partial_\mu A_\nu - \partial_\nu A_\mu + A_\mu \partial_\nu - A_\nu \partial_\mu)\partial_b \psi\right]. \quad (4.84)$$

5. Quantum Mechanics in Noncommutative Space

Recently, there is much progress in the noncommutative quantum mechanics, e.g. Landau level problem, Aharonov-Bohm effect, quantum Hall effect [57, 61, 62, 63, 64, 65, 69]. It should motivate us whether the noncommutative effect is indeed taken into account in real physics. Since the noncommutative model in the condensed matter physics is defined in terms of the model to replace a multiplication by a star-product, the noncommutative effect may be possible to observe in the nontrivial difference from the conventional effect. For example, according to the estimation of the noncommutative Aharonov-Bohm effect [63], the noncommutative parameter should be suppressed as

$$\sqrt{\theta} < 0.20 \times 10^{-7} \text{ cm}.$$

On the other hand, according to the noncommutative quantum Hall effect [66], it should be suppressed as

$$\sqrt{\theta} < 0.15 \times 10^{-7} \text{ cm}.$$

However, there is no clear evidence of the noncommutativity in any experiments until now.

In this section, we focus on the Landau level problem in the noncommutative space with its related topics. First, we review the Landau level problem in commutative space. Next, we see the noncommutative Landau problem in terms of Bopp shift formulation. Third, we reconstruct the model in gauge invariant way. Finally, I show the recent result of the noncommutative Hofstadter Butterfly diagram as well as the result of the commutative case.

5.1. Landau Level Problem in Commutative Space

5.1.1. Classical Mechanics

First of all, it should be instructive to review briefly the Landau level problem in conventional space [67]. The classical equation of motion of an electron whose charge is e and mass is m in uniform magnetic field $\boldsymbol{B} = (0, 0, B)$ is written by

$$m\frac{d\boldsymbol{v}}{dt} = -e\boldsymbol{v} \times \boldsymbol{B} = -ev_y B \boldsymbol{e}_x + ev_x B \boldsymbol{e}_y, \tag{5.1}$$

where $\boldsymbol{e}_i (i = x, y)$ is a unit vector of the i-axis. Obviously, the velocity of the electron is given by

$$\boldsymbol{v}(t) = v_0 \cos \frac{eB_0}{m} t \, \boldsymbol{e}_x - v_0 \sin \frac{eB_0}{m} t \, \boldsymbol{e}_y. \tag{5.2}$$

Furthermore, the position of the electron is given by

$$\boldsymbol{r}(t) = \left(x_0 + \frac{mv_0}{eB} \sin \frac{eB}{m} t \right) \boldsymbol{e}_x + \left(y_0 - \frac{mv_0}{eB} \sin \frac{eB}{m} t \right) \boldsymbol{e}_y. \tag{5.3}$$

This solution shows that the electron move along the circle whose radius is mv_0/eB and center is $\boldsymbol{R} = (x_0, y_0)$ in xy-plane. Note that the position of the electron is rewritten in terms of the momentum of the electron as

$$\boldsymbol{r}(t) = \left(x_0 + \frac{p_x}{eB} \right) \boldsymbol{e}_x + \left(y_0 - \frac{p_y}{eB} \right) \boldsymbol{e}_y, \tag{5.4}$$

where $\boldsymbol{p} = m\boldsymbol{v}$. It is convenient to define the relative coordinate as

$$\boldsymbol{\xi} \equiv \boldsymbol{r}(t) - \boldsymbol{R} = \frac{p_y}{eB} \boldsymbol{e}_x - \frac{p_x}{eB} \boldsymbol{e}_y. \tag{5.5}$$

5.1.2. Quantum Mechanics of Landau Problem

Now, we consider the quantum electron in the uniform magnetic field $\boldsymbol{B} = (0, 0, B)$. In this case, Hamiltonian of the electron whose charge is e and mass is m is given by

$$H = \frac{1}{2m} (\boldsymbol{p} + e\boldsymbol{A})^2, \tag{5.6}$$

where we use a little different convention of the gauge field from the previous section, which is frequently used in condensed matter physics. In this case, the gauge transformations are given by

$$\boldsymbol{A}' = \boldsymbol{A} + \nabla \chi, \qquad \psi' = e^{-ie\chi} \psi. \tag{5.7}$$

Note that there is gauge freedom in it. First, we take a symmetric gauge

$$\boldsymbol{A} = \left(-\frac{B}{2} y, \frac{B}{2} x, 0 \right). \tag{5.8}$$

In this case, the Hamiltonian becomes

$$H_S = \frac{1}{2m} \left(\Pi_x^2 + \Pi_y^2 \right), \tag{5.9}$$

where
$$\Pi_j = p_j - \frac{eB}{2}\epsilon_{jk}r_k = p_j - \frac{1}{2\ell^2}\epsilon_{jk}r_k, \qquad (j,k=1,2), \tag{5.10}$$

where $r_1 = x$, $r_2 = y$ and $\ell \equiv 1/\sqrt{eB}$. It can be easily verified that

$$[r_j, \Pi_j] = i, \qquad [\Pi_x, \Pi_y] = -i\ell^{-2}. \tag{5.11}$$

The central coordinate $\boldsymbol{R}$ of the circle and the relative coordinate $\boldsymbol{\xi}$ of the electron are defined in analog form of eq.(5.5) by

$$\boldsymbol{R} = \boldsymbol{r} - \boldsymbol{\xi} = (x - \ell^2\Pi_y,\ y + \ell^2\Pi_x), \qquad \boldsymbol{\xi} = (\ell^2\Pi_y,\ -\ell^2\Pi_x). \tag{5.12}$$

In this case, we can find the following commutation relations

$$[H, R_i] = 0, \qquad [R_x, x] = 0, \qquad [R_y, y] = 0, \qquad [R_j, \Pi_k] = 0 \tag{5.13a}$$

and

$$[R_x, y] = i\ell^2, \qquad [R_y, x] = -i\ell^2, \qquad [R_x, R_y] = i\ell^2. \tag{5.13b}$$

It is interesting that the noncommutativity about the central position of the electron emerges in terms of the constant magnetic fields.

Next, we take a Landau gauge

$$\boldsymbol{A} = (0, Bx, 0) \tag{5.14}$$

In this case, the Hamiltonian becomes

$$H_L = \frac{1}{2m}\left(\Pi_x^2 + \Pi_y^2\right) = \frac{1}{2m}\left[p_x^2 + (p_y + eBx)^2\right], \tag{5.15}$$

where

$$\Pi_x = p_x, \qquad \Pi_y = p_y + eBx. \tag{5.16}$$

Further, we have the commutation relations

$$[r_j, \Pi_j] = i, \qquad [\Pi_x, \Pi_y] = -i\ell^{-2}. \tag{5.17}$$

It is interesting to note that the system is described in terms of the harmonic oscillator. To do this, we take the canonical momentum in terms of the bosonic operators $a, a^\dagger$ as

$$\Pi_x = \frac{1}{\sqrt{2}i\ell}\left(a - a^\dagger\right), \qquad \Pi_y = \frac{1}{\sqrt{2}\ell}\left(a + a^\dagger\right) \tag{5.18a}$$

and

$$x = \ell(a - a^\dagger), \qquad y = \ell(a + a^\dagger), \tag{5.18b}$$

where

$$\left[a, a^\dagger\right] = 1. \tag{5.19}$$

It is important to note that the commutation relations,

$$[x, \Pi_x] = i, \qquad [y, \Pi_y] = i, \qquad [x, y] = 0, \qquad [\Pi_x, \Pi_y] = -i\ell^{-2}.$$

are gauge invariant. Therefore, the Hamiltonian eq.(5.9) of the electron in the uniform magnetic fields can be expressed in terms of a harmonic oscillator as

$$H = \frac{eB}{m}\left(a^\dagger a + \frac{1}{2}\right). \tag{5.20}$$

Accordingly, the energy spectrum of the electron becomes discretized spectrum due to the uniform magnetic fields. Indeed, we can observe the effect of the Landau level in experiments.

Note that each level in the energy spectrum of the Landau level depends on the external magnetic field B. Since the density of state per area becomes

$$D(E) = \frac{m}{2\pi}. \tag{5.21}$$

in two dimensions, the degeneracy of each level is given by

$$n = \frac{eB}{2\pi} = \frac{B}{\phi_0}, \tag{5.22}$$

where $\phi_0 \equiv 1/e$ is a flux quantum. Therefore, the number of flux which penetrates the area S is given by

$$N = \frac{\Phi}{\phi_0}, \tag{5.23}$$

where $\Phi = BS$ is magnetic flux.

5.1.3. Aharonov-Bohm Effect

There is an interesting phenomenon such as Aharonov-Bohm effect in the electron moving in the constant magnetic fields. The electron has nontrivial phase shift in the uniform magnetic fields, which is known as Berry phase. The Schrödinger equation of the electron $\Psi(\boldsymbol{r})$ is given by

$$\frac{1}{2m}(\boldsymbol{p} + e\boldsymbol{A})^2 \Psi(\boldsymbol{r}) = E_{\boldsymbol{k}}\Psi(\boldsymbol{r}), \qquad E_{\boldsymbol{k}} = \frac{\boldsymbol{k}^2}{2m}. \tag{5.24}$$

The formal solution of the Schrödinger equation is given by

$$\Psi(\boldsymbol{r}) = \Psi_0(\boldsymbol{r})\exp\left[-ie\int_0^r dr_j A_j\right], \tag{5.25a}$$

where

$$\Psi_0(\boldsymbol{r}) = e^{i\boldsymbol{k}\cdot\boldsymbol{r}}. \tag{5.25b}$$

Thus, the phase shift of the electron which goes around the closed loop for the area S becomes

$$\Delta\Phi_{AB} = -ie\oint dx_j A_j = -i\frac{2\pi\Phi}{\phi_0} = -2\pi N. \tag{5.26}$$

Note that it is proportional to the number of the flux which is given by eq.(5.23). Further, it is important to note that the Berry phase shift is gauge invariant. Under the gauge transformation, the shift becomes

$$\Delta\Phi'_{AB} = -ie\oint d\boldsymbol{r}\cdot\boldsymbol{A}' = -ie\oint d\boldsymbol{r}\cdot(\boldsymbol{A} + \nabla\chi).$$

According to Stokes theorem,

$$\oint \boldsymbol{F} \cdot d\boldsymbol{r} = \iint_S d\boldsymbol{S} \cdot (\nabla \times \boldsymbol{F}), \tag{5.27}$$

we have

$$\Delta\Phi'_{AB} = \Delta\Phi_{AB} - ie \oint d\boldsymbol{r} \cdot \nabla\chi = \Delta\Phi_{AB} - ie \iint d\boldsymbol{S} \cdot \nabla \times \nabla\chi.$$

Since there is well-known formulation, rot grad $\phi = 0$ for any scalar functions ϕ, we can show that the Berry phase shift is gauge invariant quantity,

$$\Delta\Phi'_{AB} = \Delta\Phi_{AB}.$$

5.2. Bopp Shift Formulation in Noncommutative Quantum Mechanics

5.2.1. Bopp Shift

Now, we consider the Landau level problem in the noncommutative space. The Hamiltonian of an electron moving under the uniform magnetic field in the noncommutative space should be given by

$$\hat{H} = \frac{1}{2m}\left(\hat{\boldsymbol{p}} + e\hat{\boldsymbol{A}}(\hat{\boldsymbol{r}})\right) \star \left(\hat{\boldsymbol{p}} + e\hat{\boldsymbol{A}}(\hat{\boldsymbol{r}})\right). \tag{5.28}$$

The noncommutative coordinate $\hat{r}$ can be written in terms of the commutative coordinate r with eq.(4.48). In this shift, the NC gauge field should be written by the commutative gauge field as

$$\hat{A}_j \to A_j - \frac{1}{2}\theta_{ab}\partial_a A_j p_b. \tag{5.29}$$

Further, due to the shift, the NC Hamiltonian can be written in terms of the commutative coordinates as

$$H_{nc} = \frac{1}{2m}\left(p_j + eA_j - \frac{e}{2}\theta_{ab}\partial_a A_j p_b\right)^2. \tag{5.30}$$

Thus, the canonical momentum Π_j in the noncommutative space becomes

$$\Pi_j = p_j + eA_j - \frac{e}{2}\theta_{ab}\partial_a A_j p_b. \tag{5.31}$$

In particular, for the symmetric gauge, the Hamiltonian eq.(5.30) becomes

$$H_S = \frac{1}{2m}\left[\left\{\left(1 - \frac{eB\theta}{4}\right)p_x - \frac{eB}{2}y\right\}^2 + \left\{\left(1 - \frac{eB\theta}{4}\right)p_y + \frac{eB}{2}x\right\}^2\right], \tag{5.32}$$

where we take $\theta_{xy} = \theta$. Moreover, the commutation relations are given by

$$[\Pi_x, \Pi_y] = -ieB\left(1 - \frac{eB\theta}{4}\right) = -i(\ell\xi_\theta)^{-2} \tag{5.33a}$$

and

$$[x, \Pi_x] = i\xi_\theta^{-2}, \quad [y, \Pi_y] = i\xi_\theta^{-2}, \quad [x, \Pi_y] = 0, \quad [y, \Pi_x] = 0, \tag{5.33b}$$

where
$$\xi_\theta^{-2} \equiv 1 - \frac{eB\theta}{4}. \tag{5.33c}$$

Similarly, we can define the center position operator R_i^θ of the noncommutative electron as
$$R_x^\theta = \xi_\theta^2 x - \ell^2 \xi_\theta^2 \Pi_y, \qquad R_y^\theta = \xi_\theta^2 y + \ell^2 \xi_\theta^2 \Pi_x. \tag{5.34}$$

Obviously, we have
$$[H, R_i^\theta] = 0. \tag{5.35}$$

Note that the Hamiltonian can be expressed in terms of a harmonic oscillator [65] as
$$H = \frac{eB}{m}\xi_\theta^{-2}\left(a^\dagger a + \frac{1}{2}\right), \qquad [a, a^\dagger] = 1, \tag{5.36}$$

where
$$\Pi_x = \frac{\xi_\theta^{-1}}{i}\sqrt{\frac{eB}{2}}\left(a - a^\dagger\right), \qquad \Pi_y = \xi_\theta^{-1}\sqrt{\frac{eB}{2}}\left(a + a^\dagger\right).$$

Therefore, the energy spectrum of the electron in the noncommutative space is discretized and the Landau level like spectrum also emerges in noncommutative space. Here, the gap of the energy spectrum is given by
$$E = \frac{eB}{m}\xi_\theta^{-2}. \tag{5.37}$$

Next, we take the Landau gauge. In this case, the Hamiltonian becomes
$$H_L = \frac{1}{2m}\left[p_x^2 + \left\{\left(1 - \frac{eB\theta}{2}\right)p_y + eBx\right\}^2\right]. \tag{5.38}$$

On the other hand, the canonical momentum Π_i becomes
$$\Pi_x = p_x, \qquad \Pi_y = \left(1 - \frac{eB\theta}{2}\right)p_y + eBx \tag{5.39}$$

Further, the commutations relations are given by
$$[\Pi_x, \Pi_y] = -ieB = -i\ell^{-2} \tag{5.40a}$$

as well as
$$[x, \Pi_x] = i, \qquad [y, \Pi_y] = i\xi_\theta'^{-2}, \qquad [x, \Pi_y] = 0, \qquad [y, \Pi_x] = 0, \tag{5.40b}$$

where
$$\xi_\theta'^{-2} \equiv 1 - \frac{eB\theta}{2}. \tag{5.40c}$$

Note that eq.(5.40a) is equivalent to the conventional commutation relation. Therefore, The noncommutative effect disappear from the energy spectrum of the noncommutative Landau level problem if we take the Landau gauge. Obviously, it is inconsistent with eq.(5.33a). This is led from the fact that the commutation relation of eq.(5.33a) is not star-gauge invariant.

5.2.2. Gauge Choice and Gauge Dependence in Bopp Shift Formulation

The wave function Ψ_{nc} of a noncommutative electron should satisfy the eigenvalue equation,

$$\hat{H}\Psi_{nc}(\boldsymbol{r}) = E\Psi_{nc}(\boldsymbol{r}). \tag{5.41}$$

If we assume the Bopp shift as well as a standard quantization procedure,

$$p_j = -i\partial_j = -i\frac{\partial}{\partial r_j} = -i\nabla_j, \tag{5.42}$$

we have the Schrödinger equation of the electron in the noncommutative space as

$$\frac{1}{2m}\left(-i\frac{\partial}{\partial r_j} + eA_j + i\frac{e}{2}\theta_{ab}\partial_a A_j \frac{\partial}{\partial r_b}\right)^2 \Psi_{nc}(\boldsymbol{r}) = E_{\boldsymbol{k}}\Psi_{nc}(\boldsymbol{r}), \tag{5.43}$$

where $E_{\boldsymbol{k}} = \boldsymbol{k}^2/2m$. The canonical momentum $\widetilde{\Pi}_j$ in the noncommutative space should be given by

$$\widetilde{\Pi}_j = p_j + eA_j - \frac{e}{2}\theta_{ab}\partial_a A_j p_b. \tag{5.44}$$

Thus, we can find that

$$\left[x, \widetilde{\Pi}_x\right] = i\left(1 + \frac{e}{2}\theta_{xy}\partial_y A_x\right), \qquad \left[y, \widetilde{\Pi}_y\right] = i\left(1 - \frac{e}{2}\theta_{xy}\partial_x A_y\right), \tag{5.45a}$$

$$\left[x, \widetilde{\Pi}_y\right] = i\frac{e}{2}\theta_{xy}\partial_y A_y, \qquad \left[y, \widetilde{\Pi}_x\right] = -i\frac{e}{2}\theta_{xy}\partial_x A_x \tag{5.45b}$$

and

$$\left[\widetilde{\Pi}_x, \widetilde{\Pi}_y\right] = -ief_{xy} + \frac{ie}{2}\theta_{ab}\left(\partial_a f_{ab}\right)p_b + ie^2\theta_{ab}\partial_a A_x \partial_b A_y, \tag{5.45c}$$

where $f_{ab} = \partial_a A_b - \partial_b A_a$. Note that eq.(5.45) is not star-gauge invariant. Here, we assume

$$[r_i, p_i] = i, \qquad [x, y] = 0, \qquad [p_x, p_y] = 0.$$

Thus, the Bopp shift formulation of the U(1) gauge theory should violate star-gauge invariance.

5.3. Star-Gauge Formulation in Noncommutative Landau Level Problem

To preserve the star-gauge invariance, the noncommutative Schrödinger equation of the electron in the noncommutative gauge theory can be written by [69]

$$-\frac{1}{2m}\left(\boldsymbol{D} \star (\boldsymbol{D} \star \psi)\right) = E_0 \psi. \tag{5.46}$$

We can straightforwardly prove the gauge invariance of the equation since we can find

$$-\frac{1}{2m}e^{-ie\chi} \star \left(\boldsymbol{D} \star (\boldsymbol{D} \star \psi)\right) = E_0 e^{-ie\chi} \star \psi$$

under the star-gauge transformations

$$\psi' = e^{-ie\chi} \star \psi, \qquad \hat{A}'_\mu = \hat{A}_\mu + \partial_\mu \chi.$$

Note that eq.(5.46) can be denoted shortly in terms of the diamond operator $\diamond$ as

$$\hat{H} \diamond \hat{\psi} = E_0 \hat{\psi}, \tag{5.47a}$$

where

$$H = \frac{1}{2m} \hat{\Pi} \star \hat{\Pi}, \qquad \Pi = p + e\boldsymbol{A}. \tag{5.47b}$$

where $p = -i\nabla$. Here, we use the fact that the canonical momentum should be related to the covariant derivative in terms of

$$\hat{\Pi} = -i\hat{D}. \tag{5.48}$$

Thus, eq.(5.46) or eq.(5.47) is a star-gauge invariant Schrödinger equation in the noncommutative space,

On the other hand, we have already discussed the star-gauge invariance about the non-commutative covariant derivative in the section 4.4.2.. From eq.(4.84), we can find

$$\left[\hat{\Pi}_x, \hat{\Pi}_y\right] \diamond \hat{\psi}_s = -i(\ell\xi_\theta)^{-2} \hat{\psi}_s, \tag{5.49}$$

for the symmetric gauge and

$$\left[\hat{\Pi}_x, \hat{\Pi}_y\right]_\star \diamond \hat{\psi}_l = -ieB\psi_l - \frac{eB}{2} \theta_{xy} \partial_x \partial_y \psi_l \tag{5.50}$$

for the Landau gauge. For the Landau gauge, there is a nontrivial term which depends on the derivative of the wave function. However, we have already seen that eq.(4.81) is star-gauge invariant. Therefore, eq.(5.49) is valid at any gauge choices.

It is interesting to note that eq.(5.49) is equivalent to eq.(5.33a) which is determined by the Bopp shift formulation in the symmetric gauge. Therefore, as long as we take a symmetric gauge, the Bopp shift formulation may have good answers.

5.4. Hofstadter Butterfly Diagram in Noncommutative Space

The Hofstadter butterfly diagram [68] illustrates the relationship between the energy spectrum of the lattice Hamiltonian of an electron and the flux $\phi = eB$. Thus, the diagram shows the band structure of an electron in the magnetic fields. One interesting aspect of the diagram is that there are q band at $\phi = p/q$ where p, q are prime numbers. As seen from Figure.3, it can be found a fractal structure in the diagram [67, 68]. Note, further, that in the weak magnetic limit $\phi \to 0$, the spectrum is composed of straight lines. Thus, it gives the Landau level in the continuous limit.

As seen before, the Bopp shift result should be valid as long as we take a symmetric gauge. From eq.(5.32), the corresponding Hamiltonian can be expressed in terms of the harmonic oscillator

$$H = \frac{eB}{m} \xi_\theta^{-2} \left(a^\dagger a + \frac{1}{2}\right), \qquad \left[a, a^\dagger\right] = 1. \tag{5.51}$$

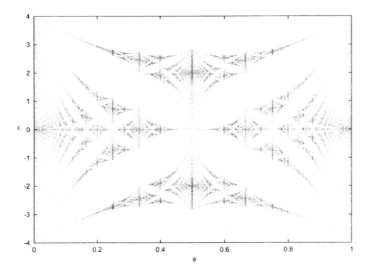

Figure 3. Hofstadter butterfly diagram in the commutative space

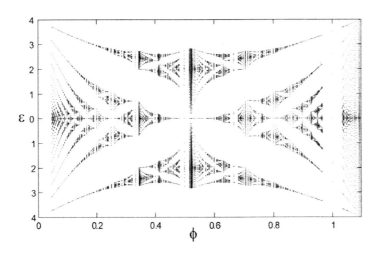

Figure 4. Noncommutative Hofstadter butterfly diagram at $\theta = 1/3$ which corresponds to the one period of commutative diagram.

Thus, the gap of the energy spectrum is given by

$$E = \frac{eB}{m}\xi_\theta^{-2}. \tag{5.52}$$

On the other hand, the commutation relation of the momentum is

$$[D_x, D_y]_\star \diamond \psi = -ieB\left(1 - \frac{eB}{4}\theta\right)\psi. \tag{5.53}$$

Consequently, the Landau level in the noncommutative space is different from the commutative case by a factor of ξ_θ^{-2}. In this respect, the Hofstadter butterfly diagram in the noncommutative space is straightforwardly calculated. The results are shown in Figure.4 and Figure.5 [69].

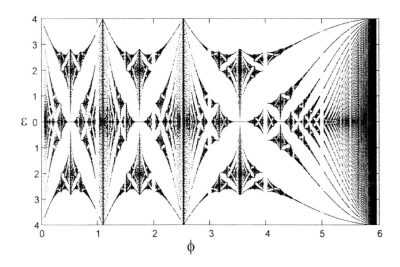

Figure 5. Noncommutative Hofstadter butterfly diagram at $\theta = 1/3$.

6. Conclusions

I presented the review of string theory and its related topics. In addition, the noncommutative space was discussed. In this chapter, I focused on the nonperturbative phenomena such as gauge/gravity correspondence, the confinement of quarks, Landau level problem in the noncommutative space and so forth. The main results in this chapter are only theoretical predictions. Thus, we need the experimental validations whether we capture essences of nature properly in terms of presented approach. Unfortunately, in the present circumstance, there are no clear evidences of which the string theory and the noncommutative space work for resolving the mysteries of modern physics.

The confiment phenomenon in QCD is most challenging issue for particle physicists. We can find the confinement phase in terms of the conjecture of AdS/CFT correspondence. Strictly speaking, we assume the duality between the gauge theories and the gravity solutions in general to explain the confinement. However, it is improtant to note that there is no proof of the dulities. Further, it should be afraid that the Wilson loop is not well-established. Accordingly, we only found that it is indeed difficult problems to solve the confinement properly.

Since the QCD is a gauge theory, the gauge symmetry should be a fundamental role in the confinement phenomenon. In particular, the non-abelian gauge symmetry plays a importtant role in the confinement. As far as we adopt gauge principle, we cannot take advantage of the isolated quarks since the isolated quarks are not gauge invariant aspects. Therefore, the confinement problem may be equivalent to why the non-abelian (color) gauge symmetry is not broken spontaneously.

References

[1] K. Nishijima *Fields and Particles*; W.A. Benjamin Inc., 1980.

[2] Fujita, T. *Symmetry and its breaking in quantum field theory*; Nova Science Publisher Inc., 2006.

[3] Fujita, T.: Kanemaki, S.; Oshima, S. (2005). *Gauge non-invariance of quark-quark interactions*. hep-ph/0511326.

[4] Fujita, T.; Hiramoto, M.; Takahashi, H. In *Focus on Boson Research*; Ling, A.V.; ed.; Nova Science Publisher Inc; 2005.

[5] Fujita, T.; Hiramoto, M.; Homma, T.; Takahashi, H. *J. Phys. Soc. Japan* 2005, 74, 1143-1149.

[6] Fujita, T.; Hiramoto, M.; Homma, T.; Matsumoto, M.; Takahashi, H. (2005). *Re-interpretation of spontaneous symmetry breaking in quantum field theory* . hep-th/0510151.

[7] Wilson, K.G. *Phys. Rev. D* 1974, 10, 2445-2459.

[8] Creutz, M. *Quarks, Gluons and Lattices*; Cambridge Monographs On Mathematical Physics; Cambridge, Uk: Univ. Pr.; 1983.

[9] Asaga, T.; Fujita, F. (2005). *No area law in QCD*. hep-th/0511172.

[10] Green, M.B.; Schwarz, J.H.; Witten, E. *Superstring theory*; Cambridge University Press, 1987.

[11] Lüst, D.; Theisen, S. *Lectures on String Theory*; Lecture Notes in Physics; Springer-Verlag Berlin: Heidelberg, 1989; 346.

[12] Polchinski, J. *String theory*; Cambridge University Press, 1998.

[13] Snyder, H.S. *Phys. Rev.* 1947, 71, 38-41.

[14] Ginsparg, P. In *Proceedings of the Les Houches school on Field, String and Critical Phenomena*; Brezin E.; Zin-Justin J.; Eds.; North Holland; 1989.

[15] Cardy, J. In *Proceedings of the Les Houches school on Field, String and Critical Phenomena*; Brezin E.; Zin-Justin J.; Eds.; North Holland; 1989.

[16] Christe, P.; Henkel, M. *Introduction to Conformal Invariance and Its Applications to Critical Phenomena*; Lecture Note in Physics; Springer-Verlag, 1993; m16.

[17] Di Francesco, P.; Mathieu, P.; Sénéchal, D. *Conformal Field Theory*; Springer: New York, 1996.

[18] Takahashi, H. *Massive Thirring Model and Finite Size Corrections*; Doctoral thesis; Physics Major, Graduate School of Science and Technology, Nihon University, 1999.

[19] Belavin, A.A.; Polyakov, A.M.; Zamolodchikov, A.B. *Nucl. Phys. B* 1984, 241, 333-380.

[20] Kac, V.G. In *Group Theoretical Methods in Physics*; Lecture Notes in Physics; Springer-Verlag Berlin: Heidelberg, 1979; Vol. 94, pp 441-445.

[21] Feigin, B.L.; Fuchs, D.B. *Funct. Anal. Appl.* 1982, 16, 114-126.

[22] Friedan, D.; Qiu, Z.; Shenker, S. *Comm. Math. Phys.* 1986, 107, 535-542.

[23] Goddard, P.; Olive, D. *Int. J. Mod. Phys. A* 1986, 1, 303-414.

[24] Thorn, C.B. In *Vertex Operators in Mathematics and Physics*; Lepowsky, J.; Mandelstam, S.; Singer, I.M.; Ed.; Springer-Verlag, 1985; 411-417.

[25] Coleman, S.R. *Comm. Math. Phys.* 1973, 31, 259-264.

[26] Yang, S.-K. *Nucl. Phys. B* 1987, 285, 183-203.

[27] Callan, C.G.; Friedan, D.; Martinec, E.J.; Perry, M.J. *Nucl. Phys. B* 1985, 262, 593-609.

[28] Callan, C.G.; Lovelance, C.; Nappi, C.R.; Yost, S.A. *Nucl. Phys. B* 1988, 308, 221-284.

[29] Giveon, A.; Porrati, M.; Rabinovici, E. *Phys. Rep.* 1994, 244, 77-202.

[30] Polchinski, J. *Phys. Rev. Lett.* 1995, 75, 4724-4727.

[31] Maldacena, J. *Int. J. Theor. Phys.* 1999, 38, 1113-1133.

[32] Aharony, O.; Gubser, S.S.; Maldacena, J.; Ooguri, H.; Oz, Y. *Phys. Rep.* 2000, 323, 183-386.

[33] Itzhaki, N.; Maldacena, J.M.; Sonnenschein, J.; Yankielowicz, S. *Phys. Rev. D* 1998, 58, 046004.

[34] 't Hooft, G. *Nucl. Phys. B* 1974, 72, 461-473.

[35] Maldacena, J. *Phys. Rev. Lett.* 1998, 80, 4859-4862.

[36] Rey, S.J.; Yee, J.T. *Eur. Phys. J. C* 2001, 22, 379-394.

[37] Dhar, A.; Kitazawa, Y. *Phys. Rev. D* 2001, 63, 125005.

[38] Dijkgraaf, R. *Nucl. Phys. B* 1999, 543, 545-571.

[39] David, J.R.; Mandal, G.; Wadia, S.R. *Phys. Rep.* 2002, 369, 549-686.

[40] Takahashi, H.; Nakajima, T.; Suzuki, K. Phys. Lett. B 2002, 546, 273-278.

[41] Alishahiha, M.; Oz, Y.; Sheikh-Jabbari, M.M. *JHEP* 1999, 11, 007.

[42] Liu, H.; Tseytlin, A.A. *Nucl. Phys. B* 1999, 553, 231-249.

[43] Ardalan, F.; Arfaei, H.; Sheikh-Jabbari, M.M. *JHEP* 1999, 02, 016.

[44] Maldacena, J.M.; Russo, J.G. *JHEP* 1999, 09, 025.

[45] Seiberg, N.; Witten, E. *JHEP* 1999, 09, 032.

[46] Russo, J.G.; Sheikh-Jabbari, M.M. *JHEP* 2000, 07, 052.

[47] Dhar, A.; Mandal, G.; Wadia, S.R.; Yogendrana, K.P. *Nucl. Phys. B* 2000, 575, 177-194.

[48] Mikhailov, A. *Nucl. Phys. B* 2000, 584, 545-588.

[49] Hashimoto, A.; Itzhaki, N. *Phys. Lett. B* 1999, 465, 142-147.

[50] Minahan, J.A. *JHEP* 1999, 01, 020.

[51] Csaki, C.; Ooguri, H.; Oz, Y.; Terning, J. *JHEP* 1999, 01, 017.

[52] Nakajima, T.; Suzuki, K.; Takahashi, H. *JHEP* 2006, 01, 016.

[53] Connes, A.; Douglas, M.R.; Schwarz, A. *JHEP* 1998, 02, 003.

[54] Douglas, M.R.; Nekrasov, N.A. *Rev. Mod. Phys.* 2001, 73, 977-1029.

[55] Szabo, R.J. *Int. J. Mod. Phys. A* 2004, 19, 1837-1861.

[56] Madore, J.; Schraml, S.; Schupp, P.; Wess, J. *Eur. Phys. J. C* 2000, 16, 161-167.

[57] Falomir, H.; Gamboa, J.; Loewe, M.; Mendez, F.; Rojas, J.C. *Phys. Rev. D* 2002, 66, 045018.

[58] Curtright, T.; Fairlie, D.; Zachos, C. *Phys. Rev. D* 1998, 58, 025002.

[59] Seiberg, N; Susskind, L.; Toumbas, N. *JHEP* 2000, 06, 044

[60] Gomis, J.; Mehen, T. *Nucl. Phys. B* 2000, 591, 265-276.

[61] Gamboa, J.; Loewe, M.; Rojas, J.C. *Phys. Rev. D* 2001, 64, 067901.

[62] Chaichian, M.; Sheikh-Jabbari, M.M.; Tureanu, A. *Phys. Rev. Lett.* 2001, 86, 2716-2719.

[63] Chaichian, M.; Presnajder, P.; Sheikh-Jbbari, M.M.; Tureanu, A. *Phys. Lett. B* 2002, 527, 149-154.

[64] Horvathy, P.A. *Ann. Phys.* 2002, 299, 128-140.

[65] Dayi, O.; Jellal, A. *J .Math. Phys.* 2002, 43, 4592-4601.

[66] Colatto, L.P.; Penna, A.L.A.; Santos, W.C. *Phys. Rev. D* 2006, 73, 105007.

[67] Hatsugai, Y.; *J. Phys.: Condens. Matter* 1997, 9, 2507-2549.

[68] Hofstadter, D.R. *Phys. Rev. B*, 1976, 14, 2239-2249.

[69] Takahashi, H.; Yamanaka, M. (2006). *Hofstadter Butterfly Diagram in Noncommutative Space*. hep-th0606168.

In: String Theory Research Progress
Editor: Ferenc N. Balogh, pp. 81-131

ISBN 1-60456-075-6
© 2008 Nova Science Publishers, Inc.

Chapter 2

IN SEARCH OF THE (MINIMAL SUPERSYMMETRIC) STANDARD MODEL STRING

Gerald B. Cleaver[*]
Center for Astrophysics, Space Physics & Engineering Research
Department of Physics, Baylor University, Waco, TX 76798-7316

Abstract

This chapter summarizes several developments in string-derived (Minimal Supersymmetric) Standard Models. Part one reviews the first string model containing solely the three generations of the Minimal Supersymmetric Standard Model and a single pair of Higgs as the matter in the observable sector of the low energy effective field theory. This model was constructed by Cleaver, Faraggi, and Nanopoulos in the $\mathbb{Z}_2 \otimes \mathbb{Z}_2$ free fermionic formulation of weak coupled heterotic strings. Part two examines a representative collection of string/brane-derived MSSMs that followed. These additional models were obtained from various construction methods, including weak coupled $\mathbb{Z}_6$ heterotic orbifolds, strong coupled heterotic on elliptically fibered Calabi-Yau's, Type IIB orientifolds with magnetic charged branes, and Type IIA orientifolds with intersecting branes (duals of the Type IIB). Phenomenology of the models is compared.

1. Introduction: The Mission of String Phenomenology

Opponents of string theory [1, 2, 3] have suggested that research in string theory [*] may be moving beyond the bounds of science, especially with the rise of the theorized string landscape containing an estimated $10^{100 \text{ to } 500}$ vacua, among which exactly one vacua corresponds to our visible universe. The claim is that "what passes for the most advanced theory in particle science these days is not really science" [2]. Instead, the opposition argues that string theory has "gone off into a metaphysical wonderland" producing "models with no testable consequences," rather than models "with testable and falsifiable consequences." Richter, for example, maintains that real progress in physics comes when "why" questions

---

[*]E-mail address: gerald_cleaver@baylor.edu

[*]The term *string theory* denotes in this chapter its manifestation as a whole, from the collection of 10-dimensional string theories pre-second string revolution to its current all encompassing, but still incompletely understood, 11-dimensional M-theory.

turn into "how" questions; he views string theory as trespassing into metaphysical speculation rather than answering the how questions. For Richter, "why" answers only provide vague, unscientific answers, whereas "how" answers reveal a clear, physical process [2].

Alternately, many within the string community contend that the role of string theory is, indeed, to answer the "why" questions. What's at odds are the meanings attached to "why" and "how". Munoz [4], for example, maintains that while the standard model of particle physics "adequately describes how the elementary particles behave," it is insufficient "to explain *why* a particular set of elementary particles, each with its own particular properties, exists." In this context, answering the "why" is "precisely one of the purposes of string theory." Munoz "why" is Richter's "how," which shows the importance of precisions of terms.

Before string theory can truly provide the "why" to the standard model (SM), two prerequisites must be satisfied: First, string theory must be shown to contain (at least) one, or a set of, vacua within the landscape that exactly reproduce all known features of the standard model. Second, string theory must then explain why that particular vacuum, or one from among the set, are realized in the observable universe. The first of these requirements may be viewed as the (primary) mission of *string phenomenology* [4].

In this chapter we review and summarize developments in the search for the SM, and for its supersymmetric extension, the minimal supersymmetric standard model (MSSM). While the claim that "no [string theory] solution that looks like our universe has been found" [2, 4] is correct (and acknowledged by both critics and proponents of string theory [4]), much progress has been made in this pursuit since the first string revolution and especially since the second string revolution. In this chapter we examine several representative advancements in the search for the SM and MSSM from string theory and some predictions such models might have.

We begin in section 2 with a review of the very first string model for which the observable sector is composed of exactly the MSSM gauge particles, the three matter generations of the Minimal Supersymmetric Standard Model, and a pair of Higgs, without *any* unwanted MSSM-charged exotics. It is a weak coupling free fermionic heterotic $\mathbf{T}^6/(\mathbb{Z}_2 \otimes \mathbb{Z}_2)$ model constructed by Cleaver, Faraggi, and Nanopoulos [5, 6, 7, 8, 9].

While the first string-based MSSM model, it has been by no means been the last. Several others have appeared since [5]. If the density of MSSM-like models within the landscape is truly around $1/10^9$ as asserted in [10], then there should indeed be *many, many* more yet to come! Section three examines a representative collection of string- or brane-derived MSSMs that have followed the heterotic $\mathbf{T}^6/(\mathbb{Z}_2 \otimes \mathbb{Z}_2)$ free fermionic model. The various methods of construction of these additional MSSM models, include $\mathbb{Z}_6$ heterotic orbifolds, strong coupled heterotic with elliptically fibered Calabi-Yau, Type IIB with magnetic charged branes, and Type IIA with intersecting branes (dual of the Type IIB). Phenomenological aspects of models are compared. Section four concludes with some discussion of the goal of string phenomenology and some common features of the MSSM string models.

Before beginning a review of string-based (MS)SM models, we should first specify what is truly meant by (MS)SM for the sake of clarity. The definition of an MSSM model is critical to any claims of constructing one from string theory and was discussed in [4] by Muñoz: A true string derived (MS)SM must possess many more phenomenological properties than simply having the (MS)SM matter content of 12 quarks and 4 leptons [†] and a (pair of) Higgs doublet(s) (and corresponding supersymmetric partners), without any MSSM-charged exotics, in its low energy effective field theory (LEEFT). Obviously particles identified with the (MS)SM must have the correct mass hierarchy, a viable CKM quark mixing matrix, a realistic neutrino mass and mixing matrix, and strongly suppressed proton decay. The correct gauge coupling strengths must also result from it. Further, potential dangers associated with string theory must also be prevented, such as hidden sector (kinetic) mixing via extra $U(1)$ charges. For MSSM candidates, viable non-perturbative supersymmetry breaking methods must ultimately be found and supersymmetric particle mass predictions must correspond to (hoped-for) LHC measurements. Ultimately the stringy characteristics of the model must be understood sufficiently well to also explain the value of the cosmological constant.

No model has yet been produced that contains all of the necessary properties. Which, then, of these characteristics have been found in a model? How deep does the phenomenology go beyond the initial matter requirement? Following the first string revolution of Schwarz and Green in 1984, numerous string models with a *net* number of three generations were constructed from Calabi-Yau compactification of the heterotic string. The vast majority of these models contain numerous extra vector-like pairs of generations. Further, even those with exactly three generations (and no extra vector-like generation pairs) contain many dozens of MSSM exotics and numerous unconstrained moduli. Eventually several models were constructed using free fermionic and orbifold compactifications that contained exactly three net MSSM generations [11, 12, 13, 14]. Nevertheless, dozens of MSSM-charged exotics are endemic to these also [15].

Proof of the existence of at least one string model containing in its observable sector solely the MSSM matter, and absent any MSSM-charge exotics, waited many years [5]. The first model with this property was found by systematic analysis of the D and F-flat directions of a free fermionic model constructed much earlier [11]. The model of [11] contains several dozen MSSM-charged exotics that where shown in [5] to gain string-scale mass along a highly constrained set of flat directions [6, 7, 8, 9]. Since construction of this first string model with exactly the MSSM matter content in its observable sector, several similar models have been constructed from alternate compactifications. In the next section we start of our reviews with the very first string model with the MSSM content.

2. First MSSM Spectrum

2.1. Heterotic Free Fermionic Models

The first model to contain exactly the MSSM gauge group, three generations of MSSM matter models, a pair of Higgs doublets, and no exotic MSSM-charged states in the observable sector is a heterotic model constructed in the free fermionic formulation [16, 17, 18, 19] corresponding to $\mathbb{Z}_2 \times \mathbb{Z}_2$ orbifold models with nontrivial Wilson lines and background fields. In the free fermionic formulation, each of the six compactified directions is replaced by two real worldsheet fermions. The related left moving bosonic modes X_i, for $i = 1..., 6$

[†]We assume the existence of the left-handed neutrino singlet, and as such each generation forms a **16** representation (rep) of $SO(10)$

are replaced by real fermion modes (y_i, w_i) and the right moving bosonic modes $\overline{X}_i$ are replaced by corresponding $(\overline{y}_i, \overline{w}_i)$. For the supersymmetric left-moving sector, each real worldsheet boson X_i is accompanied by a real worldsheet fermion x_i. Thus, the supersymmetric left-moving compactified modes are represented by a total of 18 real worldsheet fermions, arranged in six sets of three, (x_i, y_i, w_i). (Consistency with the supersymmetry sector requires the boundary conditions of x_{2I-1} and x_{2I} match, pairing these up to form a complex fermion $\chi_I = x_{2I-1} + i x_{2I}$, $I = 1, ..., 3$.)

In light cone gauge the uncompactified left-moving degrees of freedom are two space-time bosons X_μ and the associated real fermions ψ_μ, for $\mu = 1, 2$. For the bosonic right moving sector, the six real worldsheet bosonic modes $\overline{X}_i$ are unpartnered. Thus in the free fermionic formulation there are just the six pairs of real fermions, $(\overline{y}_i, \overline{w}_i)$ from the compactified space. The $\overline{X}_\mu$ are likewise unpaired.

For heterotic $E_8 \times E_8$ strings, the degrees of freedom for the observable sector E_8 are represented by five complex fermions $\overline{\psi}_j$, $j = 1, ..., 5$, and three complex fermions $\overline{\eta}_k$, $k = 1, ..., 3$, (in anticipation of $E_8 \to SO(10) \times SO(6)$), while the 8 hidden sector bosonic degrees of freedom for the hidden sector E_8 are represented by 8 complex worldsheet fermions, $\overline{\phi}_m$, $m = 1, ..., 8$. Thus, in light-cone gauge heterotic free fermionic models contain 20 left-moving real fermionic degrees of freedom and 44 right-moving real fermionic degrees of freedom.

A free fermionic heterotic string model is specified by two objects [16, 17, 18, 19]. The first is a p-dimensional basis set of free fermionic boundary vectors $\{\mathbf{V}_i, i = 1, ..., p\}$. Each vector $\mathbf{V}_i$ has 64 components, $-1 < V_i^m \leq 1$, $m = 1, ..., 64$, with the first 20 components specifying boundary conditions for the 20 real free fermions representing worldsheet degrees of freedom for the left-moving supersymmetric string, and the latter 44 components specifying boundary conditions for the 44 real free fermions representing worldsheet degrees of freedom for the right-moving bosonic string. (Components of $\mathbf{V}_i$ for complex fermions are double-counted.) Modular invariance dictates the basis vectors, $\mathbf{V}_i$, span a finite additive group $\Xi = \{\sum_{n_i=0}^{N_i-1} \sum_{i=1}^{p} n_i \mathbf{V}_i\}$, with N_i the lowest positive integer such that $N_i \mathbf{V}_i = \mathbf{0}$ (mod 2). $\mathbf{V}_1 = \mathbf{1}$ (a 64-component unit vector) must be present in all models. In a given sector $\boldsymbol{\alpha} \equiv a_i \mathbf{V}_i$ (mod 2) $\in \Xi$, with $a_i \in \{0, 1, ..., N_i - 1\}$, a generic worldsheet fermion, denoted by f_m, transforms as $f_m \to -\exp\{\pi \alpha_m\} f_m$ around non-contractible loops on the worldsheet. Boundary vector components for real fermions are thus limited to be either 0 or 1, whereas boundary vector components for complex fermions can be rational.

The second object necessary to define a free fermionic model (up to vacuum expectation values (VEVs) of fields in the LEEFT) is a $p \times p$-dimensional matrix $\mathbf{k}$ of rational numbers $-1 < k_{i,j} \leq 1$, $i, j = 1, ..., p$, that determine the GSO operators for physical states. The $k_{i,j}$ are related to the phase weights $C\binom{\mathbf{V}_i}{\mathbf{V}_j}$ in the one-loop partition function Z:

$$C\binom{\mathbf{V}_i}{\mathbf{V}_j} = (-1)^{s_i + s_j} \exp(\pi i k_{j,i} - \tfrac{1}{2} \mathbf{V}_i \cdot \mathbf{V}_j), \tag{2.1}$$

where s_i is the 4-dimensional spacetime component of $\mathbf{V}_i$. The inner product of boundary (or charge) vectors is lorentzian, taken as left-movers minus right movers. Contributions to inner products from real fermion boundary components are weighted by a factor of $\tfrac{1}{2}$ compared to contributions from complex fermion boundary components.

The phase weights $C\binom{\alpha}{\beta}$ for general sectors

$$\alpha = \sum_{j=1}^{p} a_j \mathbf{V}_j \in \Xi, \quad \beta = \sum_{i=1}^{p} b_i \mathbf{V}_i \in \Xi \qquad (2.2)$$

can be expressed in terms of the components in the $p \times p$-dimensional matrix $\mathbf{k}$ for the basis vectors:

$$C\binom{\alpha}{\beta} = (-1)^{s_\alpha + s_\beta} \exp\{\pi i \sum_{i,j} b_i (k_{i,j} - \tfrac{1}{2} \mathbf{V}_i \cdot \mathbf{V}_j) a_j\}. \qquad (2.3)$$

Modular invariance simultaneously imposes constraints on the basis vectors $\mathbf{V}_i$ and on components of the GSO projection matrix $\mathbf{k}$:

$$k_{i,j} + k_{j,i} = \tfrac{1}{2} \mathbf{V}_i \cdot \mathbf{V}_j \pmod{2} \qquad (2.4)$$
$$N_j k_{i,j} = 0 \pmod{2} \qquad (2.5)$$
$$k_{i,i} + k_{i,1} = -s_i + \tfrac{1}{4} \mathbf{V}_i \cdot \mathbf{V}_i \pmod{2}. \qquad (2.6)$$

The dependence upon the $k_{i,j}$ can be removed from equations (2.4-2.6), after appropriate integer multiplication, to yield three constraints on the $\mathbf{V}_i$:

$$N_{i,j} \mathbf{V}_i \cdot \mathbf{V}_j = 0 \pmod{4} \qquad (2.7)$$
$$N_i \mathbf{V}_i \cdot \mathbf{V}_i = 0 \pmod{8} \qquad (2.8)$$

The number of real fermions simultaneously periodic
for any three basis vectors is even. (2.9)

$N_{i,j}$ is the lowest common multiple of N_i and N_j. (2.9) still applies when two or more of the basis vectors are identical. Thus, each basis vector must have an even number of real periodic fermions.

The physical massless states in the Hilbert space of a given sector $\alpha \in \Xi$, are obtained by acting on the vacuum with bosonic and fermionic operators and by applying generalized GSO projections. The $U(1)$ charges for the Cartan generators of the unbroken gauge group are in one to one correspondence with the $U(1)$ currents $f_m^* f_m$ for each complex fermion f_m, and are given by:

$$Q_m^\alpha = \frac{1}{2} \alpha_m + F_m^\alpha, \qquad (2.10)$$

where α_m is the boundary condition of the worldsheet fermion f_m in the sector α, and F_m^α is a fermion number operator counting each mode of f_m once and of f_m^* minus once. Pseudo-charges for non-chiral (i.e., with both left- and right-moving components) real Ising fermions f_m can be similarly defined, with F_m counting each real mode f once.

For periodic fermions, $\alpha_m = 1$, the vacuum is a spinor representation of the Clifford algebra of the corresponding zero modes. For each periodic complex fermion f_m there are two degenerate vacua $|+\rangle, |-\rangle$, annihilated by the zero modes $(f_m)_0$ and $(f_m^*)_0$ and with fermion numbers $F_m^\alpha = 0, -1$, respectively.

The boundary basis vectors $\mathbf{V}_j$ generate the set of GSO projection operators for physical states from all sectors $\alpha \in \Xi$. In a given sector α, the surviving states are those that satisfy the GSO equations imposed by all $\mathbf{V}_j$ and determined by the $k_{j,i}$'s:

$$\mathbf{V}_j \cdot \mathbf{F}^\alpha = \left(\sum_i k_{j,i} a_i\right) + s_j - \tfrac{1}{2} \mathbf{V}_j \cdot \alpha \pmod{2}, \tag{2.11}$$

or, equivalently,

$$\mathbf{V}_j \cdot \mathbf{Q}^\alpha = \left(\sum_i k_{j,i} a_i\right) + s_j \pmod{2}. \tag{2.12}$$

For a given set of basis vectors, the independent GSO matrix components are $k_{1,1}$ and $k_{i,j}$, for $i > j$. This GSO projection constraint, when combined with equations (2.4-2.6) form the free fermionic re-expression of the even, self-dual modular invariance constraints for bosonic lattice models. The masses, $m^2 = m_L^2 + m_R^2$ (with $m_L^2 = m_R^2$) of physical states can also be expressed as a simple function of the charge vector,

$$\alpha' m_L^2 = -\tfrac{1}{2} + \tfrac{1}{2} \mathbf{Q}_L^\alpha \cdot \mathbf{Q}_L^\alpha \tag{2.13}$$

$$\alpha' m_R^2 = -1 + \tfrac{1}{2} \mathbf{Q}_R^\alpha \cdot \mathbf{Q}_R^\alpha. \tag{2.14}$$

The superpotential for the physical states can be determined to all order. Couplings are computable for any order in the free fermionic models, using conformal field theory vertex operators. The coupling constant can be expressed in terms of an n-point string amplitude A_n. This amplitude is proportional to a world-sheet integral I_{n-3} of the correlators of the n vertex operators V_i for the fields in the superpotential terms [20],

$$A_n = \frac{g}{\sqrt{2}} (\sqrt{8/\pi})^{n-3} C_{n-3} I_{n-3} / (M_{str})^{n-3}. \tag{2.15}$$

The integral has the form,

$$I_{n-3} = \int d^2 z_3 \cdots d^2 z_{n-1} \, \langle V_1^f(\infty) V_2^f(1) V_3^b(z_3) \cdots V_{n-1}^b(z_{n-1}) V_n^b(0) \rangle \tag{2.16}$$

$$= \int d^2 z_3 \cdots d^2 z_{n-1} \, f_{n-3}(z_1 = \infty, z_2 = 1, z_3, \cdots, z_{n-1}, z_n = 0), \tag{2.17}$$

where z_i is the worldsheet coordinate of the fermion (boson) vertex operator V_i^f (V_i^b) of the ith string state. C_{n-3} is an $\mathcal{O}(1)$ coefficient that includes renormalization factors in the operator product expansion of the string vertex operators and target space gauge group Clebsch–Gordon coefficients. $SL(2, C)$ invariance is used to fix the location of three of the vertex operators at $z = z_\infty, 1, 0$. When n_v of the fields $\prod_{i=1}^{l} \mathbf{X}_i$ take on VEVs, $\langle \prod_{i=1}^{n_v} \mathbf{X}_i \rangle$, then the coupling constant for the effective $n_e = (n - n_v)$-th order term becomes $A'_{n_e} \equiv A_n \langle \prod_{i=1}^{n_v} \mathbf{X}_i \rangle$.

In order to simply determine whether couplings are non-zero, worldsheet selection rules (in addition to the demand of gauge invariance) predicting nonvanishing correlators were formulated in [20, 21]. With n designating the total number of fields in a term, these rules are economically summarized as: [6]

1. Ramond fields must be distributed equally, mod 2, among all categories, *and*

2. For $n = 3$, either :

   (a) There is 1 field from each R category.

   (b) There is 1 field from each NS category.

   (c) There are $2R$ and $1NS$ in a single category.

3. For $n > 3$:

   (a) There must be at least $4R$ fields.

   (b) All R fields may not exist in a single category.

   (c) If $R = 4$, then only permutations of $(2_R, 2_R, n - 4_{NS})$ are allowed.

   (d) If $R > 4$, then no NS are allowed in the maximal R category (if one exists).

2.2. $SO(10)$ $\mathbb{Z}_2 \times \mathbb{Z}_2$ (NAHE based) Compactifications

In the NAHE class of realistic free fermionic models, the boundary condition basis is divided into two subsets. The first is the NAHE set [22], which contains five boundary condition basis vectors denoted $\{\mathbf{1}, \mathbf{S}, \mathbf{b}_1, \mathbf{b}_2, \mathbf{b}_3\}$. With '**0**' indicating Neveu-Schwarz boundary conditions and '**1**' indicating Ramond boundary conditions, these vectors are as follows:

	ψ^μ	χ^{12}	χ^{34}	χ^{56}	$\bar{\psi}^{1,...,5}$	$\bar{\eta}^1$	$\bar{\eta}^2$	$\bar{\eta}^3$	$\bar{\phi}^{1,...,8}$
1	1	1	1	1	1,...,1	1	1	1	1,...,1
S	1	1	1	1	0,...,0	0	0	0	0,...,0
$\mathbf{b}_1$	1	1	0	0	1,...,1	1	0	0	0,...,0
$\mathbf{b}_2$	1	0	1	0	1,...,1	0	1	0	0,...,0
$\mathbf{b}_3$	1	0	0	1	1,...,1	0	0	1	0,...,0

	$y^{3,...,6}$	$\bar{y}^{3,...,6}$	$y^{1,2}, \omega^{5,6}$	$\bar{y}^{1,2}, \bar{\omega}^{5,6}$	$\omega^{1,...,4}$	$\bar{\omega}^{1,...,4}$
1	1,...,1	1,...,1	1,...,1	1,...,1	1,...,1	1,...,1
S	0,...,0	0,...,0	0,...,0	0,...,0	0,...,0	0,...,0
$\mathbf{b}_1$	1,...,1	1,...,1	0,...,0	0,...,0	0,...,0	0,...,0
$\mathbf{b}_2$	0,...,0	0,...,0	1,...,1	1,...,1	0,...,0	0,...,0
$\mathbf{b}_3$	0,...,0	0,...,0	0,...,0	0,...,0	1,...,1	1,...,1

(2.18)

with the following choice of phases which define how the generalized GSO projections are to be performed in each sector of the theory:

$$C\begin{pmatrix}\mathbf{b}_i\\ \mathbf{b}_j\end{pmatrix} = C\begin{pmatrix}\mathbf{b}_i\\ \mathbf{S}\end{pmatrix} = C\begin{pmatrix}\mathbf{1}\\ \mathbf{1}\end{pmatrix} = -1. \quad (2.19)$$

The remaining projection phases can be determined from those above through the self-consistency constraints. The precise rules governing the choices of such vectors and phases, as well as the procedures for generating the corresponding spacetime particle spectrum, are given in refs. [18, 19].

Without the $\mathbf{b}_i$ sectors, the NAHE sector corresponds to $\mathbf{T}^6$ torus compactification at the self-dual radius. The addition of the sectors $\mathbf{b}_1$, $\mathbf{b}_2$ and $\mathbf{b}_3$ produce an effective $\mathbf{T}^6/(\mathbb{Z}_2 \times \mathbb{Z}_2)$ compactification. $\mathbf{b}_1$, $\mathbf{b}_2$ and $\mathbf{b}_3$ correspond to the three twisted sectors of the $\mathbb{Z}_2 \times \mathbb{Z}_2$ orbifold model: $\mathbf{b}_1$ provides the $\mathbb{Z}_2^a$ twisted sector, $\mathbf{b}_2$ provides the $\mathbb{Z}_2^b$ twisted sector, and $\mathbf{b}_3$ provides the $\mathbb{Z}_2^a \otimes \mathbb{Z}_2^b$ twisted sector.

After imposing the NAHE set, the resulting model has gauge group $SO(10) \times SO(6)^3 \times E_8$ and $N = 1$ spacetime supersymmetry. The model contains 48 multiplets in the **16** representation of $SO(10)$, 16 from each twisted sector $\mathbf{b}_1$, $\mathbf{b}_2$ and $\mathbf{b}_3$.

In addition to the spin 2 multiplets and the spacetime vector bosons, the untwisted sector produces six multiplets in the vectorial **10** representation of $SO(10)$ and a number of $SO(10) \times E_8$ singlets. As can be seen from Table (2.18), the model at this stage possesses a cyclic permutation symmetry among the basis vectors $\mathbf{b}_1$, $\mathbf{b}_2$ and $\mathbf{b}_3$, which is also respected by the massless spectrum.

2.3. Minimal Heterotic Superstring Standard Model

The second stage in the construction of these NAHE-based free fermionic models consists of adding three additional basis vectors, denoted $\{\alpha, \beta, \gamma\}$ to the above NAHE set. These three additional basis vectors correspond to "Wilson lines" in the orbifold construction.

The allowed fermion boundary conditions in these additional basis vectors are also constrained by the string consistency constraints, and must preserve modular invariance and worldsheet supersymmetry. The choice of these additional basis vectors nevertheless distinguishes between different models and determines their low-energy properties. For example, three additional vectors are needed to reduce the number of massless generations to three, one from each sector $\mathbf{b}_1$, $\mathbf{b}_2$, and $\mathbf{b}_3$, and the choice of their boundary conditions for the internal fermions $\{y, \omega | \bar{y}, \bar{\omega}\}^{1,\cdots,6}$ also determines the Higgs doublet-triplet splitting and the Yukawa couplings. These low-energy phenomenological requirements therefore impose strong constraints [12, 13, 14] on the possible assignment of boundary conditions to the set of internal worldsheet fermions $\{y, \omega | \bar{y}, \bar{\omega}\}^{1,\cdots,6}$.

The additional sectors corresponding to the first MSSM model are:

	ψ^μ	χ^{12}	χ^{34}	χ^{56}	$\bar{\psi}^{1,\cdots,5}$	$\bar{\eta}^1$	$\bar{\eta}^2$	$\bar{\eta}^3$	$\bar{\phi}^{1,\cdots,8}$
α	1	1	0	0	1 1 1 1 1	1	0	0	0 0 0 0 0 0 0 0
β	1	0	0	1	1 1 1 0 0	1	0	1	1 1 1 1 0 0 0 0
γ	1	0	1	0	$\frac{1}{2}\,\frac{1}{2}\,\frac{1}{2}\,\frac{1}{2}\,\frac{1}{2}$	$\frac{1}{2}$	$\frac{1}{2}$	$\frac{1}{2}$	$\frac{1}{2}$ 0 1 1 $\frac{1}{2}\,\frac{1}{2}\,\frac{1}{2}$ 1

	$y^3y^6\ y^4\bar{y}^4\ y^5\bar{y}^5\ \bar{y}^3\bar{y}^6$	$y^1\omega^6\ y^2\bar{y}^2\ \omega^5\bar{\omega}^5\ \bar{y}^1\bar{\omega}^6$	$\omega^1\omega^3\ \omega^2\bar{\omega}^2\ \omega^4\bar{\omega}^4\ \bar{\omega}^1\bar{\omega}^3$
α	1 0 0 1	0 0 1 0	0 0 1 0
β	0 0 0 1	0 1 0 1	1 0 1 0
γ	0 0 1 1	1 0 0 1	0 1 0 0

(2.20)

with corresponding generalized GSO coefficients:

$$C\begin{pmatrix}\alpha\\ \mathbf{b}_j, \beta\end{pmatrix} = -C\begin{pmatrix}\alpha\\ \mathbf{1}\end{pmatrix} = -C\begin{pmatrix}\beta\\ \mathbf{1}\end{pmatrix} = C\begin{pmatrix}\beta\\ \mathbf{b}_j\end{pmatrix} =$$

$$-C\begin{pmatrix}\beta\\\gamma\end{pmatrix} = C\begin{pmatrix}\gamma\\\mathbf{b}_2\end{pmatrix} = -C\begin{pmatrix}\gamma\\\mathbf{b}_1,\mathbf{b}_3,\alpha,\gamma\end{pmatrix} = -1 \qquad (2.21)$$

($j = 1, 2, 3$), (The remaining GSO coefficients are specified by modular invariance and spacetime supersymmetry.)

The full massless spectrum of the model, together with the quantum numbers under the right-moving gauge group, are given in ref. [11].

2.4. Initial Gauge Group

Prior to any scalar fields receiving non-zero VEVS, the observable gauge group consists of the universal $SO(10)$ subgroup, $SU(3)_C \times SU(2)_L \times U(1)_C \times U(1)_L$, generated by the five complex worldsheet fermions $\bar{\psi}^{1,\cdots,5}$, and six observable horizontal, flavor-dependent, Abelian symmetries $U(1)_{1,\cdots,6}$, generated by $\{\bar{\eta}^1, \bar{\eta}^2, \bar{\eta}^3, \bar{y}^3\bar{y}^6, \bar{y}^1\bar{\omega}^6, \bar{\omega}^1\bar{\omega}^3\}$, respectively. The hidden sector gauge group is the E_8 subgroup of $(SO(4) \sim SU(2) \times SU(2)) \times SU(3) \times U(1)^4$, generated by $\bar{\phi}^{1,\cdots,8}$.

The weak hypercharge is given by

$$U(1)_Y = \frac{1}{3}U(1)_C \pm \frac{1}{2}U(1)_L, \qquad (2.22)$$

which has the standard effective level k_1 of $5/3$, necessary for MSSM unification at M_U.[‡] The "+" sign was chosen for the hypercharge definition in [11]. [5] showed that the choice of the sign in eq. (2.22) has interesting consequences in terms of the decoupling of the exotic fractionally charged states. The alternate sign combination of $U(1)_C$ and $U(1)_L$ is orthogonal to $U(1)_Y$ and denoted by $U(1)_{Z'}$. Cancellation of the Fayet-Iliopoulos (FI) term by directions that are D-flat for all of the non-anomalous $U(1)$ requires that at least one of the $U(1)_Y$ and $U(1)_{Z'}$ be broken. Therefore, viable phenomenology forced $SU(3)_C \times SU(2)_L \times U(1)_Y$ to be the unbroken $SO(10)$ subgroup below the string scale. This is an interesting example of how string dynamics may force the $SO(10)$ subgroup below the string scale to coincide with the Standard Model gauge group.

2.5. Massless Matter

The full massless spectrum of the model, together with the quantum numbers under the right-moving gauge group, were first presented in ref. [11]. In the model, each of the three sectors $\mathbf{b}_1$, $\mathbf{b}_2$, and $\mathbf{b}_3$ produce one generation in the 16 representation, (Q_i, u_i^c, d_i^c, L_i, e_i^c, N_i^c), of $SO(10)$ decomposed under $SU(3)_C \times SU(2)_L \times U(1)_C \times U(1)_L$, with charges under the horizontal symmetries.

In addition to the gravity and gauge multiplets and several singlets, the untwisted Neveu-Schwarz (NS) sector produces three pairs of electroweak scalar doublets

---

[‡]The sign ambiguity in eq. (2.22) can be understood in terms of the two alternative embeddings of $SU(5)$ within $SO(10)$ that produce either the standard or flipped $SU(5)$. Switching signs in (2.22) flips the representations, $(e_L^c, u_L^c, h) \leftrightarrow (N_L^c, d_L^c, \bar{h})$. In the case of $SU(5)$ string GUT models, only the "−" (i.e., flipped version) is allowed, since there are no massless matter adjoint representations, which are needed to break the non-Abelian gauge symmetry of the unflipped $SU(5)$, but are not needed for the flipped version. For MSSM-like strings, either choice of sign is allowed since the GUT non-Abelian symmetry is broken directly at the string level.

$\{h_1, h_2, h_3, \bar{h}_1, \bar{h}_2, \bar{h}_3\}$. Each NS electroweak doublet set $(h_i, \bar{h}_i)$ may be viewed as a pair of Higgs with the potential to give renormalizable (near EW scale) mass to the corresponding $\mathbf{b}_i$-generation of MSSM matter. Thus, to reproduce the MSSM states and generate a viable three generation mass hierarchy, two out of three of these Higgs pairs must become massive near the string/FI scale. The twisted sector provides some additional $SU(3)_C \times SU(2)_L$ exotics: one $SU(3)_C$ triplet/antitriplet pair $\{H_{33}, H_{40}\}$; one $SU(2)_L$ up-like doublet, H_{34}, and one down-like doublet, H_{41}; and two pairs of vector-like $SU(2)_L$ doublets, $\{V_{45}, V_{46}\}$ and $\{V_{51}, V_{52}\}$, with fractional electric charges $Q_e = \pm\frac{1}{2}$. $h_4 \equiv H_{41}$ and $\bar{h}_4 \equiv H_{34}$ play the role of a fourth pair of MSSM Higgs. Hence, all exotics form vector-like pairs with regard to MSSM-charges, a generic requirement for their decoupling.

Besides the anti-electrons e_i^c and neutrino singlets N_i^c, the model contains another 57 non-Abelian singlets. 16 of these carry electric charge and 41 do not. The set of 16 are twisted sector states,[§] eight of which carry $Q_e = \frac{1}{2}$, $\{H_3^s, H_5^s, H_7^s, H_9^s, V_{41}^s, V_{43}^s, V_{47}^s, V_{49}^s\}$, and another eight of which carry $Q_e = -\frac{1}{2}$, $\{H_4^s, H_6^s, H_8^s, H_{10}^s, V_{42}^s, V_{44}^s, V_{48}^s, V_{50}^s\}$.

Three of the 41 $Q_e = 0$ states, $\{\Phi_1, \Phi_2, \Phi_3\}$, are the completely uncharged moduli from the NS sector. Another fourteen of these singlets form vector-like pairs, $(\Phi_{12}, \overline{\Phi}_{12})$, $(\Phi_{23}, \overline{\Phi}_{23})$, $(\Phi_{13}, \overline{\Phi}_{13})$, $(\Phi_{56}, \overline{\Phi}_{56})$, $(\Phi'_{56}, \overline{\Phi}'_{56})$, $(\Phi_4, \overline{\Phi}_4)$, $(\Phi'_4, \overline{\Phi}'_4)$, possessing charges of equal magnitude, but opposite sign, for all local Abelian symmetries. The remaining 24 $Q_e = 0$ singlets, $\{H_{15}^s, H_{16}^s, H_{17}^s, H_{18}^s, H_{19}^s, H_{20}^s, H_{21}^s, H_{22}^s, H_{29}^s, H_{30}^s, H_{31}^s, H_{32}^s, H_{36}^s, H_{37}^s, H_{38}^s, H_{39}^s,$ and $\{V_1^s, V_2^s, V_{11}^s, V_{12}^s, V_{21}^s, V_{22}^s, V_{31}^s, V_{32}^s\}$, are twisted sector states carrying both observable and hidden sector Abelian charges.

The model contains 34 hidden sector non-Abelian states, all of which also carry both observable and hidden $U(1)_i$ charges: Five of these are $SU(3)_H$ triplets, $\{H_{42}, V_4, V_{14}, V_{24}, V_{34}\}$, while another five are antitriplets, $\{H_{35}, V_3, V_{13}, V_{23}, V_{33}\}$. The remaining hidden sector states are 12 $SU(2)_H$ doublets, $\{H_1, H_2, H_{23}, H_{26}, V_5, V_7, V_{15}, V_{17}, V_{25}, V_{27}, V_{39}, V_{40}\}$ and a corresponding 12 $SU(2)'_H$ doublets, $\{H_{11}, H_{13}, H_{25}, H_{28}, V_9, V_{10}, V_{19}, V_{20}, V_{29}, V_{30}, V_{35}, V_{37}\}$. The only hidden sector NA states with non-zero Q_e (in half-integer units) are the four of the hidden sector doublets, H_1, H_2, H_{11}, and H_{13}.

For a string derived MSSM to result from this model, the several exotic MSSM-charged states must be eliminated from the LEEFT. Along with three linearly independent combinations of the h_i, and of the $\bar{h}_i$, for $i = 1, ..., 4$, the 26 states, $\{H_{33}, H_{40}, V_{45}, V_{46}, V_{51}, V_{52}, H_1, H_2, H_{11}, H_{12}\}$, $\{H_3^s, H_5^s, H_7^s, H_9^s, V_{41}^s, V_{43}^s, V_{47}^s, V_{49}^s\}$, and $\{H_4^s, H_6^s, H_8^s, H_{10}^s, V_{42}^s, V_{44}^s, V_{48}^s, V_{50}^s\}$ must be removed.

Examination of the MSSM-charged state superpotential shows that three out of four of each of the h_i and $\bar{h}_i$ Higgs, and *all* of the 26 states above can be decoupled from the LEEFT via the terms,

$$\Phi_{12} h_1 \bar{h}_2 + \Phi_{23} h_3 \bar{h}_2 + H_{31}^s h_2 H_{34} + H_{38}^s \bar{h}_3 H_{41} +$$
$$\Phi_4 [V_{45} V_{46} + H_1 H_2] + \overline{\Phi}_4 [H_3^s H_4^s + H_5^s H_6^s + V_{41}^s V_{42}^s + V_{43}^s V_{44}^s] + \quad (2.23)$$

[§]Vector-like representations of the hidden sector are denoted by a "V", while chiral representations are denoted by a "H". A superscript "s" indicates a non-Abelian singlet.

$$\Phi'_4[V_{51}V_{52} + H^s_7 H^s_8 + H^s_9 H^s_{10}] + \overline{\Phi}'_4[V^s_{47}V^s_{48} + V^s_{49}V^s_{50} + H_{11}H_{13}] +$$
$$\Phi_{23} H^s_{31} H^s_{38}[H_{33}H_{40} + H_{34}H_{41}].$$

This occurs when all states in the set

$$\{\Phi_4, \overline{\Phi}_4, \Phi'_4, \overline{\Phi}'_4, \Phi_{12}, \Phi_{23}, H^s_{31}, H^s_{38}\}, \tag{2.24}$$

take on near string scale VEVs through FI term anomaly cancellation. All but one of the dominant terms in (2.23) are of third order and will result in unsuppressed FI scale masses, while the remaining term is of fifth order, giving a mass suppression of $\frac{1}{10}$ for H_{33} and H_{40}.

2.6. Anomalous $U(1)$

All known quasi-realistic chiral three generation $SU(3)_C \times SU(2)_L \times U(1)_Y$ heterotic models, of lattice, orbifold, or free fermionic construction, contain an anomalous local $U(1)_A$ [23, 24]. Anomaly cancellation provides a means by which VEVs naturally appear. While non-perturbatively chosen, some perturbative possibilities may exist that provide the needed effective mass terms to eliminate all unwanted MSSM exotics from models generically containing them.

An anomalous $U(1)_A$ has non-zero trace of its charge over the massless states of the LEEFT,

$$\text{Tr}\, Q^{(A)} \neq 0. \tag{2.25}$$

String models often appear to have not just one, but several anomalous Abelian symmetries $U(1)_{A,i}$ ($i = 1$ to n), each with $\text{Tr}\, Q^{(A)}_i \neq 0$. However, there is always a rotation that places the entire anomaly into a single $U(1)_A$, uniquely defined by

$$U(1)_A \equiv c_A \sum_{i=1}^{n} \{\text{Tr}\, Q^{(A)}_i\} U(1)_{A,i}, \tag{2.26}$$

with c_A a normalization coefficient. Then $n - 1$ traceless $U(1)'_j$ are formed from linear combinations of the n $U(1)_{A,i}$ that are orthogonal to $U(1)_A$.

Prior to rotating the anomaly into a single $U(1)_A$, six of the FNY model's twelve $U(1)$ symmetries are anomalous: $\text{Tr}\, U_1 = -24$, $\text{Tr}\, U_2 = -30$, $\text{Tr}\, U_3 = 18$, $\text{Tr}\, U_5 = 6$, $\text{Tr}\, U_6 = 6$ and $\text{Tr}\, U_8 = 12$. Thus, the total anomaly can be rotated into a single $U(1)_A$ defined by

$$U_A \equiv -4U_1 - 5U_2 + 3U_3 + U_5 + U_6 + 2U_8. \tag{2.27}$$

Five mutually orthogonal U'_s, for $s = 1, ..., 5$ are then formed that are all traceless and orthogonal to U_A. A set of vacuum expectations values (VEVs) will automatically appear in any string model with an anomalous $U(1)_A$ as a result of the Green-Schwarz-Dine-Seiberg-Witten anomaly cancellation mechanism [25, 26]. Following the anomaly rotation of eq. (2.26), the universal Green-Schwarz (GS) relations,

$$\frac{1}{k_m k_A^{1/2}} \text{Tr}\, T(R) Q_A = \frac{1}{3k_A^{3/2}} \text{Tr}\, Q_A^3 = \frac{1}{k_i k_A^{1/2}} \text{Tr}\, Q_i^2 Q_A = \frac{1}{24 k_A^{1/2}} \text{Tr}\, Q_A$$

$$\equiv 8\pi^2 \delta_{\text{GS}}, \tag{2.28}$$

$$\frac{1}{k_m k_i^{1/2}} \operatorname*{Tr}_{G_m} T(R) Q_i = \frac{1}{3k_i^{3/2}} \operatorname{Tr} Q_i^3 = \frac{1}{k_A k_i^{1/2}} \operatorname{Tr} Q_A^2 Q_i = \frac{1}{(k_i k_j k_A)^{1/2}} \operatorname{Tr} Q_i Q_{j \neq i} Q_A$$

$$= \frac{1}{24 k_i^{1/2}} \operatorname{Tr} Q_i = 0, \tag{2.29}$$

where k_m is the level of the non-Abelian gauge group G_m and $2T(R)$ is the index of the representation R of G_m, defined by

$$\operatorname{Tr} T_a^{(R)} T_b^{(R)} = T(R) \delta_{ab}, \tag{2.30}$$

removes all Abelian triangle anomalies except those involving either one or three U_A gauge bosons.¶

The standard anomaly cancellation mechanism breaks U_A and, in the process, generates a FI D-term,

$$\epsilon \equiv \frac{g_s^2 M_P^2}{192\pi^2} \operatorname{Tr} Q^{(A)}, \tag{2.31}$$

where g_s is the string coupling and M_P is the reduced Planck mass, $M_P \equiv M_{Planck}/\sqrt{8\pi} \approx 2.4 \times 10^{18}$.

2.7. Flat Direction Constraints

Spacetime supersymmetry is broken in a model when the expectation value of the scalar potential,

$$V(\varphi) = \tfrac{1}{2} \sum_\alpha g_\alpha^2 D_a^\alpha D_a^\alpha + \sum_i |F_{\varphi_i}|^2, \tag{2.32}$$

becomes non-zero. The D-term contributions in (2.32) have the form,

$$D_a^\alpha \equiv \sum_m \varphi_m^\dagger T_a^\alpha \varphi_m, \tag{2.33}$$

with T_a^α a matrix generator of the gauge group g_α for the representation φ_m, while the F-term contributions are,

$$F_{\Phi_m} \equiv \frac{\partial W}{\partial \Phi_m}. \tag{2.34}$$

The φ_m are the scalar field superpartners of the chiral spin-$\tfrac{1}{2}$ fermions ψ_m, which together form a superfield Φ_m. Since all of the D and F contributions to (2.32) are positive semidefinite, each must have a zero expectation value for supersymmetry to remain unbroken.

For an Abelian gauge group, the D-term (2.33) simplifies to

$$D^i \equiv \sum_m Q_m^{(i)} |\varphi_m|^2 \tag{2.35}$$

¶The GS relations are a by-product of modular invariance constraints.

where $Q_m^{(i)}$ is the $U(1)_i$ charge of φ_m. When an Abelian symmetry is anomalous, the associated D-term acquires the FI term (2.31),

$$D^{(A)} \equiv \sum_m Q_m^{(A)} |\varphi_m|^2 + \epsilon \,. \tag{2.36}$$

g_s is the string coupling and M_P is the reduced Planck mass, $M_P \equiv M_{Planck}/\sqrt{8\pi} \approx 2.4 \times 10^{18}$ GeV.

The FI term breaks supersymmetry near the string scale, $V \sim g_s^2 \epsilon^2$, unless its can be cancelled by a set of scalar VEVs, $\{\langle \varphi_{m'} \rangle\}$, carrying anomalous charges $Q_{m'}^{(A)}$,

$$\langle D^{(A)} \rangle = \sum_{m'} Q_{m'}^{(A)} |\langle \varphi_{m'} \rangle|^2 + \epsilon = 0 \,. \tag{2.37}$$

To maintain supersymmetry, a set of anomaly-cancelling VEVs must simultaneously be D-flat for all additional Abelian and the non-Abelian gauge groups,

$$\langle D^{i,\alpha} \rangle = 0 \,. \tag{2.38}$$

A non-trivial superpotential W also imposes numerous constraints on allowed sets of anomaly-cancelling VEVs, through the F-terms in (2.32). F-flatness (and thereby supersymmetry) can be broken through an nth-order W term containing Φ_m when all of the additional fields in the term acquire VEVs,

$$\langle F_{\Phi_m} \rangle \sim \langle \frac{\partial W}{\partial \Phi_m} \rangle \sim \lambda_n \langle \varphi \rangle^2 \left(\frac{\langle \varphi \rangle}{M_{str}}\right)^{n-3} \,, \tag{2.39}$$

where φ denotes a generic scalar VEV. If Φ_m additionally has a VEV, then supersymmetry can be broken simply by $\langle W \rangle \neq 0$.

F-flatness must be retained up to an order in the superpotential that is consistent with observable sector supersymmetry being maintained down to near the electroweak (EW) scale. However, it may in fact be *desirable* to allow such a term to escape at some elevated order, since it is known that supersymmetry does *not* survive down to 'everyday' energies. Depending on the string coupling strength, F-flatness cannot be broken by terms below eighteenth to twentieth order. As coupling strength increases, so does the required order of flatness.

Generically, there are many more D-flat directions that are simultaneously F-flat to a given order in the superpotential for the LEEFT of a string model than for the field-theoretic counterpart. In particular, there are usually several D-flat directions that are F-flat to all order in a string model, but only flat to some finite, often low, order in the corresponding field-theoretic model. This may be attributed to the string worldsheet selection rules, which impose strong constraints on allowed superpotential terms beyond gauge invariance.

2.8. MSSM Flat Directions

The existence of an all-order D- and F-flat direction containing VEVs for all of the fields in (2.24) was proved in [5]. This $U(1)_A$ anomaly-cancelling flat direction provided the *first* known example of a superstring derived model in which, of all the $SU(3)_C \times SU(2)_L \times U(1)_Y$-charged states, only the MSSM spectrum remains light below the string scale.

From the first stage of a systematic study of D- and F-flat directions, a total of four all-order flat directions formed solely of non-Abelian singlet fields with zero hypercharge were found [6]. The scale of the overall VEV for all of these directions was computed to be $|<\alpha>| \sim 10^{17}$ GeV. Other than these four, no other directions were found to be F-flat past 12th order. The resulting phenomenology of the all-order flat directions was presented in [7]. For these directions renormalizable mass terms appeared for one complete set of up-, down-, and electron-like fields and their conjugates. However, the apparent top and bottom quarks did not appear in the same $SU(2)_L$ doublet. Effectively, these flat directions gave the strange quark a heavier mass than the bottom quark. This inverted mass effect was a result of the field Φ_{12} receiving a VEV in all of the above direction. A search for MSSM-producing singlet flat directions that did not contain $\langle \Phi_{12} \rangle$ was then performed. None were found. This, in and of itself, suggested the need for non-Abelian VEVs in more phenomenologically appealing flat directions. Too few first and second generation down and electron mass terms implied similarly.

Thus, in the second stage of the study, hidden sector non-Abelian fields (with zero hypercharge) were also allowed to take on VEVs. This led to discovery of additional all-order flat directions containing non-Abelian VEVs [8] with a similar overall VEV scale $\sim 10^{17}$ GeV. Their phenomenology was explored in [9]. Several phenomenological aspects improved when non-Abelian VEVs appeared.

All MSSM flat directions were found via a computer search that generated combinations of maximally orthogonal D-flat basis directions [27]. *Maximally orthogonal* means that each of the basis directions contained at least one VEV unique to itself. Thus, unless a basis direction's defining VEV was vector-like, the basis vector could only appear in a physical flat direction multiplied by positive (real) weights. The physical D-flat directions generated where required to minimally contain VEVs for the set of states, $\{\Phi_4, \overline{\Phi}_4, \Phi'_4, \overline{\Phi}'_4, \Phi_{12}, \Phi_{23}, H^s_{31}, H^s_{38}\}$ necessary for decoupling of all 32 SM-charged MSSM exotics, comprised of the three extra pairs of Higgs doublets and the 28 MSSM-charged exotics. The all-order flat directions were found to eliminate no less than seven of the extra non-anomalous $U(1)'$, thereby greatly reducing the horizontal symmetries.

Stringent F-flatness [6] was demanded of the all singlet flat directions. This requires that the expectation value of each component of a given F-term vanish, rather than allowing elimination of an F-term via cancellation between expectation values of two or more components. Such stringent demands are clearly not necessary for F-flatness. Total absence of all individual non-zero VEV terms can be relaxed: collections of such terms appear without breaking F-flatness, so long as the terms separately cancel among themselves in each $\langle F_{\Phi_m} \rangle$ and in $\langle W \rangle$. However, even when supersymmetry is retained at a given order in the superpotential via cancellation between several terms in a specific F_{Φ_m}, supersymmetry could well be broken at a slightly higher order. Thus, stringent flatness enables all-order flat directions to be found without demanding precise fine-tuning among the field VEVs.

Non-Abelian VEVs offer one solution to the stringent F-flatness issue. Because non-Abelian fields contain more than one field component, *self-cancellation* of a dangerous F-term can sometimes occur along non-Abelian directions. That is, for some directions it may be possible to maintain "stringent" F-flatness even when dangerous F-breaking terms appear in the stringy superpotential. Since Abelian D-flatness constraints limit only VEV magnitudes, the gauge freedom of each group remains (phase freedom, in particular, is ubiquitous) with which to attempt a cancellation between terms (whilst retaining consistency with non-Abelian D-flatness). However, it can often be the case that only a single term from W becomes an offender in a given F-term. If a contraction of non-Abelian fields (bearing multiple field components) is present it may be possible to effect a self-cancellation that is still, in some sense, stringently flat.

Self-cancellation was first demonstrated in [8]. In the model, eighth order terms containing non-Abelian fields posed a threat to F-flatness of two non-Abelian directions. It was showed that a set of non-Abelian VEVs exist that is consistent with D-flat constraints and by which self-cancellation of the respective eighth order terms can occur. That is, for each specific set of non-Abelian VEVs imposed by D-flatness constraints, the expectation value of the dangerous F-term is zero. Hence, the "dangerous" superpotential terms posed no problem and two directions became flat to all finite order.

The non-Abelian fields taking on VEVs were the doublets of the two hidden sector $SU(2)$ gauge symmetries, with doublet fields of only one of the $SU(2)_H$ taking on VEVs in a given flat directions. Each flat direction with field VEVs charged under one of the $SU(2)_H$ was matched by a corresponding flat direction with isomorphic field VEVs charged under the other $SU(2)_H$. Example directions wherein self-cancellation was not possible were compared to examples where self-cancellation occurred. Rules for $SU(2)$ self-cancellation of dangerous F-terms were developed. As examples, one direction involving just the $SU(2)_H$ doublet fields $\{H_{23}, H_{26}, V_{40}\}$ was compared to another containing both $SU(2)_H$ and $SU(2)'_H$ doublets: $\{H_{23}, V_{40}, H_{28}, V_{37}\}$.

In generic non-Abelian flat directions, the norms of the VEVs of all fields are fixed by Abelian D-term cancellation, whereas the signs of the VEV components of a non-Abelian field are fixed by non-diagonal mixing of the VEVs in the corresponding non-Abelian D-terms (2.33). For the first direction, Abelian D-term cancellation required the ratio of the norms-squared of the H_{23}, H_{26}, V_{40} VEVs to be $1:1:2$, while the non-Abelian D-term cancellation required

$$\langle D^{SU(2)_H} \rangle = \langle H_{23}^\dagger T^{SU(2)} H_{23} + H_{26}^\dagger T^{SU(2)} H_{26} + V_{40}^\dagger T^{SU(2)} V_{40} \rangle = 0 , \quad (2.40)$$

where

$$T^{SU(2)} \equiv \sum_{a=1}^{3} T_a^{SU(2)} = \begin{pmatrix} 1 & 1-i \\ 1+i & -1 \end{pmatrix} . \quad (2.41)$$

The only solutions (up to a $\alpha \leftrightarrow -\alpha$ transformation) to (2.40) are

$$\langle H_{23} \rangle = \begin{pmatrix} \alpha \\ -\alpha \end{pmatrix}, \quad \langle H_{26} \rangle = \begin{pmatrix} \alpha \\ -\alpha \end{pmatrix} \quad \langle V_{40} \rangle = \begin{pmatrix} \sqrt{2}\alpha \\ \sqrt{2}\alpha \end{pmatrix}, \quad (2.42)$$

and

$$\langle H_{23} \rangle = \begin{pmatrix} \alpha \\ \alpha \end{pmatrix}, \quad \langle H_{26} \rangle = \begin{pmatrix} \alpha \\ \alpha \end{pmatrix} \quad \langle V_{40} \rangle = \begin{pmatrix} \sqrt{2}\alpha \\ -\sqrt{2}\alpha \end{pmatrix}. \quad (2.43)$$

A ninth-order superpotential term jeopardizes flatness of this non-Abelian D-flat direction via,

$$\langle F_{V_{39}} \rangle \equiv \langle \frac{\partial W}{\partial V_{39}} \rangle \quad (2.44)$$

$$\propto \langle \Phi_{23}\overline{\Phi}_{56}\Phi'_4 H^s_{31} H^s_{38}\rangle \langle H_{23}\cdot H_{26} V_{40} + H_{23} H_{26}\cdot V_{40} + H_{26} H_{23}\cdot V_{40}\rangle. \tag{2.45}$$

Self-cancellation of this F-term could occur if the non-Abelian VEVs resulted in

$$\langle H_{23}\cdot H_{26} V_{40} + H_{23} H_{26}\cdot V_{40} + H_{26} H_{23}\cdot V_{40}\rangle = 0. \tag{2.46}$$

However, neither (2.42) nor (2.43) are solution to this.

The contrasting, self-cancelling flat direction contains the non-Abelian VEVs of H_{23}, V_{40}, H_{28}, V_{37} with matching magnitudes. The $SU(2)_H$ D-term,

$$\langle D^{SU(2)_H}\rangle = \langle H^\dagger_{23} T^{SU(2)} H_{23} + V^\dagger_{40} T^{SU(2)} V_{40}\rangle = 0 \tag{2.47}$$

has the two solutions

$$\langle H_{23}\rangle = \begin{pmatrix} \alpha \\ -\alpha \end{pmatrix},\quad \langle V_{40}\rangle = \begin{pmatrix} \alpha \\ \alpha \end{pmatrix}, \tag{2.48}$$

and

$$\langle H_{23}\rangle = \begin{pmatrix} \alpha \\ \alpha \end{pmatrix},\quad \langle V_{40}\rangle = \begin{pmatrix} \alpha \\ -\alpha \end{pmatrix}. \tag{2.49}$$

(The $SU(2)'_H$ D-term solutions for H_{28} and V_{37} have parallel form.)

Flatness of this direction was threatened by an eighth-order superpotential term through

$$\langle F_{V_{35}}\rangle \equiv \langle \frac{\partial W}{\partial V_{35}}\rangle \tag{2.50}$$

$$\propto \langle \Phi_{23}\overline{\Phi}_{56} H^s_{31} H^s_{38}\rangle \langle H_{23}\cdot V_{40}\rangle \langle H_{28}\rangle. \tag{2.51}$$

Either set of $SU(2)_H$ VEVs (2.48) or (2.49) results in $\langle H_{23}\cdot V_{40}\rangle = 0$, which made this non-Abelian direction flat to all finite order!

2.9. Higgs μ & Generation Mass Terms

The all-order flat directions giving mass to all MSSM charged exotics were also shown to generate string scale mass for all but one linear combination of the electroweak Higgs doublets h_i and $\bar{h}_i$, for $i = 1, ..., 4$ [9]. Effective Higgs mass "μ-terms" take the form of one h_i field and one $\bar{h}_i$ plus one or more factors of field VEVs. Collectively, they may be expressed in matrix form as the scalar contraction $h_i M_{ij} \bar{h}_j$ with eigenstates h and $\bar{h}$ formed from linear combinations of the h_i and $\bar{h}_i$ respectively,[||]

$$h = \frac{1}{n_h}\sum_{i=1}^{4} c_i h_i;\qquad \bar{h} = \frac{1}{n_{\bar{h}}}\sum_{i=1}^{4} \bar{c}_i \bar{h}_i, \tag{2.52}$$

with normalization factors $n_h = \sqrt{\sum_i (c_i)^2}$, and $n_{\bar{h}} = \sqrt{\sum_i (\bar{c}_i)^2}$. These combinations will then in turn establish the quark and lepton mass matrices.

[||] The possibility of linear combinations of MSSM doublets forming the physical Higgs is a feature generic to realistic free fermionic models.

All MSSM-producing flat directions were shown to necessarily contain Φ_{23}, H_{31}^s, and H_{38}^s VEVs. Together these three VEVs produce four (linearly dependent) terms in the Higgs mass matrix: $h_3\bar{h}_2\langle\Phi_{23}\rangle$, $h_2\bar{h}_4\langle H_{31}\rangle$, $h_4\bar{h}_3\langle H_{38}\rangle$, and $h_4\bar{h}_4\langle H_{31}\rangle$. When these are the only non-zero terms in the matrix, the massless Higgs eigenstates simply correspond to $c_1 = \bar{c}_1 = 1$ and $c_j = \bar{c}_j = 0$ for $j = 2, 3, 4$. In this case all possible quark and lepton mass terms containing h_j for $j \in \{2, 3, 4\}$, decouple from the low energy MSSM effective field theory. However, when one or more of the c_j or $\bar{c}_j$ are non-zero, then some of these terms are not excluded and provide addition quark and lepton mass terms. In such terms, the Higgs components can be replaced by their corresponding Higgs eigenstates along with a weight factor,

$$h_i \to \frac{c_i}{n_h} h; \qquad \bar{h}_i \to \frac{\bar{c}_i}{n_{\bar{h}}} \bar{h}. \qquad (2.53)$$

Thus, in string models such as this, two effects can contribute to inter-generational (and intra-generational) mass hierarchies: generic suppression factors of $\frac{\langle\phi\rangle}{M_P}$ in non-renormalizable effective mass terms and $\frac{c_i}{n_h}$ or $\frac{\bar{c}_i}{n_{\bar{h}}}$ suppression factors. This means a hierarchy of values among the c_i and/or among the $\bar{c}_i$ holds the possibility of producing viable inter-generational $m_t : m_c : m_u \sim 1 : 7 \times 10^{-3} : 3 \times 10^{-5}$ mass ratios even when all of the quark and lepton mass terms are of renormalizable or very low non-renormalizable order. More than one generation of such low order terms necessitates a hierarchy among the c_i and $\bar{c}_i$.[*]

For any MSSM all-order flat direction, at least one generation must receive mass from renormalizable terms. The up- and down-like renormalizable terms are

$$\bar{h}_1 Q_1 u_1^c, \quad h_2 Q_2 d_2^c, \quad h_3 Q_3 d_3^c. \qquad (2.54)$$

Since $\bar{h}_1$ is either the only component or a primary component in $\bar{h}$, the top quark is necessarily contained in Q_1 and u_1^c is the primary component of the left-handed anti-top mass eigenstate. Thus, the bottom quark must be the second component of Q_1. Since there are no renormalizable $h_i Q_1 d_m^c$ terms in eq. (2.54), a bottom quark mass that is hierarchically larger than the strange and down quark masses requires that

$$\frac{|c_{j=2,3}|}{n_h} \ll 1. \qquad (2.55)$$

Non-zero $c_{2,3}$ satisfying eq. (2.55) could, perhaps, yield viable strange or down mass terms.

The first possible bottom mass term appears at fourth order, $h_4 Q_1 d_3^c H_{21}^s$. Realization of the bottom mass via this term requires h to contain a component of h_4 and for H_{21}^s to acquire a VEV. Of all flat directions, only two, here denoted *FDNA1* and *FDNA2* give H_{21}^s a VEV. Of these two, only *FDNA2* embeds part of h_4 in h.

The physical mass ratio of the top and bottom quark is of order $\sim 3 \times 10^{-2}$. In free fermionic models there is no significant suppression to the effective third order superpotential coupling constant, $\lambda_3^{\text{eff}} = \lambda_4 \langle\phi\rangle$, originating from a fourth order term [29]. Hence, a

---

[*]Generational hierarchy via suppression factor in Higgs components was first used in free fermionic models of the flipped $SU(5)$ class [28].

reasonable top to bottom mass ratio would imply

$$\frac{|c_2|}{n_h}, \frac{|c_3|}{n_h} \ll \frac{|c_4|}{n_h} \sim 10^{-2 \text{ to } -3} \qquad (2.56)$$

when $\frac{|\bar{c}_1|}{n_{\bar{h}}} \sim 1$ and $\langle h \rangle \sim \langle \bar{h} \rangle$. However, [9] showed that $\frac{|c_4|}{n_h} \gtrsim 10^{-3}$ value cannot be realized along any of the flat directions explored.

The next possible higher order bottom mass terms do not occur until sixth order, all of which contain h_j, where $j \in \{2,3,4\}$. Beyond fourth order a suppression factor of $\frac{1}{10}$ per order is generally assumed [30]. Thus, a sixth order down mass term would imply $\frac{|c_j|}{n_h} \sim 1$, where $j \in \{2,3,4\}$ as appropriate, when $\frac{|\bar{c}_1|}{n_{\bar{h}}} \sim 1$. However, none of the flat directions transform any of such sixth order terms into mass terms [9].

If not sixth order, then seventh order is the highest order that could provide a sufficiently large bottom mass. There are no such seventh order terms containing h_1. However, h_2 is in 15 of these terms, one of which becomes a bottom mass term for *FDNA2*. No seventh order terms containing h_3 or h_4 become μ terms for any of the flat directions. Therefore, the only possible bottom quark mass terms resulting from the flat directions explored in [9] are the fourth order h_4 term and the seventh order h_2 terms. However, no flat directions contain the VEVs required to transform any of the seventh or eighth order terms into effective mass terms.

While several flat directions can generate a particular ninth order Higgs μ term, $h_1 \bar{h}_3 \langle N_3^c \Phi_4' H_{15}^s H_{30}^s H_{31}^s H_{28} \cdot V_{37} \rangle$, *FDNA2* is the only one that simultaneously generates the fourth order bottom mass term. Thus, *FDNA2* was singled out for detailed analysis.

[7, 8] showed that a small $\overline{\Phi}_{12} \ll$ FI-scale VEV produces superior quark and lepton mass matrix phenomenology. However, since $\overline{\Phi}_{12}$ does not acquire a VEV in any flat direction found, $\langle \overline{\Phi}_{12} \rangle \neq 0$ was only allowed a second order effect at or below the MSSM unification scale. Thus, in the Higgs mass matrix, $h_2 \bar{h}_1 \langle \overline{\Phi}_{12} \rangle$ was allowed to provide significantly suppressed mass term. When $\langle \overline{\Phi}_{12} \rangle \sim 10^{-4}$ was assumed, the numeric form of the *FDNA2* Higgs doublet mass matrix becomes (in string-scale mass units)

$$M_{h_i,\bar{h}_j} \sim \begin{pmatrix} 0 & 0 & 10^{-5} & 0 \\ 10^{-4} & 0 & 0 & 1 \\ 0 & 2 & 0 & 0 \\ 0 & 0 & 1 & 10^{-1} \end{pmatrix}. \qquad (2.57)$$

The massless Higgs eigenstates produced (by diagonalizing $M^\dagger M$) are of order[*]

$$h = h_1 + 10^{-7} h_2 - 10^{-5} h_4, \qquad (2.58)$$
$$\bar{h} = \bar{h}_1 + 10^{-6} \bar{h}_3 - 10^{-4} \bar{h}_4. \qquad (2.59)$$

These Higgs eigenstates provide examples of how several orders of magnitude mass suppression factors can appear in the low order terms of a Minimal Standard Heterotic String Model (MSHSM). Here, specifically, the h_4 coefficient in (2.58) can provide 10^{-5} mass suppression for one down-quark generation and one electron generation. When

---

[*] In the limit of $\langle \overline{\Phi}_{12} \rangle = 0$, the h eigenstate reduces to $\bar{h}_1$.

rewritten in terms of the Higgs mass eigenstates, $h_4(Q_i d_j^c + L_i e_j^c)$ contains a factor of $10^{-5}h(Q_i d_j^c + L_i e_j^c)$. Similarly, the h_2 coefficient can provide 10^{-7} suppression. Further, the $\bar{h}_4$ coefficient in (2.59) can provide 10^{-4} up like-quark mass suppression and the $\bar{h}_3$ coefficient a corresponding 10^{-6} suppression.

Under the assumption of the above Higgs (near) massless eigenstates, the quark and lepton mass matrices were calculated up to ninth order, the level at which suppression in the coupling (assumed here to be $\sim 10^{-5}$) is comparable to that coming out of eqs. (2.58, 2.59). The up-quark mass matrix contains only a single term (corresponding to the top mass) when $\langle\bar{\Phi}_{12}\rangle = 0$, but develops an interesting texture when $\langle\bar{\Phi}_{12}\rangle \sim 10^{-4}$ FI-scale. To leading order, the general form (in top quark mass units) is

$$M_{Q,u^c} = \begin{pmatrix} \bar{h}_1 & .1\bar{h}_3 + 10^{-3}\bar{h}_4 & 0 \\ 10^{-2}\bar{h}_3 + 10^{-4}\bar{h}_4 & 0 & 0 \\ 0 & 0 & \bar{h}_4 \end{pmatrix} \sim \begin{pmatrix} 1 & 10^{-6} & 0 \\ 10^{-7} & 0 & 0 \\ 0 & 0 & 10^{-4} \end{pmatrix}. \quad (2.60)$$

The up-like mass eigenvalues are 1, 10^{-4}, and 10^{-13}. A more realistic structure would appear for a $\bar{h}_4$ suppression factor of 10^{-2} rather than 10^{-4}.

The corresponding down-quark mass matrix has the form

$$M_{Q,d^c} = \begin{pmatrix} 0 & 10^{-3}h_2 + 10^{-5}h_4 & h_4 \\ 10^{-5}h_2 & h_2 + .1h_4 & 10^{-5}h_2 \\ 0 & 10^{-4}h_2 & 0 \end{pmatrix} \sim \begin{pmatrix} 0 & 10^{-9} & 10^{-5} \\ 10^{-11} & 10^{-6} & 10^{-11} \\ 0 & 10^{-10} & 0 \end{pmatrix}. \quad (2.61)$$

The resulting down-quark mass eigenvalues are 10^{-5}, 10^{-6}, and 10^{-15} (in top quark mass units). This provides quasi-realistic down and strange masses, but lacks a bottom mass. Unfortunately, the down-like quark eigenstate corresponding to a 10^{-5} mass is in Q_1, making it the bottom quark. The second and third generation masses would be more viable if the h_4 suppression factor were 10^{-2} instead.

The electron mass matrix takes the form

$$M_{L,e^c} = \begin{pmatrix} 0 & 10^{-4}h_2 & 0 \\ 10^{-4}h_2 & h_2 + .1h_4 & 10^{-4}h_2 \\ .1h_4 & 10^{-5}h_2 & 0 \end{pmatrix} \sim \begin{pmatrix} 0 & 10^{-10} & 0 \\ 10^{-10} & 10^{-6} & 10^{-10} \\ 10^{-5} & 10^{-11} & 0 \end{pmatrix}. \quad (2.62)$$

The three corresponding electron-like mass eigenvalues 10^{-5}, 10^{-6}, and 10^{-14}. As with the down-like quark masses, more viable second and third generation electron masses would appear if the h_4 suppression factor was only 10^{-2}.

2.10. Hidden Sector Condensation

The free fermionic MSHSM model proved that non-Abelian VEVs can yield a three generation MSSM model with no exotics while maintaining supersymmetry at the string/FI scale. Supersymmetry must ultimately be broken slightly above the electroweak scale, though. Along some of the flat directions (both Abelian and non-Abelian), this model showed qualitatively how supersymmetry may be broken dynamically by hidden sector field condensation. Two of the non-Abelian flat directions break both of the hidden sector $SU(2)_H$

and $SU(2)'_H$ gauge symmetries, but leave untouched the hidden sector $SU(3)_H$. Thus, condensates of $SU(3)_H$ fields can initiate supersymmetry breaking [31].

The set of nontrivial $SU(3)_H$ fields is composed of five triplets, $\{H_{42}, V_4, V_{14}, V_{24}, V_{34}\}$, and five corresponding anti-triplets, $\{H_{35}, V_3, V_{13}, V_{23}, V_{24}\}$. Along these two non-Abelian flat directions, singlet VEVs give unsuppressed FI-scale mass to two triplet/antitriplet pairs via trilinear superpotential terms,[†]

$$\langle \Phi_{12} \rangle V_{23} V_{24} + \langle \Phi_{23} \rangle V_{33} V_{34} \tag{2.63}$$

and slightly suppressed mass to another triplet/anti-triplet pair via a fifth order term,

$$\langle \overline{\Phi}_{56} H^s_{31} H^s_{38} \rangle H_{42} H_{35} . \tag{2.64}$$

and a significantly suppressed mass to a fourth pair via a tenth order term,

$$\langle \Phi_{23} \overline{\Phi}_{56} H^s_{31} H^s_{38} H_{23} V_{40} H_{28} V_{37} \rangle V_4 V_3 . \tag{2.65}$$

Before supersymmetry breaking, the last triplet/antitriplet pair, V_{14}/V_{13}, remain massless to all finite order.

Consider a generic $SU(N_c)$ gauge group containing N_f flavors of matter states in vector-like pairings $T_i \bar{T}_i$, $i = 1, \ldots N_f$. When $N_f < N_c$, the gauge coupling g_s, though weak at the string scale M_{str}, becomes strong at a condensation scale defined by

$$\Lambda = M_P e^{8\pi^2/\beta g_s^2} , \tag{2.66}$$

where the β-function is given by,

$$\beta = -3N_c + N_f . \tag{2.67}$$

The N_f flavors counted are only those that ultimately receive masses $m \ll \Lambda$. Thus, in this model $N_c = 3$ and $N_f = 1$ (counting only the vector-pair, V_{14} and V_{13}), which corresponds to $\beta = -8$ and results in an $SU(3)_H$ condensation scale

$$\Lambda = e^{-20} M_P 10^{10} \text{ GeV}. \tag{2.68}$$

At this condensation scale Λ, the matter degrees of freedom are best described in terms of the composite "meson" fields, $T_i \bar{T}_i$. (Here the meson field is $V_{14} V_{13}$.) Minimizing the scalar potential of the meson field induces a VEV of magnitude,

$$\langle V_{14} V_{13} \rangle = \Lambda^3 \left(\frac{m}{\Lambda} \right)^{N_f/N_c} \frac{1}{m} . \tag{2.69}$$

This results in an expectation value of

$$\langle W \rangle = N_c \Lambda^3 \left(\frac{m}{\Lambda} \right)^{N_f/N_c} \tag{2.70}$$

for the non-perturbative superpotential.

[†]These mass terms even occur for the simplest Abelian flat directions.

Supergravity models are defined in terms of two functions, the Kähler function, $G = K + \ln|W|^2$, where K is the Kähler potential and W the superpotential, and the gauge kinetic function f. These functions determine the supergravity interactions and the soft-supersymmetry breaking parameters that arise after spontaneous breaking of supergravity, which is parameterized by the gravitino mass $m_{3/2}$. The gravitino mass appears as a function of K and W,

$$m_{3/2} = \langle e^{K/2} W \rangle. \tag{2.71}$$

Thus,

$$m_{3/2} \sim \langle e^{K/2} \rangle \langle W \rangle \sim \langle e^{K/2} \rangle N_c \Lambda^3 \left(\frac{m}{\Lambda}\right)^{N_f/N_c}. \tag{2.72}$$

Restoring proper mass units explicitly gives,

$$m_{3/2} \sim \langle e^{K/2} \rangle N_c \left(\frac{\Lambda}{M_P}\right)^3 \left(\frac{m}{\Lambda}\right)^{N_f/N_c} M_P. \tag{2.73}$$

Hence the meson field $V_{14}V_{13}$ acquires a mass of at least the supersymmetry breaking scale. It was assumed in [9] that $m_{V_{14}V_{13}} \approx 1$ TeV. The resulting gravitino mass is

$$\begin{aligned} m_{3/2} &\sim \langle e^{K/2} \rangle \left(\frac{7 \times 10^9 \, \text{GeV}}{2.4 \times 10^{18} \, \text{GeV}}\right)^3 \left(\frac{1000 \, \text{GeV}}{7 \times 10^9 \, \text{GeV}}\right)^{1/3} 2.4 \times 10^{18} \, \text{GeV} \\ &\approx \langle e^{K/2} \rangle \, 0.3 \, \text{eV}. \end{aligned} \tag{2.74}$$

In standard supergravity scenarios, one generally obtains soft supergravity breaking parameters, such as scalar and gaugino masses and scalar interaction, that are comparable to the gravitino mass: $m_o, m_{1/2}, A_o \sim m_{3/2}$. A gravitino mass of the order of the supersymmetry breaking scale would require $\langle e^{K/2} \rangle \sim 10^{12}$ or $\langle K \rangle \sim 55$. On the other hand, for a viable model, $\langle e^{K/2} \rangle \sim \mathcal{O}(1)$ would necessitate a decoupling of local supersymmetry breaking (parametrized by $m_{3/2}$) from global supersymmetry breaking (parametrized by $m_o, m_{1/2}$). This is possible in the context of no-scale supergravity [32], endemic to weak coupled string models.

In specific types of no-scale supergravity, the scalar mass m_o and the scalar coupling A_o have null values thanks to the associated form of the Kähler potential. Furthermore, the gaugino mass can go as a power of the gravitino mass, $m_{1/2} \sim \left(\frac{m_{3/2}}{M_P}\right)^{1-\frac{2}{3}q} M_P$, for the standard no-scale form of G and a non-minimal gauge kinetic function $f \sim e^{-Az^q}$, where z is a hidden sector moduli field [33]. A gravitino mass in the range 10^{-5} eV $\lesssim m_{3/2} \lesssim 10^3$ eV is consistent with the phenomenological requirement of $m_{1/2} \sim 100$ GeV for $\frac{3}{4} \gtrsim q \gtrsim \frac{1}{2}$.

2.11. First MSHSM Model Summary and Inspired Research

The NAHE-based free fermionic heterotic model presented initially in [5] was the first to succeed at removing *all* MSSM-charged exotic states from the LEEFT. This model has since received the designation of Minimal Standard Heterotic String Model. The existence of the first Minimal Standard Heterotic String Models, which contain solely the three generations of MSSM quarks and leptons and a pair of Higgs doublets as the massless SM-charged states in the LEEFT, was a significant discovery. The MSHSM offered the first potential realizations of possible equivalence, in the strong coupling limit, between the string scale and the minimal supersymmetric standard model unification scale $M_U \approx 2.5 \times 10^{16}$ GeV. This requires that the observable gauge group just below the string scale should be $SU(3)_C \times SU(2)_L \times U(1)_Y$ with charged spectrum consisting solely of the three MSSM generations and a pair of Higgs doublets.

In the MSHSM, masses of the exotic states were driven to be above the MSSM unification at around $\frac{1}{10}$ FI scale by scalars taking on D- and F-flat VEVs to cancel an anomalous $U(1)_A$ (which are endemic to realistic free fermionic heterotic MSSM-like models. The first class of flat directions explored for this model involved only non-Abelian singlet fields [5, 6, 7]. The MSHSM flat direction search was expanded in [8, 9] to include hidden sector non-Abelian field VEVs. This provided for improved phenomenology, showing that quasi-realistic patterns to quark and charged-lepton mass matrices can appear. This was a result of both the new non-Abelian VEVs and the related structure of the physical Higgs doublets h and $\bar{h}$. In the more realistic free fermionic heterotic models, the physical Higgs can each contain up to four components with weights vastly differing by several orders of magnitude. These components generically have generation-dependent coupling strengths. Thus, mass suppression factors for the first and second quark and lepton generations can appear even at very low order as a result of the different weights of the Higgs components.

In this model the top quark can receive a viable, unsuppressed mass (given realistic Higgs VEVs), while masses for the bottom quark, most second generation and some first generation quarks and leptons were too small. This resulted from too small weight factors for one h component and for one $\bar{h}$ component. Phenomenology would be significantly improved if the h_4 and $\bar{h}_4$ weights in h and $\bar{h}$ were larger by a factor of 100 than their respective values of 10^{-5} and 10^{-4} in the best non-Abelian flat direction.

In this model the emergence of new techniques for the removal of dangerous terms from $\langle W \rangle$ and from $\langle F \rangle$ was also observed. Flatness of four non-Abelian directions was lifted to all order by the vanishing of terms with more than two non-Abelian fields. Non-Abelian self-cancellation within single terms was developed in the MSHSM as a promising tool for extending the order to which a non-Abelian direction is safe.

The flat directions of this MSHSM present some interesting phenomenological features such as multi-component physical Higgs that couple differently to given quarks and leptons. Exploration of flat direction phenomenology for this model demonstrated that non-Abelian VEVs are necessary (but perhaps not sufficient) for viable LEEFT MSSM phenomenology. This is in agreement with similar evidence presented suggesting this might be true as well for all MSHSM $\mathbb{Z}_2 \times \mathbb{Z}_2$ models. Nonetheless, the stringent flat F- and D-flat directions producing this MSHSM do not themselves lead to viable quark and lepton mass matrices. This implies significant worth to exploring the generic properties of non-Abelian flat directions in $\mathbb{Z}_2 \times \mathbb{Z}_2$ models that contain exactly the MSSM three generations and two Higgs doublets as the only MSSM-charged fields in the LEEFT.

One direction suggested is non-stringent MSHSM directions flat to a finite order due to cancellation between various components in an F-term. While the absence of any non-zero terms from within $\langle F_{\Phi_m} \rangle$ and $\langle W \rangle$ is clearly sufficient for F-flatness along a given

D-flat direction, such stringent demands was not necessary. Total absence of these terms can be relaxed, so long as they appear in collections which cancel among themselves in each $\langle F_{\Phi_m} \rangle$ and in $\langle W \rangle$. It is desirable to examine the mechanisms of such cancellations as they can allow additional flexibility for the tailoring of phenomenologically viable particle properties while leaving SUSY inviolate.

This insufficiency in phenomenology from stringent flat directions (Abelian or otherwise) for this model, and from non-Abelian singlet vacua for a range of other models inspired further analysis of non-Abelian flat direction technology and self-cancellation from a geometrical framework [34]. The geometrical perspective facilitates manipulations of non-Abelian VEVs, such as treating superpotential contractions with multiple pairings, and examining the new possibility of self-cancellation between elements of a single term. When expressed in geometric language, the process of describing valid solutions, or compatibilities between the F and D conditions, can become more accessible and intuitive. This could provide further assistance in closing the gap between string model building and low energy experimental evidence [34].

From the geometric pint of view, [34] showed that non-Abelian D-term flatness translates into the imperative that the adjoint space representation of all expectation values form a closed vector sum. Furthermore, the possibility emerges that a single seemingly dangerous F-term might experience a self-cancellation among its components. In [34] it was examined whether this geometric language can provide an intuitive and immediate recognition of when the D and F conditions are simultaneously compatible, as well as a powerful tool for their comprehensive classification. Some initial success to this process was found. Geometric interpretation of F and D flat directions was made concrete by an examination of the specific cases of $SU(2)$ and $SO(2n)$ symmetries. As a rank 1 group with a number of generators equal to its fundamental dimension 3, $SU(2)$ represents the simplest specific case on which to initiate discussion. On the other hand, $SO(2n)$ introduces new complications by way of higher rank groups and an adjoint space of dimension greater than the fundamental.[‡] A geometric interpretation for simultaneity of D- and F-flatness for $SU(2)$-charged fields was introduced.

Solutions to non-Abelian D-flatness were found to not necessarily be associated with gauge invariant superpotential terms. A counter example was presented in [34] for which the VEVs within a single **6** provide $SO(6)$ D-flatness. The constraint equations for non-trivial D-flatness were developed for any number of **6**'s.[§]

The discovery of an MSHSM in the neighborhood of the string/M-theory parameter space allowing free-fermionic description strongly suggested a search for more phenomeno-

---

[‡]As an interesting aside, note that although the fields under consideration were all spacetime scalars, superpotential terms can inherit an induced symmetry property from the analytic rotationally invariant contraction form of the group under study. $SU(2)$ will have a "fermionic" nature, with an antisymmetric contraction, while that of $SO(2n)$ will be symmetric, or "bosonic".

[§]Development of systematic methods for geometrical analysis of the landscape of D- and F-flat directions begun in [34] offers a possibility for further understanding of the geometry of brane-anti-brane systems. This is related to the connection between 4D supergravity D-terms of a string and a $D_{3+q} - \bar{D}_{3+q}$ wrapped brane/antibrane system. In this association, the energy of a $D_{3+q}/\bar{D}_{3+q}$ system appears as an FI D-term. An open string tachyon connecting brane and anti-brane is revealed as an FI-cancelling Higgs field, and a D_{1+q}-brane produced in an annihilation between a D_{3+q}-brane and a $\bar{D}_{3+q}$-anti-brane is construed to be a D-term string [35].

logically realistic MSHSMs in the free fermion region. The concrete results, obtained in the analysis of a specific model, highlighted the underlying, phenomenologically successful, structure generated by the NAHE set and promoted further investigation of the string vacuum in the vicinity of this model. That is, it warranted further investigation of $\mathbb{Z}_2 \times \mathbb{Z}_2$ models in the vicinity of the self-dual radius in the Narain moduli space.

One variation among MSHSM models in this neighborhood is the number of pairs of Higgs doublets. Investigation in this direction were recently conducted in [36], wherein the removal of some (all) of the three or four extra pairs of Higgs doublets through modifications to free fermion boundary conditions was studied. A general mechanism was developed that achieves Higgs reduction through asymmetric boundary conditions between the left-moving (y^i, w^i) and right-moving $(\bar{y}^i, \bar{w}^i)$ internal fermions for the 6 compactified dimensions [36]. By this, the number of pairs of Higgs doublets was reduced to one. However a correlation was found between the reduction in Higgs doublets and the size of the flat direction moduli space. The change in boundary conditions substantially reduces the number of non-hypercharged scalar field singlets. This vast reduction eliminated any flat directions that could simultaneously cancel the anomalous $U(1)$ FI term while retaining the MSSM gauge group. Thus, the only stable supersymmetric vacua in the model of [36] destroys the MSSM gauge group. If this pattern holds for all MSHSM models in the neighborhood, then the physical Higgs must be a composite state with different coupling strengths to each generation. This result could lead to some interesting phenomenological predictions for LHC physics.

3. Subsequent String-Derived MSSM Models ...

Following the construction of the first string-derived model containing exactly the MSSM states in the observable sector, several other string models with this feature have been generated from different heterotic compactifications and from Type IIA & IIB theories. In this section a representative, but nonetheless incomplete, set of such models are reviewed. The discussions are arranged by model class and compactification method, rather than chronologically. The methods of construction and phenomenological features of these models are summarized. For further details of these example models, the original papers should be consulted.

3.1. From Heterotic Orbifolds

An additional $E_8 \times E_8$ heterotic model with solely the MSSM spectrum in the observable sector of the LEEFT was recently constructed using a $\mathbb{Z}'_6 = \mathbb{Z}_3 \otimes \mathbb{Z}_2$ orbifold alternative to $\mathbb{Z}_2 \otimes \mathbb{Z}_2$ [37, 38]. In the $\mathbb{Z}'_6$ model the quarks and leptons appear as three **16**'s of $SO(10)$, two of which are localized at fixed points of the orbifold and one for which some of the quarks and leptons are distributed across the bulk space (untwisted sector) and some are localized in twisted sectors. Like the free fermionic model above, this model initially contains MSSM-charged exotics. This model specifically contains 4 $SU(3)_C$ **3/3̄** pairs, 5 pairs of $SU(2)$ doublets with $\pm\frac{1}{2}$ hypercharge, and 7 extra pairs of doublets with zero hypercharge. Flat directions, formed from a set of 69 non-Abelian singlets without hypercharge, are argued to exist that may give near string-scale mass to all MSSM-charged exotics (depending on the severity of mass suppression for 7th and 8th order terms), thereby decoupling the exotics from the low energy effective field theory.

3.1.1. Orbifold Construction

In the orbifold construction, all internal degrees of freedom are bosonized. For the left-moving supersymmetric sector, the 6 compactified bosonic modes X_L^i, $i = 1$ to 6, are complexified into $Z_L^j = \frac{1}{\sqrt{2}}(X_L^{2j-1} + iX_L^{2j})$, $j = 1$ to 3. The accompanying six fermionic worldsheet modes x^i are likewise paired into three complex fermions $\chi^j = \frac{1}{\sqrt{2}}\{x^{2j-1} + ix^{2j}\}$, $j = 1, ..., 3\}$, which are then replaced by a real bosonic mode ϕ^j via the relation $\chi^j = \exp(-2i\phi^j)$. The single left-moving complex spacetime fermion ψ^0 in lightcone gauge is also replaced by a real bosonic mode ϕ^0. The corresponding right-moving modes are the compactified complex $Z_R^j = \frac{1}{\sqrt{2}}(X_R^{2j-1} + iX_R^{2j})$ and the 16 real bosonic modes $\bar{X}^I$, $I = 1, ..., 16$, for which the momenta are on the $E_8 \otimes E_8$ root lattice.

The compactification lattice factorizes as $\mathbf{T}^6 = \mathbf{T}^2 \otimes \mathbf{T}^2 \otimes \mathbf{T}^2$, in parallel to the free fermionic models. However, in the orbifold model the torus is $\mathbf{T}^6 = R^6/\Lambda_{G_2 \otimes SU(3) \otimes SO(4)}$, where $\Lambda_{G_2 \otimes SU(3) \otimes SO(4)}$ is the lattice for the $G_2 \otimes SU(3) \otimes SO(4)$ algebra. Therefore, complex coordinates z^i on the torus are identified when they differ by a lattice vector, $\mathbf{z} \sim \mathbf{z} + 2\pi \mathbf{l}$ (where $\mathbf{z}$ is a vector with three complex components z_j) and $\mathbf{l} = m_a \mathbf{e}_a$, with $\mathbf{e}_a$ the basis vectors of the three lattice planes and $m_a \in \mathbb{Z}$. This lattice has a $\mathbb{Z}_6' \equiv \mathbb{Z}_3 \otimes \mathbb{Z}_2$ discrete symmetry, $\mathbf{z} \to \theta \mathbf{z}$, $\theta_j^i \equiv \exp(2\pi i v_6^i)\delta_j^i$, for $i,j = 1, 2, 3$, with $\theta^6 = \mathbf{1}$, and $6v_6^i = 0$ mod 1. $N = 1$ supersymmetry is maintained if the $\mathbb{Z}_6$ twist is a discrete symmetry within $SU(3) \in SO(6)$, which further requires that $\sum_i v_6^i = 0$ mod 1. The combination of lattice translations and twists of $\mathbf{z}$ form the space group $\$$ whose elements are $(\theta^k, \mathbf{l})$, for $k = 0, ..., 5$. Points on the orbifolded torus $\mathbf{T}^6/\mathbb{Z}_6'$ (alternately $\mathbb{R}^6/\$$) are then identified with $\mathbf{z} \sim \theta^k \mathbf{z} + 2\pi \mathbf{l}$.

The $\mathbb{Z}_6' = \mathbb{Z}_3 \otimes \mathbb{Z}_2$ twist acts on the $\mathbf{Z}$ and χ fields as $\mathbf{Z} \equiv \mathbf{Z}_L + \mathbf{Z}_R \to \theta \mathbf{Z}$ and $\chi \to \theta \chi$ and therefore on the ϕ fields as $\phi \to \phi - \pi \mathbf{v}_6$. In travelling around non-contractible loops in the σ direction, the fields transform as

$$\mathbf{Z}(\sigma + 2\pi) = \mathbf{Z}(\sigma) + 2\pi m_a \mathbf{e}_a, \quad (3.1)$$
$$\chi(\sigma + 2\pi) = \pm \chi(\sigma), \quad (3.2)$$

(where $+$ is for a Ramond fermion and $-$ is for a Neveu-Schwarz fermion) in the untwisted sectors and

$$\mathbf{Z}(\sigma + 2\pi) = \theta^k \mathbf{Z}(\sigma) + 2\pi m_a \mathbf{e}_a, \quad (3.3)$$
$$\chi(\sigma + 2\pi) = \pm \theta^k \chi(\sigma), \quad (3.4)$$

in the kth twisted sector. Thus,

$$\phi(\sigma + 2\pi) = \phi(\sigma) - \pi k \mathbf{v}_6 \quad (3.5)$$

in the kth twisted sector. Twisted strings are localized at the fixed points.

Orbifolding of the model is obtained by simultaneously modding the $E_8 \otimes E_8$ torus by a correlated $\mathbb{Z}_6'$ twist, as required for modular invariance. The bosonic fields on the lattice transform as $\bar{X}^I \to \bar{X}^I + \pi k V_6^I$. Discrete Wilson lines add shifts, $\bar{X}^I \to \bar{X}^I + \pi n_l W_l^I$, with integer n_l, to accompany torus lattice translations.¶ Thus, the twists and lattice shifts are embedded into the gauge degrees of freedom $\bar{X}^I$, as $(\theta^k, m_a \mathbf{e}_a) \to (\mathbf{1}, k\mathbf{V}_6 + m_a \mathbf{W}_{la})$.

Modular invariance requires that both $6\mathbf{V}_6$ and $n\mathbf{W}_l$, with $n \leq N$ the order of the Wilson line, must lie on the $E_8 \otimes E_8$ lattice. In addition, the modular invariance rules, parallel to (2.4-2.6), are

$$\tfrac{1}{2}(\mathbf{V}_6^2 - \mathbf{v}_6^2) = 0 \bmod 1, \qquad (3.6)$$

$$(\mathbf{V}_6 \cdot \mathbf{W}_n) = 0 \bmod 1, \qquad (3.7)$$

$$(\mathbf{W}_n \cdot \mathbf{W}_m) = 0 \bmod 1, \; (W_n \neq W_m) \qquad (3.8)$$

$$\tfrac{1}{2}\mathbf{W}_n^2 = 0 \bmod 1, \qquad (3.9)$$

which can be nicely unified [37, 38] as

$$\tfrac{1}{2}\left[(r\mathbf{V}_6 + m_a \mathbf{W}_{na})^2 - r\mathbf{v}_6^2\right] = 0 \bmod 1. \qquad (3.10)$$

The orbifold expression for the generalized GSO projections specifying the physical states in the untwisted sector is

$$\mathbf{v}_6 \cdot \mathbf{q} - \mathbf{V}_6 \cdot \mathbf{p} = 0 \bmod 1; \; \mathbf{W}_6 \cdot \mathbf{p} = 0 \bmod 1, \qquad (3.11)$$

where the components of vectors $(\mathbf{v}_6, \mathbf{V}_6, \mathbf{W}_n)$ are the (v_6^j, V_6^I, W_n^I). This generalizes for the twisted sectors as

$$\bar{k}\mathbf{v}_6 \cdot (\bar{\mathbf{N}}_f - \bar{\mathbf{N}}_f^*) - \bar{k}\mathbf{v}_6 \cdot (\mathbf{q} + k\mathbf{v}_6)$$
$$+ (\bar{k}\mathbf{V}_6 + \bar{m}_a \mathbf{W}_{na}) \cdot (\mathbf{p} + \bar{k}\mathbf{V}_6 + \bar{m}_a \mathbf{W}_{na}) = 0 \bmod 1 \qquad (3.12)$$

for all $\mathbf{W}_n$ and $(\bar{k}, \bar{m}_a)$ depending on conjugacy class (with modifications for non-prime orbifolds). The components of $\mathbf{q}$ are q^j, the momenta of the left-moving bosons ϕ^j on the $\mathbf{T}^6$ torus, the components of $\mathbf{p}$ are p^I, the momenta of the right-moving $\bar{X}^I$ bosons on the $E_8 \otimes E_8$ lattice, and $\mathbf{V}_f^I \equiv k V_6^I$.

In the untwisted sector the mass of the physical states are

$$\alpha' m_L^2 = -\tfrac{1}{2} + \tfrac{1}{2}\mathbf{q}^2 + N + N^* \qquad (3.13)$$

$$\alpha' m_R^2 = -1 + \tfrac{1}{2}\mathbf{p}^2 + \bar{N} + \bar{N}^*, \qquad (3.14)$$

for which $N^{(*)} = \sum_j N^{(*)j}$, where $N^{(*)j}$ are the sums of the eigenvalues of the number operators for the complex left-moving $Z_L^{(*)j}$ modes, respectfully, and $\bar{N}^{(*)}$ are the corresponding for the complex right-moving modes. The generalization for the twisted sectors is,

$$\alpha' m_L^2 = -\tfrac{1}{2} + \tfrac{1}{2}(\mathbf{q} + k\mathbf{v}_6)^2$$
$$+ \tfrac{1}{2}\left[k\mathbf{v}_6 \cdot (\mathbf{1} - k\mathbf{v}_6) + k\mathbf{v}_6 \cdot \mathbf{N}_f - k\mathbf{v}_6 \cdot \mathbf{N}_f^*\right] \qquad (3.15)$$

$$\alpha' m_R^2 = -1 + \tfrac{1}{2}(\mathbf{p} + \mathbf{V}_f)^2$$
$$+ \tfrac{1}{2}\left[k\mathbf{v}_6 \cdot (\mathbf{1} - kv_{6j}) + k\mathbf{v}_6 \cdot \bar{\mathbf{N}}_f - k\mathbf{v}_6 \cdot \bar{\mathbf{N}}_f^*\right]. \qquad (3.16)$$

¶In a free fermion model, the corresponding complex fermion field is $\bar{\lambda}^I = \pm \exp(2\pi i \bar{X}^I)$ and transforms as $\bar{\lambda}^I(\sigma + 2\pi) \to \pm \Theta \bar{\lambda}^I(\sigma)$, where $\Theta^I = \exp(2\pi i k V_6^I)\delta_J^I$ without Wilson lines and as $\bar{\lambda}^I(\sigma + 2\pi) \to \pm \Theta \bar{\lambda}^I(\sigma)$, where $\Theta^I = \exp(2\pi i \{k V_6^I + m_a W_{la}^I\})\delta_J^I$ with Wilson lines.

At each fixed point on the orbifold, determined by the combination $\mathbf{V}_f$, $m_a\mathbf{W}_{na}$, and $k\mathbf{v}_n$, $E_8 \otimes E_8$ is broken to a subgroup. For the $G_2 \otimes SU(3) \otimes SO(4)$ torus, the twist vector is chosen to be $\mathbf{v}_6 = (\frac{1}{6}, \frac{1}{3}, -\frac{1}{2})$. For the associated order six $E_8 \otimes E_8$ twist, only two choices guarantee complete **16**'s of $SO(10)$ in the twisted sector:

$$\mathbf{V}_6 = (\tfrac{1}{2}, \tfrac{1}{2}, \tfrac{1}{3}, 0, 0, 0, 0, 0)(\tfrac{1}{3}, 0, 0, 0, 0, 0, 0, 0), \tag{3.17}$$

$$\mathbf{V}'_6 = (\tfrac{1}{3}, \tfrac{1}{3}, \tfrac{1}{3}, 0, 0, 0, 0, 0)(\tfrac{1}{6}, \tfrac{1}{6}, 0, 0, 0, 0, 0, 0), \tag{3.18}$$

with the first being a $\mathbb{Z}'_6 \sim \mathbb{Z}_3 \otimes \mathbb{Z}_2$ twist and the second a non-factorizable $\mathbb{Z}_6$ twist. The orbifold model uses $\mathbf{V}_6$ [37, 38].

The simplest choice of additional Wilson lines leads to three equivalent fixed points with local $SO(10)$ symmetry, arranging for one generation per fixed point. This would be analogous to the free fermionic model. However, it was found in [37, 38] that three fixed points always produced chiral, rather than vector-like, exotic MSSM states. These chiral exotic could only get EW mass, and therefore not be decoupled from the low energy effective field theory. Thus, two Wilson lines,

$$\mathbf{W}_2 = (\tfrac{1}{2}, 0, \tfrac{1}{2}, \tfrac{1}{2}, \tfrac{1}{2}, 0, 0, 0)(-\tfrac{3}{4}, \tfrac{1}{4}, \tfrac{1}{4}, -\tfrac{1}{4}, \tfrac{1}{4}, \tfrac{1}{4}, \tfrac{1}{4}, -\tfrac{1}{4}) \tag{3.19}$$

$$\mathbf{W}_3 = (\tfrac{1}{3}, 0, 0, \tfrac{1}{3}, \tfrac{1}{3}, \tfrac{1}{3}, \tfrac{1}{3}, \tfrac{1}{3})(1, \tfrac{1}{3}, \tfrac{1}{3}, \tfrac{1}{3}, 0, 0, 0, 0), \tag{3.20}$$

were chosen that produce two generations at fixed points and a third spread between in the bulk U and additional fixed points in the twisted (fixed point) sectors of T_2 and T_4. For these Wilson lines, the additional exotic MSSM states are all vector-like [37, 38].

Since the $\mathbf{V}$ twist is of order 6, the untwisted sector, U, is accompanied by five twisted sectors, T_k, $k = 1$ to 5. The combination of orbifold and Wilson lines generate 2 equivalent sets of twisted sector fixed points. The local non-Abelian observable gauge groups at one set of the 6 fixed points are $SO(10) \otimes SO(4)$, $SO(12)$, $SU(7)$, $SO(8) \otimes SO(6)$, $SO(8)' \otimes SO(6)'$ and $SO(8)'' \otimes SO(6)''$, respectively. The MSSM gauge group is realized as the intersection of these groups.[||] The net gauge group, as common factor of all gauge groups at the 6 fixed points, is

$$[SU(3)_C \otimes SU(2)_L \otimes U(1)_Y \otimes \prod_{i=1}^{4} U(1)_i] \otimes [SU(4)_H \otimes SU(2)_H \otimes \prod_{i=5}^{8} U(1)_i]. \tag{3.21}$$

The hypercharge has the standard $SO(10)$ embedding,

$$U(1)_Y \equiv U(1)_1 = (0, 0, 0, \tfrac{1}{2}, \tfrac{1}{2}, -\tfrac{1}{3}, -\tfrac{1}{3}, -\tfrac{1}{3})(0, 0, 0, 0, 0, 0, 0, 0) \tag{3.22}$$

(following the notation of the $\mathbf{V}$ and $\mathbf{W}$), another aspect in common with the free fermionic model. The remaining eight local Abelian symmetries are

$$U(1)_2 = (1, 0, 0, 0, 0, 0, 0, 0)(0, 0, 0, 0, 0, 0, 0, 0) \tag{3.23}$$

[||] Notice that, as expected, formation of the net orbifold gauge group parallels that of free fermionic language. In the free fermion case, the twisted sectors (Wilson lines) each act to reduce the untwisted sector gauge group and the net gauge group is the intersection of all twisted sector gauge reductions. However, free fermionic formulation does not yield to fixed point interpretation.

$$U(1)_3 = (0,1,0,0,0,0,0,0)(0,0,0,0,0,0,0,0) \qquad (3.24)$$
$$U(1)_4 = (0,0,1,0,0,0,0,0)(0,0,0,0,0,0,0,0) \qquad (3.25)$$
$$U(1)_5 = (0,0,0,1,1,1,1,1)(0,0,0,0,0,0,0,0) \qquad (3.26)$$
$$U(1)_6 = (0,0,0,0,0,0,0,0)(1,0,0,0,0,0,0,0) \qquad (3.27)$$
$$U(1)_7 = (0,0,0,0,0,0,0,0)(0,1,1,0,0,0,0,0) \qquad (3.28)$$
$$U(1)_8 = (0,0,0,0,0,0,0,0)(0,0,0,1,0,0,0,0) \qquad (3.29)$$
$$U(1)_9 = (0,0,0,0,0,0,0,0)(0,0,0,0,-1,-1,-1,1) \qquad (3.30)$$

$U(1)_2$ though $U(1)_8$ are all anomalous. As with all weak coupled heterotic models, the anomaly can be rotated into a single generator of the form

$$U(1)_A = \sum_{i=2}^{8} \operatorname{tr} Q_i \, U(1)_i \qquad (3.31)$$

with $\operatorname{tr} Q_A = 88$. A set of mutually orthogonal, anomaly-free $U(1)'_k$, for $k = 2$ to 7, are formed from linear combinations of $U(1)_i$, for $i = 2$ to 8, orthogonal to $U(1)_A$. As with the free fermionic model, the anomaly is eliminated by the Green-Schwarz-Dine-Seiberg-Witten anomaly cancellation mechanism [25, 26]. A non-perturbatively determined set of vacuum expectation values (VEVs) will turn on to cancel the resulting FI D-term $\sim \operatorname{tr} Q_A$ with a contribution of opposite sign.

The cause of the anomaly, the massless matter, is found in the twisted sectors as well as the untwisted. Twisted matter, located at fixed points, appears in complete multiplet representations of the local gauge group, while untwisted bulk matter only appears in the gauge group formed from the intersection of all local gauge groups. Located at each of two equivalent fixed points on the $SO(4)$ plane in the twist sector T_1 is a complete MSSM generation. Each generation forms a complete 16 rep of the untwisted sector $SO(10)$. Then, as a result of the Wilson lines, MSSM-charged matter from sectors U, T_2, and T_4 combine to form an additional complete 16. The remaining matter (Higgs, MSSM exotics and hidden sector matter) appear in vector-like pairs. Since the Higgs originate in the untwisted sector, they naturally appear as MSSM reps, rather than as $SO(10)$ reps, providing a natural solution to the normal Higgs doublet/triplet splitting issue.

As in the free fermionic model, the string selection rules play an important role in matter Yukawa couplings. The orbifold worldsheet selections rules allow, at third order, only UUU, $T_1T_2T_3$, $T_1T_1T_4$, UT_2T_4, and UT_3T_3 non-zero sector couplings. Of these, the four twisted sector couplings are prohibited for MSSM states either by gauge symmetry or lack of MSSM states in the T_3 sector. For the heaviest generation, the quark doublet and anti-up quark originate in the untwisted sector, whereas the anti-down quark is from a twisted sector [37, 38]. This provides an unsuppressed top quark mass and suppressed higher order masses for all other MSSM matter, offering a possible explanation for the top to bottom mass hierarchy. Generational mass hierarchy is also afforded. Both are re-occurring themes in MSSM string models.

The exotic down-like and lepton-like states yield an interesting feature for the model. These states mix with the MSSM states in mass matrices and the physical states contain both. The 4 d-like quarks pair with all 7 $\bar{d}$-like quarks in a potentially rank 4 mass matrix, given sufficient non-zero VEVs from among the 69 non-Abelian singlet, hypercharge-free s_m^o ($m = 1$ to 69) fields. The mass terms are formed by factors of 1 to 6 s_m^o VEVs. The 8 lepton doublet fields with $-\frac{1}{2}$ hypercharge appear with the 5 antilepton doublet fields in a potentially rank 5 mass matrix. The lepton mass terms are similarly formed by products of 1 to 6 s_m^o VEVs. The 8 remaining lepton doublets with zero hypercharge appear in their own potentially rank 4 mass matrix. The 16 vector-like pairs of non-Abelian singlet with $\pm\frac{1}{2}$ hypercharge similarly appear in their own potentially rank 8 mass matrix. The VEVs required for mass terms in all four matrices sufficient to remove 4 pairs of exotic quarks, 4 pairs of exotic lepton/anti-lepton hypercharged doublets (one of the pairs must be a pair of EW Higgs), 4 pairs of hypercharge-free lepton/anti-lepton doublets, and 16 pairs of hypercharged non-Abelian singlets, correspond to D-flat directions [37, 38].

Suppression of R-violating terms, in particular $\bar{u}\bar{d}\bar{d}$, imposes constraints on the allowed exotic down quark content of the physical states, thereby restricting the phenomenologically viable flat directions. Also, for generic non-Abelian singlet flat directions, all five pairs of non-generational hypercharged $SU(2)_L$ doublets acquire near FI scale mass, prohibiting any EW Higgs. Hence, fine tuning must be called upon to keep some VEVs far below the FI scale to allow for one Higgs doublet pair.

Like the free fermionic MSSM, the orbifold model has a large perturbative vacuum degeneracy that maintains supersymmetry after cancellation of the $U(1)_A$, while reducing gauge symmetries and making various combinations of states massive. In this model, a parameter space of VEVs from the set s_m^o was found that is simultaneously (i) D- and F-flat to all order, (ii) supplies near string-scale mass to most (or possibly all as claimed in [38]) of the MSSM-charged exotics, and (iii) breaks all of the Abelian gauge factors other than hypercharge. In this subspace of flat directions the gauge group is reduced to

$$[SU(3)_C \otimes SU(2)_L \otimes U(1)_Y] \otimes [SU(4)_H \otimes SU(2)_H], \qquad (3.32)$$

completely separating the observable and hidden sectors.

When only the s_m^o uncharged under a non-anomalous extended $U(1)_{B-L}$, defined as

$$U(1)_1 = (0, 1, 1, 0, 0, -\tfrac{2}{3}, -\tfrac{2}{3}, -\tfrac{2}{3})(\tfrac{1}{2}, \tfrac{1}{2}, \tfrac{1}{2}, -\tfrac{1}{2}, 0, 0, 0, 0), \qquad (3.33)$$

and some $SU(2)_H$ doublets uncharged under $U(1)_{B-L}$ are allowed to receive VEVs, the gauge group is alternately reduced to

$$[SU(3)_C \otimes SU(2)_L \otimes U(1)_Y \otimes U(1)_{B-L}] \otimes [SU(4)_H]. \qquad (3.34)$$

(The standard $U(1)_{B-L} \in SO(10)$ is anomalous and required hidden sector extension to remove the anomaly.) The fields with VEVs were shown to form D-flat monomials, but F-flatness was not required [38]. Rather, it was assumed F-flatness can be satisfied by specific linear combinations of D-flat directions up to a sufficiently high order. Thus, the phenomenology associated with these VEVS is not necessarily as realistic as that for the prior parameter space. With the reduced set of singlet VEVs, decoupling of the MSSM exotic states require mass terms up to 11th order in the superpotential. These masses are likely highly suppressed, with a value far below the MSSM unification scale. Therefore the related exotics likely do not decouple. The advantage to keeping a gauged $B - L$ symmetry is the removal of renormalizable R-parity violating couplings that otherwise lead to strong proton decay. When this $U(1)_{B-L}$ is retained, the up quarks are heavier than the down quarks for each generation, which are also heavier then the electrons. The up quark generational mass hierarchy is on the order of 1: 10^{-6}: 10^{-6}. The down-quarks and for the electron quarks masses for the lighter two generations are also of the same order.

All order flatness for the initial collection of non-Abelian singlet directions was proved using the orbifold parallels to the NSR free fermionic selections rules. First, these rules prohibit self-coupling among the non-Abelian singlets in sectors U, T_2, and T_4. Additionally, singlets from these sectors can only couple to two or more states in $T_{1,3}$. Thus, any D-flat directions involving only VEVs of U, T_2, and T_4 non-Abelian singlets is guaranteed to be F-flat to all order. F-flat directions were formed as 39 independent linear combinations of the D-flat directions containing only s_m^o from U, T_2, and T_4. This set of flat directions allowed decoupling of many, *but not all*, of the MSSM exotics [37, 38]. An increase in parameter space of all order D- and F- flat directions space, via hidden sector non-Abelian fields, that might induce mass to all MSSM exotics is reportedly under investigation, as are finite order D- and F-flat directions consistent with EW breaking [38].

Critically, [38] reminds us that flat directions are not always necessary for decoupling of exotics. Rather, isolated special points generically exist in the VEV parameter space that are not located along flat directions, but for which all D- and F-terms are nonetheless zero. [38] calls upon the proof by Wess and Bagger [39] that D-terms do not actually increase (except by one for FI term cancellation) the number of constraints for supersymmetric flat directions beyond the F-term constraints.** Thus, the system of D- and F-equations is not over constraining. Once a solution to $F_m = 0$, for all fields s_m^o, is found (to a given order), complexified gauge transformations of the fields s_m^0, that continue to provide a $F_m = 0$ solution, can be performed that simultaneously arrange for D-flatness. Thus, since the $F_m = 0$ equations impose m (non-linear) constraints on m fields, there should be at least one non-trivial solution for any set of fields s_m^o. A parallel proof for non-Abelian field VEVs also exists.†† This reduction in apparent total constraints is possible because the F-term equations constrain gauge invariant polynomials, which also correspond to non-anomalous D-flat directions [38]. Hence [38] argues that since supersymmetric field configurations in generic string models form low dimensional manifolds or points, all of the MSSM singlets should generally attain non-zero VEVs. Systematic surveys of these low dimensional manifolds for free fermionic models indicates the overall FI VEV scale is around $\frac{1}{10}$ of the string scale, and therefore is around 10^{17} GeV. This argument nevertheless does not prove that *all* of the exotics will necessarily obtain mass. At even the special points some VEVs required for certain masses may be prohibited by supersymmetry.

The exotic states are given mass through superpotential terms (consistent with orbifold string selection rules [40, 41]) involving up to six singlet VEVs (i.e., up to 8th order in the superpotential.* The set of required s_m^o singlet VEVs was shown to be provided by

---

**FI term cancellation requires the existence of one monomial that is D-flat for all non-anomalous symmetries, but that carries the opposite sign to the FI-term in for the anomalous $U(1)_A$, imposing an additional constraint.

††Complications to these proofs do arise when different scalar fields possess the same gauge charges.

*In [37, 38] it is argued that the high order of some of these mass terms does not necessarily imply mass suppression, citing combinatorial effects in the coupling. However, for free fermionic models, the mass suppression factor is estimated to be $\sim \frac{1}{10}$ per order [30] beginning at 5th order [29] Thus, if free fermionic patterns in mass suppression continue for orbifolds, then it should be expected that orbifold masses involving more than

D-flat directions. As discussed above, simultaneous F-flatness of a linear combination of the D-flat directions is then required.

On par with the MSHSM free fermionic model, this orbifold model admits spontaneous supersymmetry breaking via hidden sector gaugino condensation $<\lambda\lambda>$ from the $SU(4)$. When the **6**'s and the $4/\bar{4}$'s of the hidden $SU(4)$ receive flat direction mass through allowed couplings, the condensation scale for $SU(4)$ gauginos is in the range of 10^{11} to 10^{13} GeV. Supersymmetry breaking then results after dilaton stabilization. As with the free fermionic model, stabilization must be assumed to result from non-perturbative corrections to the Kähler potential, since the multiple gaugino racetrack solution is not possible with the hidden $SU(2)$. The latter either does not condense or does so at too low of an energy. [38] assumes a dilaton non-perturbative correction of the form

$$K = -\ln(S + S^*) + \Delta K_{np}, \qquad (3.35)$$

where $\mathrm{Re} < S > \sim 2$. Supersymmetry is spontaneously broken by the dilaton F-term,

$$F_S \sim \frac{<\lambda\lambda>}{M_P} \qquad (3.36)$$

The resulting soft SUSY-breaking terms from dilaton stabilization are universal with the only independent parameter being the gravitino mass $m_{3/2}$. The universal gaugino and scalar masses are then $\sqrt{3}m_{3/2}$ and $m_{3/2}$, respectfully.

A nice feature of the orbifold model is possible mitigation of the factor-of-20 difference between the MSSM unification scale $\sim 2.5 \times 10^{16}$ GeV and the weak coupled heterotic string scale $\sim 5 \times 10^{17}$ GeV. One or two large orbifolded radii at the MSSM unification length scale are consistent with perturbativity and could lower the string scale down toward the MSSM scale. This is one aspect not possible from the free fermionic perspective, since the free fermionic formulation effectively fixes compactification at the self-dual radius, i.e., at the string length scale.[†]

In their search for MSSM heterotic orbifolds, the authors of [37, 38] reported finding roughly 10^4 $\mathbb{Z}_6' = \mathbb{Z}_3 \otimes \mathbb{Z}_2$ orbifold models with the SM gauge group. Of these, approximately 100 have exactly 3 MSSM matter generations plus vector-like MSSM matter. The model reviewed above was the only one for which flat directions exist whereby all MSSM exotics can be made massive and, thereby, decouple from the LEEFT. Thus, within the context of weak coupled heterotic strings a total of two MSSM models with no exotics have been identified. A vast array of MSSM heterotic models without exotics likely remains yet undiscovered in the weak coupling domain.

3.2. Heterotic Elliptically Fibered Calabi-Yau

The free fermionic and orbifold models are not the only heterotic MSSM's that have been found. An equal number of strong coupled heterotic models have also been constructed.

---

four VEVs should also fall below the MSSM unification scale and, therefore, do not actually decouple from the model.

[†]Alternatively, weak coupled free fermionic heterotic strings may offer a resolution [42, 43, 44] to the factor-of-20 difference through "optical unification" [45] effects, whereby an apparent MSSM unification scale below the string scale is guaranteed by the existence of certain classes of intermediate scale MSSM exotics.

(Strong coupling is implied by the presence of D-branes that provide needed anomaly cancellation.) Another $E_8 \otimes E_8$ heterotic model containing solely the MSSM spectrum in the observable sector was constructed by compactification on an elliptically fibered Calabi-Yau three-fold $X = \tilde{X}/(\mathbb{Z}_3 \otimes \mathbb{Z}_3)$ containing a $SU(4)$ gauge instanton and a $\mathbb{Z}_3 \otimes \mathbb{Z}_3$ Wilson line [46].* The observable sector E_8 is spontaneously broken by the $SU(4)$ gauge instanton into $SO(10)^\dagger$ which is further broken to $SU(3)_C \otimes SU(2)_L \otimes U(1)_Y \otimes U(1)_{B-L}$ by the Wilson line.* The Calabi-Yau X, with $\mathbb{Z}_3 \otimes \mathbb{Z}_3$ fundamental group, was first investigated in [47], while the $SU(4)$ instanton was obtained in [48, 49, 50] as a connection on a holomorphic vector bundle. This model was produced as a slight variation from another in [48, 49, 50], with the MSSM-charged content of the latter varying by only an extra Higgs pair. Both models contain an additional 19 uncharged moduli, formed of 3 complex structure moduli, 3 Kähler moduli, and 13 vector bundle moduli.

The renormalizable MSSM Yukawa couplings were computed in [51]. Each superfield in the model is associated with a $\bar{\partial}$-closed $(0, 1)$ form Φ_i taking values in some bundle over X. The tree-level (classical) couplings $\lambda_{i,j,k}$ can be expressed in the large-volume limit as

$$\lambda_{i,j,k} \sim \int_X \Omega \wedge \Phi_i \wedge \Phi_j \wedge \Phi_k, \qquad (3.37)$$

where Ω is the Calabi-Yau's holomorphic $(3,0)$-form. (3.37) provides the unique way of generating a complex number from 3 superfields. Rather than computing the numeric value of each λ, general rules were constructed for non-zero values by evaluating the corresponding triplet products of cohomology classes for two matter fields and a Higgs. The resulting Calabi-Yau worldsheet selection rules provide a cubic texture with only non-zero couplings for interactions of the first generation with the second and third generations. Thus, two of the three generations receive electroweak scale mass for their quarks and leptons, while masslessness of one generation can receive correction from higher order, non-renormalizable terms. Couplings for non-zero moduli-dependent μ-terms for the Higgs were similarly computed in [52]. Selection rules were found to severely limit the number of moduli that can couple to a Higgs pair. For the two Higgs pair model of [48, 49, 50] only four of the nineteen moduli can thereby produce renormalizable μ-terms when they acquire non-zero VEVS. No third order mu-terms are allowed for the single Higgs pair in [46]. Thus in the minimal model, the Higgs mass is naturally suppressed. The coupling coefficients for both matter and Higgs cubic terms are not expected to be constant over the moduli space. Instead, they likely depend on the moduli [52].

A highly non-trivial consistency requirement of this model is the slope-stability of the $SU(4)$ vector bundle V, which is necessary for $N = 1$ SUSY. In [53] the vector bundle V was proven to be slope-stable for any Kähler class ω in a maximum dimensional (i.e., 3-dimensional) subcone of the full Kähler cone. A prerequisite for slope stability is the

---

*Interestingly, the MSSM-charged content of this model is claimed to not vary between weak and strong string coupling strengths.

†The reduction is actually to Spin(10), the universal covering group of $SO(10)$, but from hereon this distinction will not be made.

*The $U(1)_{B-L}$, which prohibits $\Delta L = 1$ and $\Delta B = 1$ dimension four nucleon decay terms must be broken above the electroweak scale.

"Bogomoli inequality" for non-trivial V,

$$\int_X \omega \wedge c_2(V) > 0. \tag{3.38}$$

The viable hidden sector content of the model is so far undetermined–the possible range of hidden sector holomorphic vector bundles V' has not yet been completely identified [46]. Any V' must also be slope-stable, and thereby also satisfy (3.38) for the same Kähler classes for which V is slope-stable. Further, the topology of V' must satisfy the anomaly cancellation condition (relating it to the topology of the observable vector bundle V),

$$c_2(V) + c_2(V') = c_2(TX) + [\mathcal{W}] - [\overline{\mathcal{W}}], \tag{3.39}$$

where c_2 is the second Chern class, TX denotes the tangent bundle of the Calabi-Yau three-fold X, and $[\mathcal{W}]$ and $[\overline{\mathcal{W}}]$ are the Poincare dual of the curves on which the brane and anti-branes are wrapped (if either one or both are present). The percentage of the subcone of V stability for which V' is also stable and can provide anomaly cancellation is claimed to be large [46].

A trivial vector-bundle V', (with $c_2(V') = 0$), corresponding to an unbroken hidden sector E_8 gauge symmetry is one slope-stable choice. In this case (3.39) becomes

$$c_2(V) = c_2(TX) + [\mathcal{W}] - [\overline{\mathcal{W}}], \tag{3.40}$$

For this, [54] investigated the general form of the potential when the combined anomaly from a non-trivial slope-stable V and X can be cancelled by the curve of a single five-brane rather than a brane/anti-brane combination. Their findings indicated "that for a natural range of parameters, the potential energy function of the moduli fields has a minimum which fixes the values of all moduli."* Further, the minimum of the potential energy is negative and typically of order $-10^{-16} M_{\text{Pl}}^4$, giving the theory a large negative cosmological constant. They also showed that at the potential minimum the Kähler covariant derivatives vanish for all moduli fields. Since the moduli are uncharged, D-flatness is also retained. Therefore, supersymmetry remains unbroken in the anti-deSitter vacuum. All of this remains true even in the case of hidden sector gaugino condensation. It was argued that the pattern found for the example case continued for non-trivial hidden sector vector bundles V'.

In [54] the addition of an anti-brane was shown to shift the minimum of the potential by an amount proportional to

$$\mathcal{J} = c_2(V) - c_2(TX) + [\mathcal{W}] + [\overline{\mathcal{W}}], \tag{3.41}$$

which by anomaly cancellation is

$$\mathcal{J} = 2[\overline{\mathcal{W}}] > 0. \tag{3.42}$$

For a viable range of the moduli, the minimum of the potential can be shifted to be positive and on the scale of the cosmological constant. The up-lifted vacuum becomes meta-stable, with a very long lifetime.

---

*Simultaneous stability of the $SU(4)$ vector bundle and anomaly cancellation with a trivial hidden sector vector bundle has been challenged in [55, 56].

The same effect was found when a hidden $SU(4)$ vector bundle V' is chosen [54]. Without an anti-brane the potential minimum is highly negative, but this can be uplifted by an anti-brane. Further, when $SU(4)$ was chosen as the hidden sector vector bundle for the MSSM model described above, a consistent configuration was claimed with the addition of a 5-brane/anti-5-brane pair for anomaly cancellation. (Since $V' = V$, slope stability was assured for the hidden sector.) With this, the total gauge group is $[SU(3)_C \otimes SU(2)_L \otimes U(1)_Y \otimes U(1)_{B-L}] \otimes SO(10)_H$.

In [57, 58] the aspects of dynamical SUSY breaking process for the $SO(N_c = 10)$ hidden symmetry was analyzed in further detail, based on recent work in [59] regarding four dimensional $N = 1$ theories with both supersymmetric and non-supersymmetric vacua. For SUSY $SO(10)$ moduli space, both stable supersymmetric vacua and meta-stable non-supersymmetric vacua exist for an even number N_f of massive **10** matter reps in the free-magnetic range,

$$N_c - 4 = 6 \leq N_f \leq \frac{3}{2}(N_c - 2) = 12. \tag{3.43}$$

At certain points in the $SO(10)$ moduli space some or all of these massive reps can become massless. In the neighborhood of the moduli space of these points, the matter reps gain slight masses and supersymmetry is also broken. The meta-stable non-supersymmetric vacua are long lived in the limit

$$\sqrt{\frac{m}{\Lambda}} \ll 1 \tag{3.44}$$

where m is the typical mass scale of the **10** reps and Λ is the strong-coupling scale. [57, 58] show that $SU(4)$ hidden sector bundle moduli can be consistently chosen to provide for long-lived meta-stable vacua.

The stability and/or anomaly cancellation of the observable sector $SU(4)$ vector bundle, with trivial hidden sector vector bundle, has been claimed as likely, but is nonetheless uncertain. An alternate model, with slightly modified elliptically fibered Calabi-Yau three-fold X' and alternate $SU(5)$ vector bundle, has proven to be definitely stable. The latter model provides anomaly cancellation with a trivial hidden vector bundle [56]. The initial Calabi-Yau three-fold is the same, $\tilde{X}$, but this time $X' = \tilde{X}/\mathbb{Z}_2$ has a fundamental group $\phi_1(X') = \mathbb{Z}_2$, rather than $\phi_1(X) = \mathbb{Z}_3 \otimes \mathbb{Z}_3$. On X', slope stability for an observable $SU(5)$ vector bundle V' could be proved absolutely. The $SU(5)$ vector bundle breaks the initial E_8 to the commutant of $SU(5)$, which is also $SU(5)$. A $\mathbb{Z}_2$ Wilson line, allowed by the fundamental group, then breaks $SU(5)$ to $SU(3)_C \times SU(2)_L \times U(1)_L$ [56].

The particle spectrum is again given by the decomposition of the adjoint rep of E_8 under this breaking pattern. The result is exactly three generations of MSSM matter, 0, 1, or 2 pairs of Higgs doublets (depending on location in the moduli space), and no MSSM exotics (other than perhaps one extra Higgs pair). Additionally, the particle spectrum includes at least 87 uncharged moduli [56]. If a trivial hidden sector vector bundle is chosen (thereby giving E_8 local hidden symmetry), anomaly cancellation proves possible with addition of a single $M5$ brane wrapping an effective curve $[W]$ specified by $c_2(TX') - c_2(V') = [W]$ [56]. The presence of the $M5$ brane places this model in the strong coupling region.

Trilinear Yukawa couplings for this model were analyzed in [60]. Non-zero couplings were computed for all three generations of up-quarks. Depending on the location in the moduli space, these couplings may provide a realistic mass hierarchy. Parallel non-zero down-quark and electron couplings were not found. All R-parity violating terms, including B and L violating terms leading to proton decay, were found to vanish. μ mass parameters for the Higgs and neutrino mass terms were provided by vector bundle moduli [56].

An enlarged class of $E_8 \otimes E_8$ heterotic strings on elliptically fibered Calabi-Yau manifolds X with vector bundles was introduced in [61]. These models have structure group $U(N) \sim SU(N) \otimes U(1)$ and M5-branes. This construction gives rise to GUT models containing $U(1)$ factors like flipped $SU(5)$ or directly to the MSSM, even on simply connected Calabi-Yau manifolds. MSSM-like models were constructed in which the only chiral states are the MSSM states, but nevertheless contain vector-like MSSM exotics. In the example given, the vector exotics were undetermined, as were the Yukawa couplings for all states. The MSSM-like models constructed continue to possess the most attractive features of flipped $SU(5)$ such as doublet-triple splitting and proton stability.

In the prior two elliptically-fibered models, the observable sector vector bundles $V = SU(4)$ or $SU(5)$ break E_8 to the respective commutants $SO(10)$ or $SU(5)$. Reduction to the MSSM then depends on Wilson lines, since adjoint or higher rep Higgs are not possible for the level-1 Kač-Moody algebras from which most heterotic gauge groups are derived. The Wilson lines require Calabi-Yau three-folds with non-zero first fundamental group $\pi(CY_3)$, provided in the two prior models by freely-acting $\mathbb{Z}_3 \otimes \mathbb{Z}_3$ or $\mathbb{Z}_2$ orbifoldings of a simply connected Calabi-Yau three-fold. By using $U(N)$ vector bundles that break E_8 to either flipped $SU(5)$ (for $N = 4$) or directly to the MSSM (for $N = 5$) the need for a non-trivial first fundamental group is eliminated. This allows a larger number of geometric backgrounds.

A vector bundle in this new class of models takes the form $W = V \oplus L^{-1}$ with V being either a $U(4)$ or $U(5)$ vector bundle in the observable E_8 and L^{-1} a line bundle included such that $c_1(W) = 0$. W is embedded into $SU(6) \in E_8$ such that the commutant is $SU(3)_C \otimes SU(2)_L \otimes U(1)_1$ ($U(1)_1 \neq U(1)_Y$). The $U(1)$ bundle in V is embedded in $SO(10)$ as $Q_1 = (1, 1, 1, 1, -5)$. The surviving $U(1)_1$ does not remain massless unless it is also embedded in the hidden sector, which is performed by embedding the line bundle also into the hidden E_8 (thereby bringing part of it into the observable sector). The line bundle breaks the hidden E_8 into $E_7 \otimes U(1)_2$ and the surviving $U(1)$ becomes

$$U(1)_Y = \tfrac{1}{3}(U(1)_1 + 3U(1)_2). \tag{3.45}$$

$U(1)_Y$ was shown to remain massless as long as

$$\int_X c_1(L)c_2(V) = 0; \quad \int_{\Gamma_a} c_1(L) = 0. \tag{3.46}$$

where Γ_a is the internal two-cycle wrapped by N_a five branes.

As a result of the non-standard $U(1)_Y$ embedding, the tree level relations among the MSSM gauge couplings at the unification scale are changed. Rather than $\alpha_{MSSM} = \tfrac{5}{3}\alpha_Y$, this embedding yields $\alpha_{MSSM} = \tfrac{8}{3}\alpha_Y$. However, it was reported that through threshold corrections, a co-dimension one hypersurface in the Kähler moduli space allows MSSM gauge coupling unification. Relatedly, the non-standard hypercharge embedding may lead to some falsifiable predictions. Since some "hidden" matter now carries hypercharge, it becomes coupled to MSSM matter. Sufficiently low mass hidden matter could therefore have detectable effects at LHC.

The resulting MSSM-charged chiral massless spectrum contains

$$g = \tfrac{1}{2} \int_X c_3(V) \tag{3.47}$$

generations of 15-plets that do not include the neutrino singlet. A few $g = 3$ generation models were obtained. As a result of the hidden sector contribution to hypercharge, the anti-electrons carry both observable and hidden sector E_8 charge unless

$$\int_X c_3(L) = 0. \tag{3.48}$$

The MSSM-charged spectrum also contains an undetermined number of vector-like pairs of exotics. However, exact MSSM models of this class without exotics may very well exist. Exploration of the MSSM-like models from vector bundles on either non-simply or simply connected Calabi-Yau threefolds has just begun.

3.3. Type IIB Magnetic Charged Branes

Realization of gauge groups and matter reps from D-brane stacks in compactified Type IIB models (with odd dimension branes) or Type IIA models (with even dimension branes) provides another bottom-up route to (MS)SM-like models from string theory. A supersymmetric Type IIB model with magnetized $D3$, $D5$, $D7$, and $D9$ branes is T-dual to a Type IIA intersecting $D6$ brane model. Relatedly, the model building rules for Type IIB and IIA models are very similar [62, 63, 64, 65].

The starting point in these models is two stacks, a, with $N_a = 3$ D-branes of the same type and similarly b, with $N_b = 2$, which generate a $U(3)_a \otimes U(2)_b$ symmetry. Along with $U(1)_a \in U(3)_a$ and $U(1)_b \in U(3)_b$, additional single (unstacked) branes $U(1)_i$ provide for $U(1)_Y$ and charge cancellation requirements. The $U(3)_a = SU(3)_C \times U(1)_a$ ($U(2)_b = SU(2)_L \times U(1)_b$) gauge generators are open strings with both ends bound on the a (b) stack. Quark doublet MSSM matter appears in bi-fundamental representations $(\mathbf{N}_a, \bar{\mathbf{N}}_b) = (\mathbf{3}, \bar{\mathbf{2}})$ as open strings at the intersections of a and b with one end bound on the a stack and the other end on the b stack. The $\mathbf{3}$ has charge $Q_a = +1$ of $U(1)_a$, while the $\bar{\mathbf{2}}$ has charge $Q_b = -1$ of $U(1)_b$. Exotic $(\bar{\mathbf{3}}, \mathbf{1})$ quarks, $(\mathbf{3}, \mathbf{1})$ anti-quarks, and $(\mathbf{1}, \mathbf{2})$ leptons & Higgs can appear when open strings have one end attached to the a (b) stack and the other end attached to an additional single $U(1)$ brane. The multiplicity of given chiral bi-fundamental reps is specified by the intersection numbers between the two related stacks.

In orientifold compactifications, the a and b stacks become paired with image stacks a' and b'. Additional quarks appear in $(\mathbf{N}_a, \bar{\mathbf{N}}_b) = (\mathbf{3}, \mathbf{2})$ reps at the intersections of a and b' as open strings with one end bound on the a stack and the other end on the b' stack. The $\mathbf{2}$ has charge $Q_b = +1$ of $U(1)_b$. Open strings can also be bound between stacks and their mirrors, producing exotic matter in symmetric (adjoint) reps, that is $\mathbf{8}$ for $SU(3)_C$ or $\mathbf{3}$ for $SU(2)_L$ and antisymmetric representations, i.e., anti-quarks ($\bar{\mathbf{3}}$ for $SU(3)_C$ or (anti)-lepton singlets $\bar{\mathbf{1}}$ for $SU(2)_L$. Thus, models with only the MSSM content cannot produce symmetric reps from the $SU(3)_C$ or $SU(2)_C$ stacks and must not produce more than three anti-symmetric $SU(3)$ representations.

Exact (MS)SM spectra place constraints on the intersection numbers. If the number of intersections of a with b is p and the intersections of a with b' is q, then exactly 3 quark doublets requires $p + q = 3$. The six anti-quark singlets can arise from a combination of r antisymmetric intersections of a with its image a' and $6 - r$ intersections of a with $U(1)$ from single D-brane stacks c, d, etc. The three quark doublets carry $Q_a = 2(+1)$, the r antisymmetric states carry $Q_a = 1 + 1$, and the $6 - r$ antiquarks carry $Q_a = -1$. Assuming no exotics requires, $6 + 2r - (6 - r) = 0$ for tadpole cancellation for Q_a. Hence $r = 0$ and all antiquarks must come from open strings connecting a stack and single branes.

Similarly, for t copies of $(\mathbf{1}, \mathbf{2})$ and u copies of $(\mathbf{1}, \bar{\mathbf{2}})$, with $t + u = 3$ ($t, u \geq 0$), and 6 leptons singlets (three of which carry hypercharge), s of which are from antisymmetric intersections of b and b' and $6 - s$ from intersections between generic c and d singlet branes, tadpole cancellation requires $t - (3 - t) + 2s - 3p + 3(3 - p) = 0$, or equivalently, $t + s - 3p = -3$. Solutions exist for all values of $0 \leq p \leq 3$, which is thus a requirement for the (MS)MS exact spectra from orientifolds [62, 63, 66]. $s = 3$ antisymmetric singlets with $t = 0$, $s = 4$ antisymmetric singlets with $t = 2$, and $s = 6$ antisymmetric singlets with $t = 0$ allows for $(p, q) = (3, 0)$ or $(0, 3)$. All other viable values of s and t require $(p, q) = (2, 1)$ or $(1, 2)$ (for fixed definition of chirality, otherwise $\pm$ sign changes allowed). However, the antisymmetric states of $SU(2)$ do not have the MSSM Yukawa couplings to the Higgs [62, 63].

Type IIB and Type IIA have produced models with exactly the (MS)SM states, i.e., with no MSSM-charged exotics, that also yield somewhat realistic MSSM Yukawa terms. The most successful Type IIB models have been compactified on $\mathbf{T}^6/(\mathbb{Z}_2 \times \mathbb{Z}_2)$, with $\mathbf{T}^6 = \mathbf{T}^2 \times \mathbf{T}^2 \times \mathbf{T}^2$ [67, 68]. This is similar to the compactification for the free fermionic MSSM-spectrum model, except that orientifolding replaces orbifolding. The $\mathbb{Z}_2 \times \mathbb{Z}_2$ generators are θ and ω, where $\theta : (z_1, z_2, z_3) \to (-z_1, -z_2, z_3)$ and $\omega : (z_1, z_2, z_3) \to (z_1, -z_2, -z_3)$. The additional orientifold modding is performed by the operator product ΩR where $R : (z_1, z_2, z_3) \to (-z_1, -z_2, -z_3)$ and Ω is worldsheet parity. After orientifolding, these models contain 64 $O3$-planes and 4 $O7_i$ planes, with i denoting the ith $\mathbf{T}_i^2$ on which the plane is localized at a $\mathbb{Z}_2$ fixed point and wrapped around the other two $\mathbf{T}_{j \neq i}^2$.

Crosscap tadpoles are produced in these models via non-trivial contribution to the Klein bottle amplitude. These tadpoles can be cancelled by Type IIB $D(3 + 2n)$-branes, for $n = 0, 1, 2, 3$, that fill up $D = 4$ Minkowski space and wrap $2n$ cycles on the compact space. If desired, $D7$ and $D9$ branes can contain non-trivial magnetic field strength $F = dA$ in the compactified volume. These non-trivial gauge bundles generally reduce the rank of the gauge group and lead to $D = 4$ chiral fermions.

Of vital importance is that magnetic flux induces D-brane charges of lower dimension. That is, magnetic flux on $D9$ branes produces $D7$, $D5$, and $D3$ brane charges. The standard convention for magnetized D-branes was defined in [69], with topological information specified by the six integers (n_a^i, m_a^i) with n_a^i the unit of magnetic flux in the given torus and m_a^i the number of times that the D-branes wrap the ith $\mathbf{T}^2$ [70]. The magnetic field flux constraint for any brane is

$$\frac{m_a^i}{2\pi} \int_{\mathbf{T}_i^2} F_a^i = n_a^i. \tag{3.49}$$

The chiral spectrum produced by two stacks of D-branes, a and b, (and their images a'

and b') is a result of the intersection products [70],

$$I_{ab} = \prod_{i=1}^{3}(n_a^i m_b^i - m_a^i n_b^i), \quad (3.50)$$

$$I_{ab'} = -\prod_{i=1}^{3}(n_a^i m_b^i + m_a^i n_b^i), \quad (3.51)$$

$$I_{aa'} = -8\prod_{i=1}^{3}(n_a^i m_a^i), \quad (3.52)$$

$$I_{aO} = 8(-m_a^1 m_a^2 m_a^3 + m_a^1 n_a^2 n_a^3 + n_a^1 m_a^2 n_a^3 + n_a^1 n_a^2 m_a^3), \quad (3.53)$$

where subscript O denotes the contributions from intersections with the $O3$ and $O7_i$ planes.

In models of this class, the initial $U(N_a)$ from N_a stacked D-branes reduce to $U(N_a/2)$ under $\mathbb{Z}_2 \times \mathbb{Z}_2$. Further, ΩR invariance requires that to each set of topological numbers (n_a^i, m_a^i) be added its ΩR image, $(n_a^i, -m_a^i)$, for each D-brane a. The total $D9$- and $D5$-brane charge will vanish [70]. The D-branes fixed by some elements of $\mathbb{Z}_2 \otimes \mathbb{Z}_2$ and ΩR will carry $USp(N_a)$ gauge group. The complete physical spectrum must be invariant under $\mathbb{Z}_2 \otimes \mathbb{Z}_2 \otimes \Omega R$.

Cancellation of the R-R tadpole for this class of model imposes [70]

$$\sum_a N_a n_a^1 n_a^2 n_a^3 = 16 \quad (3.54)$$

$$\sum_a N_a m_a^1 m_a^2 n_a^3 = \sum_a N_a m_a^1 n_a^2 m_a^3 = \sum_a N_a n_a^1 m_a^2 m_a^3 = -16. \quad (3.55)$$

These conditions correspond to cancellation of non-Abelian triangle diagrams and of $U(1)$ anomalies. The R-R tadpole constraint is satisfied by

$$g^2 + N_f = 14, \quad (3.56)$$

where g is the number of quark and lepton generations and $8N_f$ is the number of $D3$-branes. $N = 1$ SUSY (and NS-NS tadpole cancellation) additionally requires

$$\sum_i \tan^{-1}\left(\frac{m_a^i A_i}{n_a^i}\right) = 0, \quad (3.57)$$

with A_i the area (in α' units) of the ith T^2, which is satisfied by $A_2 = A_3 = A$ and

$$\tan^{-1}(A/3) + \tan^{-1}(A/4) = \frac{\pi}{2} + \tan^{-1}(A_1/2). \quad (3.58)$$

In [70] an additional consistency, K-charge anomaly cancellation, was also examined. K-charge anomaly first arose in [71]. In addition to homological R-R charges, D-branes can also carry K-theory $\mathbb{Z}_2$ charges that are invisible to homology. K-charge anomaly had been ignored before [70] because of its automatic cancellation in simple models. However, it was found to be more severe than uncancelled NS-NS tadpoles, because there is no analogue of the Fischler-Susskind mechanism for K charge anomalies [70]. In Type I theory, non-BPS D-branes exist that carry non-trivial K-theory $\mathbb{Z}_2$ charges. Elimination of the anomaly requires that these non-BPS branes must be paired [72, 73]. Since a Type I non-BPS $D7$-brane is regarded as a $D7$-brane and its worldsheet parity image $\overline{D7}$-brane in Type IIB theory, even numbers of these brane pairs must be required in Type IIB (and its dual Type IIA). For a $\mathbb{Z}_2 \times \mathbb{Z}_2$ orientifold, this expresses itself as global cancellation of $\mathbb{Z}_2$ R-R charges carried by the $D5_i\overline{D5}_i$ and $D9_i\overline{D9}_i$ brane pairs [72, 73]. Cancellation conditions for K-charge were found stringent enough in [70] to render inconsistent seemingly consistent flux compactified Pati-Salem (PS)-like vacua presented elsewhere. In contrast, the [74, 70] models were found K anomaly-free.

In [70] a model with $SU(4) \times SU(2)_L \times SU(2)_R \times U(1)^3 \times USP(8N_f)$ gauge group (using $USP(2) \sim SU(2)$) was constructed. A generalized Greene-Schwarz mechanism breaks some of the extra $U(1)$. Further, the $USP(8N_f)$ reduces to $U(1)^{2N_f}$ as the $8N_F$ branes are moved away from an orbifold singularity, breaking the gauge group to,

$$SU(4) \times SU(2)_L \times SU(2)_R \times U(1)' \times U(1)^{2N_f}. \tag{3.59}$$

In addition to three generations of PS matter and a single Higgs bi-doublet, the model contains numerous exotic $(\mathbf{4}, \mathbf{1}, \mathbf{1})$, $(\overline{\mathbf{4}}, \mathbf{1}, \mathbf{1})$, $(\mathbf{1}, \mathbf{2}, \mathbf{1})$, and $(\mathbf{1}, \mathbf{1}, \mathbf{2})$ PS reps, along with 196 moduli uncharged under PS (many more than in generic three generation MSSM-like free-fermionic models, because the latter are fixed at the self-dual radius). The 196 moduli form a vast parameter space of $N = 1$ flat directions, which corresponds in brane language to (anti)D9 recombination. It is possible to leave a $\bar{D}$-brane with a gauge bundle that makes massive *all* of the extra $U(1)^{2N_f}$, reducing the gauge group to

$$SU(4) \times SU(2)_L \times SU(2)_R \times U(1)', \tag{3.60}$$

while making massive all but four of the PS-charged exotics, That is, all but two $(\mathbf{1}, \mathbf{2}, \mathbf{1})$, and two $(\mathbf{1}, \mathbf{1}, \mathbf{2})$ reps become massive. The standard PS reps are unaffected by the D9 recombination because the PS sector is associated with the three sets of $D7$ branes instead. The MSSM spectrum can be produced from the standard PS spectrum by higgsing.[*] The MSSM Yukawa couplings were computed in terms of theta functions, which showed that exactly one generation could receive renormalizable EW-scale mass, while the other two generations remain massless at renormalizable order [70].

A slight variation of the $D7$ stacks in [70] allows an improved model with gauge group

$$SU(3) \times SU(2)_L \times SU(2)_R \times U(1)_{B-L}. \tag{3.61}$$

The corresponding exotic MSSM-charged mass is three $(\mathbf{1}, \mathbf{1}, \mathbf{2})_{-1}$ states. A recombination process of the D-brane stacks generates VEVs for these three exotics (a frequent occurrence also in free fermionic models) and breaks

$$SU(2)_R \times U(1)_{B-L} \to U(1)_Y. \tag{3.62}$$

Following this, the MSSM-charged matter corresponds to exactly three generations *without* a neutrino singlet.

---

[*]This model was similarly produced in [75] by intersecting D-brane construction.

3.3.1. With Background Flux

Problems with the magnetic brane models above include (1) the large number of uncharged closed string moduli, producing unobserved massless fundamental scalars in the low energy effective theory, and (2) the trivial hidden sector prohibiting SUSY breaking from gaugino condensation in the strong coupling limit. These two issues can be (partially) resolved by the introduction of non-trivial R-R and NS-NS 3-form fluxes, F_3 and H_3 respectively. Attempts to stabilize string moduli though the addition of compactified background fluxes can also induce SUSY-breaking soft terms. In addition meta-stable vacua can be induced. Nevertheless, even the best models of this class have shortcomings. In particular, they generally either have exotics or lack other critical phenomenological features [76, 77, 78, 79].

A systematic approach to embedding the Standard Model in flux compactifications was first offered in [69]. Compatibility between chirality and $N = 1$ flux compactifications was first shown in [70]. Therein, example models were presented with chiral $D = 4$ flux vacua leading to MSSM-like spectrum for both $N = 1$ and $N = 0$ SUSY. Since both R-R and NS-NS tadpoles cancel in these models, the broken SUSY case was free of the usually associated stabilization problems. In the latter model SUSY breaking induces soft terms in the LEEFT. Introduction of the magnetized $D9$-branes with large negative $D3$-brane charges enabled the construction of three-family standard-like models with one, two, and three units of quantized flux (the last case being supersymmetric).

In models with flux, both R-R and NS-NS fluxes must obey the Bianchi identities,

$$dF_3 = 0; \; dH_3 = 0, \tag{3.63}$$

for the case of no discrete internal B-field. These 3-form fluxes generate a scalar potential for the dilaton and complex structure moduli, freezing them at particular values. The $D3$-brane R-R charge given by a combined $G_3 = F_3 - \tau H_3$ field, with $\tau = a + i/g_s$ axion-dilaton coupling, is

$$N_{\text{flux}} = \frac{1}{(4\pi^2\alpha')^2} \int_{M_6} H_3 \wedge \overline{F}_3 = \frac{i}{(4\pi^2\alpha')^2} \int_{M_6} \frac{G_3 \wedge \overline{G}_3}{2\text{Im}\tau} \in 64\mathbb{Z}, \tag{3.64}$$

which is a topological quantity. To satisfy Dirac quantization, N_{flux} must also be quantized in units of 64. The D3-brane tension is given by

$$T_{\text{flux}} = \frac{-1}{(4\pi^2\alpha')^2} \int_{M_6} \frac{G_3 \wedge *_6\overline{G}_3}{2\text{Im}\tau} \in 64\mathbb{Z} \geq |N_{\text{flux}}|. \tag{3.65}$$

This isn't a topological quantity, but depends on the dilaton and complex structure. The quantity

$$V_{\text{eff}} \equiv T_{\text{flux}} - N_{\text{flux}} \tag{3.66}$$

produces an effective potential for these moduli [70].

The minimum of the potential is at either $*_6\overline{G}_3 = iG_3$ (ensuring the contribution of G_3 to the R-R charge is positive), corresponding to $|N_{\text{flux}}|$ D3 branes, or $*_6\overline{G}_3 = -iG_3$, corresponding to $|N_{\text{flux}}|\,\overline{D3}$ branes. The $\overline{D3}$ fluxes will necessarily break SUSY, while the $D3$ fluxes may or may not break SUSY, depending on their class.

(3.64) modifies the $D3$-brane tadpole constraint (3.56) to

$$g^2 + N_f + \frac{1}{16}N_{\text{flux}} = 14, \qquad (3.67)$$

as a result of (3.54) being altered to

$$\sum_a N_a n_a^1 n_a^2 n_a^3 + \frac{1}{2}N_{\text{flux}} = 16. \qquad (3.68)$$

The $g = 1$ generation solution is $N_{\text{flux}} = 3 \cdot 64$ (three units of quantized flux) and $N_f = 1$ for which G_3, formed solely of $N = 1$ SUSY preserving $(2, 1)$ flux, can be chosen to fix all untwisted moduli and the dilaton. On the other hand, the $g = 3$ generation solution has $N_{\text{flux}} = 64$ and $N_f = 1$, for which G_3 contains both $N = 1$ SUSY preserving $(2, 1)$ and SUSY breaking $(0, 3)$ components. In the latter case, NS-NS tadpole cancellation remains, but a gravitino mass is generated

$$m_{3/2}^2 \sim \frac{\int |G_3 \wedge \Omega|^2}{\text{Im}\tau\, \text{Vol}(\mathcal{M}_6)^2}, \qquad (3.69)$$

along with soft terms in the low energy effective lagrangian for the MSSM. The cosmological constant is, nonetheless, kept to zero at first order due to the no-scale structure of V_{eff} [70]. The addition of discrete torsion to these models was also investigated in [70].

Models of $(\mathbf{T}^2)^3/(\mathbb{Z}_2 \otimes \mathbb{Z}_2)$ compactification class with varying fluxes were also studied intensively in [80, 81]. Issues among these continued to be realization of viable masses, mixings, and moduli stabilization. A self dual flux of the form

$$G_3 = \frac{8}{\sqrt{3}} \exp^{-\pi/6} (d\bar{z}_1 dz_2 dz_3 + dz_1 d\bar{z}_2 dz_3 + dz_1 dz_2 d\bar{z}_3) \qquad (3.70)$$

was shown to stabilize the remaining complex structure torus moduli (setting all to $e^{2\pi i/3}$), following prior stabilization of the first two moduli by NS-NS tadpole cancellation (3.58). The difficulty with this model class is cancellation of three-form flux contributions to the $D3$ charge; $D3$ charge conservation constraints are hard to satisfy in semi-realistic models [81].

Three generation MSSM-like models with one, two, and three units of flux were presented in [81]. Several models with a single unit of flux are analyzed. For two of these models the MSSM Higgs have no renormalizable couplings to the three generations of quarks and leptons. Three of these models contain four or five pairs of Higgs which have renormalizable couplings with quarks and leptons. In addition to the extra Higgs, the models also possess varying numbers of chiral MSSM-charged exotics, only a few of which can obtain large masses via coupling to the Higgs. Another of these models was the first MSSM-like model (with several exotics) containing flux to possess strong gauge dynamics, resulting in gaugino and matter condensation on $D7$-branes of the hidden sector. A two flux model was also reported that contains several MSSM-charged exotics. One pair of chiral exotics couple to the MSSM Higgs and, thus, obtains an electroweak scale mass. This model also comes with five pairs of MSSM higgs doublets with correct charges to create renormalizable Yukawa mass terms for all quarks and leptons. In the supersymmetric flux model, the MSSM Higgs doublets do not have Yukawa couplings to quarks or leptons of any generation, due to wrong charges under broken anomalous $U(1)_A$ gauge symmetries. In addition this model contains several MSSM-charged exotics.

A first-of-its-kind MSSM-like model was reported in [73]. While the model is again a $\mathbf{T}^6/(\mathbb{Z}_2 \times \mathbb{Z}_2)$ orientifold, its magnetized $D9$-branes with large negative charge where introduced into the hidden sector, rather than into the observable sector. Models with this property had not been studied previously because of the difficulty of arranging supersymmetric D-brane configurations with more than three stacks of $U(n)$ branes. These problems were overcome in [73].

The initial gauge group of this model is

$$U(4)_C \times U(2)_L \times U(2)_R. \qquad (3.71)$$

The intersection numbers were chosen to satisfy

$$I_{ab} = 3; \ I_{ac} = -3; \ I_{bc} \geq 1, \qquad (3.72)$$

which produces a model with one unit of flux. No filler branes are needed; hence, there is no USp groups. This gauge group (3.71) can be broken down to

$$SU(3)_C \times SU(2)_L \times U(1)_{I_{3R}} \times U(1)_{B-L}. \qquad (3.73)$$

by the generalized Green-Schwarz mechanism and the splittings of the $U(4)_C$ and $U(2)_R$ stacks of $D6$-branes. A significant feature of this model is that the $U(1)_{I_{3R}} \times U(1)_{B-L}$ gauge symmetry can only be broken to $U(1)_Y$ at the TeV scale by giving VEVs to the neutrino singlet scalar field or the neutral component from a $(\bar{\mathbf{4}}, \mathbf{1}, \bar{\mathbf{2}})$. As is typical, only one family gets mass because of anomalous charge constraints.

3.4. Type IIA Intersecting Branes

Type IIA intersecting $D6$-brane models are the T-duals to Type IIB with magnetic charged D-branes. As reported in [73], a large number of non-supersymmetric three family SM-like models of Type IIA origin have been constructed that satisfy the R-R tadpole cancellation conditions [82, 83, 84, 85, 86, 75]. However, generically the NS-NS tadpoles remain uncancelled in these models. A gauge hierarchy issue is also generic to this class [73].

Translation of rules for Type IIA intersecting $D6$-brane model building from Type IIB with magnetic charged branes was provided in [87, 88]. [73] applied this translation for the $\mathbf{T}^6/(\mathbb{Z}_2 \times \mathbb{Z}_2)$, with $\mathbf{T}^6 = \mathbf{T}^2 \times \mathbf{T}^2 \times \mathbf{T}^2$, compactification in particular. Related consistent Type IIA $N = 1$ supersymmetric near-MSSMs with intersecting $D6$-branes have been constructed [68, 89, 90]. Following this, a wide range of MSSM-like (but not exact), PS-like, $SU(5)$, and flipped-$SU(5)$ three generation models were discovered [91, 92, 93, 94, 95]. In the phenomenological investigations of these models [96, 97], the re-occurring concern was moduli stabilization in open and closed strings [73]. Hidden sector gaugino condensation was determined to stabilize the dilaton and complex structure in the Type IIA frame, but further stabilization remains via fluxes or the equivalent.

The Type IIB ($\mathbb{Z}_2 \times \mathbb{Z}_2$) orientifold notation applies to this dual process, except now R: $(z_1, z_2, z_3) \to (\bar{z}_1, \bar{z}_2, \bar{z}_3)$. In the $D6$ models, (n^i, m^i) are the respective brane wrapping numbers along the canonical bases of homology one-cycles $[a_i]$ and $[b_i]$, on the ith two torus $\mathbf{T}_i^2$ and the general homology class for one cycle on $\mathbf{T}_i^2$ is $n^i[a_i] + m^i[b_i]$. Hence the homology classes $[\Pi_a]$ for the three cycles wrapped by a stack of N_a $D6$ branes are

$$[\Pi_a] = \prod_{i=1}^{3}(n^i[a_i] + m^i[b_i]); \quad [\Pi_{a'}] = \prod_{i=1}^{3}(n^i[a_i] - m^i[b_i]). \tag{3.74}$$

There are also homology classes for the four $O6$-planes associated with the orientifold projections ΩR, $\Omega R\omega$, $\Omega R\Theta\omega$, and $\Omega R\Theta$ defined by $[\Pi^1] = [a_1][a_2][a_3]$, $[\Pi^2] = -[a_1][b_2][b_3]$, $[\Pi^3] = [b_1][a_2][b_3]$, and $[\Pi^3] = [b_1][b_2][a_3]$ [73].

The R-R tadpole cancellation conditions (3.75) are parallel to (3.68,3.55),

$$-N^{(1)} - \sum_a N_a n_a^1 n_a^2 n_a^3 = -N^{(2)} + \sum_\alpha N_a m_a^1 m_a^2 n_a^3 =$$
$$-N^{(3)} + \sum_a N_a m_a^1 n_a^2 m_a^3 = -N^{(4)} + \sum_a N_a n_a^1 m_a^2 m_a^3 = -16, \tag{3.75}$$

with $N^{(i)}$ denoting the number of filler branes on the top of the ith $O6$-brane. $N = 1$ SUSY survives the orientation projection if the rotation angle of any $D6$ brane with respect to the orientifold-plane is an element of $SU(3)$. This again translates into two additional constraints on triplets of wrapping numbers [73]. Last, K-theory charges must cancel in like manner as in Type IIB models.

From Type IIA with $\mathbf{T}^6/(\mathbb{Z}_2 \times \mathbb{Z}_2)$ compactification (without flux), [73] constructed a three family trinification model with gauge group $U(3)_C \times U(3)_L \times U(3)_R$, wherein all MSSM fermions and Higgs fields belong to bi-fundamental reps. While the three $SU(3)$ gauge groups contain very large R-R charges, a supersymmetric intersecting $D6$-brane trinification model was found which satisfies the R-R tadpole and K-theory cancellation conditions. The trinification gauge group was then broken down to $SU(3)_C \times SU(3)_L \times SU(3)_R$ by the generalized Green-Schwarz mechanism, which involves the untwisted R-R forms $B_2^{i=1,\ldots,4}$. These four fields couple to the $U(1)$ field strength F_a for each stack a via,

$$N_a m_a^1 n_a^2 n_a^3 \int_{M4} B_2^1 \text{tr} F_a, \quad N_a n_a^1 m_a^2 n_a^3 \int_{M4} B_2^2 \text{tr} F_a,$$
$$N_a m_a^1 n_a^2 m_a^3 \int_{M4} B_2^3 \text{tr} F_a, \quad N_a m_a^1 m_a^2 m_a^3 \int_{M4} B_2^4 \text{tr} F_a. \tag{3.76}$$

The next stage, reduction to $SU(3)_C \times SU(2)_L \times U(1)_{Y_L} \times U(1)_{I_{3R}} \times U(1)_{Y_R}$, results from $D6$ brane splittings. The SM is then obtained by a higgsing that necessarily breaks D- and F-flatness. Quark Yukawa couplings are allowed for just one EW massive generation because, while there is only one Higgs doublet pair, the Q_L^i arise from the intersections on the second two-torus and the Q_R^i arise from the intersections on the third two-torus. Lepton Yukawa couplings are forbidden by anomalous $U(1)_A$ symmetries [73]. A Type IIB dual to this model has not yet been found. While this trinification model has very large R-R charge to start with, its dual Type IIB model also has supergravity fluxes to contend with, which makes R-R tadpole cancellation even more difficult [73].[†]

[†]Thus, as a practical aspect Type IIA intersecting $D6$ models and Type IIB with magnetically charged branes, though T dual, may offer some different opportunities for (MS)SM searches.

$\mathbb{Z}'_6 = \mathbb{Z}_3 \otimes \mathbb{Z}_2$ orientifolding of Type IIA models has been investigated by Bailin and Love [62, 63]. They determined that unlike $\mathbb{Z}_6$ orientifold models to date, $\mathbb{Z}'_6$ can support $(a \cap b, a \cap b') = \pm(2,1)$ or $\pm(1,2)$ intersection numbers, which may lead to the most phenomenologically viable MSSM models without exotics. The initial torus was chosen as $\mathbf{T}^6 = \mathbf{T}^2 \times \mathbf{T}^2 \times \mathbf{T}^2$, where $\mathbf{T}^2_{1,2} = \mathbb{R}^2/\Lambda_{SU(3)}$. After $\mathbb{Z}'_6$ orbifolding, the two basis 1-cycles on each $\mathbf{T}^2$ can be arranged in one of two configurations, producing a total of eight different configurations for the $\mathbb{Z}'_6$ orientifold. Allowed intersection numbers for the eight different models with $\mathbb{Z}'_6$ orientifoldings were analyzed. By construction, there is no matter in the symmetric representations in any of the models. The models contain up to two generations of antiquarks and antileptons in the antisymmetric rep of the $SU(2)$ brane and its mirror. Unfortunately, none of the models could be enlarged to contain exactly the MSSM content and no more. This is prohibited by Abelian charge cancellations. Comparison of the possible intersection numbers for $\mathbb{Z}_6$ and $\mathbb{Z}'_6$ orbifolds, and the variation in the number of singlets within the $\mathbb{Z}'_6$ orbifolds models, showed that different orbifold point groups produce different physics, as does the same point group realized on different lattices.

4. Conclusion

The Standard Model (SM) is truly one of the most outstanding theoretical and experimental achievement of the 20th century. Nevertheless, the very nature of the Standard Model (SM) and the inconsistency of quantum gravity beg the question of what more unified, self-consistent theory hides behind both. The details of the Standard Model cannot be explained from within; nor can the infinities of quantum gravity be resolved from within. "Why" is the SM as it is? "Why" does quantum theory seem inconsistent with gravity. A deeper, underlying theory is necessary to answer these "why" questions.

String theory proposes viable answers to the questions. While the concept of a string landscape has significantly diminished hopes of locating *the* unique string vacuum that either resolves *all* the "Why's" of the SM and gravity by postdicting, in it low energy effect field theory limit, all experimentally verified SM physics or is eliminated as a viable theory by experimental disconfirmation. Instead, string theory in its present form responds to these questions not with a single unique answer, but with a collection of viable answers, all of a geometrical or topological nature. Just as a given Lie algebra has an infinite set of representations, so too the underlying physical laws within string theory may well have a vast array of different representations, of which our observable universe is but one.

Several hundred years ago the debate was over a geocentric universe verses a heliocentric universe. Eventually the heliocentric perception was supplanted by a "galacticentric" view. Then within the last century Hubble proved how limited the latter outlook was also. Perhaps string theory is taking us to the next step of realization, that our universe may be but one of many. That to understand the "why" of our universe and its laws, we must look to a multiverse beyond. Some may argue that this suggestion passes beyond science into the realm of philosophy. However, if string theory describes the nature of reality, then is it not still science? Is that not the very role of science, to uncover physical reality? String theory may well be teaching us that a "univercentric" view is as outmoded today as the geocentric view was realized to be long ago.

Whether string theory ultimately predicts but one possibility for a consistent universe (if the landscape picture eventually collapses) or a vast array, to be viable string theory must predict *at least* our observable universe as a possible outcome. To prove that string theory allows for the physics of the observable universe is the mission of string phenomenology. String phenomenologists are pursuing this investigation from many fronts. (MS)SM-like model realization from weak coupled heterotic models via free fermionic and orbifold constructions, strong coupled heterotic models on elliptically fibered Calabi-Yau's, Type IIB orientifolds with magnetic charged branes, and Type IIA orientifolds with intersecting branes was reviewed in this chapter.

A string derived (MS)SM must possess far more realistic features than just the correct gauge forces and matter states. Also required for the model are realistic gauge coupling strengths, a correct mass hierarchy, a viable CKM quark mixing matrix, a realistic neutrino mass and mixing matrix, and a severely suppressed proton decay. The hidden sector must be sufficiently hidden. MSSM candidates must also provide viable non-perturbative supersymmetry breaking that yields testable predictions for supersymmetric particle masses. Finally, the physical value of the cosmological constant must be produced. No current string-derived (MS)SM-like models are a perfect match with the (MS)SM, but significant progress had been made in the last decade.

The first step in proving that the MSSM can be realized in string theory was accomplished when Cleaver, Faraggi, and Nanopoulos constructed a string model that yields exactly the matter content of the MSSM in the observable sector, with no MSSM-charged exotics. This was shown in the context of a weak coupled heterotic model constructed in the free fermionic formalism [5, 6, 7, 8, 9]. Glimpses of other necessary phenomenology were also shown by this model. Generational mass hierarchy appears as an ubiquitous effect of vacuum expectation values of scalar fields induced via Abelian anomaly cancellation, a feature endemic to all quasi-realistic heterotic models. In this model the physical Higgs is a mixed state of pure Higgs that carry extra generational charges. The mixed states are induced by off-diagonal terms in a Higgs mass matrix that produce exactly one pair of Higgs that is massless at the string scale. The relative weights of the generational Higgs components in the physical Higgs can vary by several orders of magnitude, thereby allowing for a large mass hierarchy even with only low order couplings. The model also provides for supersymmetry breaking through hidden sector condensates.

Following the $\mathbb{Z}_2 \otimes \mathbb{Z}_2$ NAHE-based free fermionic model [5, 6, 7, 8, 9], several additional MSSM models, also without exotic MSSM-charged states, were constructed by alternate means. A representative sample of these were reviewed herein, including the heterotic $\mathbb{Z}_6' = \mathbb{Z}_3 \otimes \mathbb{Z}_2$ orbifold of [37, 38]; the heterotic elliptically fibered Calabi-Yau's of [46, 47, 48, 49, 50, 51, 52, 53, 57, 58], [56, 60], and [61]; the Type IIB magnetic charged branes of [70, 74] without flux, and those of [87], [80], [81], and [73] with flux; and the Type IIA intersecting brane models of [73] and [68, 89].

These examples showed several (differing) areas of progress toward more realistic phenomenology. For instance, some models reconfirmed the importance of a local anomalous Abelian symmetry and the related flat direction VEVs, resulting from the Green-Schwarz-Dine-Seiberg-Witten anomaly cancellation mechanism [25, 26]. These VEVs invoke many necessary features of an MSSM: observable sector gauge breaking to $SU(3)_C \otimes SU(2)_L \otimes U(1)_Y$, decoupling of MSSM exotic matter, production of effective

Higgs mu-terms, and formation of intergenerational and intragenerational mass hierarchy.

The greatest advancement was in the role of branes and antibranes, especially with regard to supersymmetry breaking and moduli stabilization. Brane-based MSSM-like models with stable supersymmetric anti-deSitter vacua have been constructed. Further, uplifting from stable anti-deSitter vacua to metastable deSitter vacua with cosmological constants at a viable scale has been crafted into MSSM-like models by the addition of antibranes. In the years ahead further realistic features of string models containing the (MS)SM gauge group, solely the three generations of (MS)SM matter and a Higgs pair will likely be found, as string phenomenologists continue to analyze the content of the string landscape.

Acknowledgments

G.C. wishes to acknowledge the authors of all papers reviewed herein and of all MSSM-string model papers as a whole. Based on length limitations, reviews herein were limited to a representative set from most construction methods.

References

[1] P. Woit, *Not Even Wrong: The Failure of String Theory and the Search for Unity in Physical Law*, (Basic Books, New York, 2006).

[2] B. Richter, *Theory in Particle Physics: Theological Speculation Versus Practical Knowledge*, Physics Today **59** #10, 8 (2006).

[3] B. Schroer, arXiv:hep-th/0611132.

[4] C. Munoz, arXiv:hep-ph/0312091.

[5] G. B. Cleaver, A. E. Faraggi and D. V. Nanopoulos, *Phys. Lett. B* **455**, 135 (1999) [arXiv:hep-ph/9811427].

[6] G. B. Cleaver, A. E. Faraggi and D. V. Nanopoulos, *Int. J. Mod. Phys. A* **16**, 425 (2001) [arXiv:hep-ph/9904301].

[7] G. B. Cleaver, A. E. Faraggi, D. V. Nanopoulos and J. W. Walker, *Nucl. Phys. B* **593**, 471 (2001) [arXiv:hep-ph/9910230].

[8] G. B. Cleaver, A. E. Faraggi, D. V. Nanopoulos and J. W. Walker, *Mod. Phys. Lett. A* **15**, 1191 (2000) [arXiv:hep-ph/0002060].

[9] G. B. Cleaver, A. E. Faraggi, D. V. Nanopoulos and J. W. Walker, *Nucl. Phys. B* **620**, 259 (2002) [arXiv:hep-ph/0104091].

[10] F. Gmeiner, R. Blumenhagen, G. Honecker, D. Lust and T. Weigand, *JHEP* **0601**, 004 (2006) [arXiv:hep-th/0510170].

[11] A. E. Faraggi, D. V. Nanopoulos and K. J. Yuan, *Nucl. Phys. B* **335**, 347 (1990).

[12] A. E. Faraggi, *Phys. Lett. B* **274**, 47 (1992).

[13] A. E. Faraggi, *Phys. Lett.* B **278**, 131 (1992).

[14] A. E. Faraggi, *Nucl. Phys.* B **387**, 239 (1992) [arXiv:hep-th/9208024].

[15] A. E. Faraggi, *Phys. Rev.* D **46**, 3204 (1992).

[16] I. Antoniadis, C. P. Bachas and C. Kounnas, *Nucl. Phys.* B **289**, 87 (1987).

[17] H. Kawai, D. C. Lewellen and S. H. H. Tye, *Phys. Rev. Lett.* **57**, 1832 (1986) [Erratum-ibid. **58**, 429 (1987)].

[18] H. Kawai, D. C. Lewellen, J. A. Schwartz and S. H. H. Tye, *Nucl. Phys.* B **299**, 431 (1988).

[19] I. Antoniadis and C. Bachas, *Nucl. Phys.* B **298**, 586 (1988).

[20] S. Kalara, J. L. Lopez and D. V. Nanopoulos, *Phys. Lett.* B **245**, 421 (1990).

[21] J. Rizos and K. Tamvakis, *Phys. Lett.* B **262**, 227 (1991).

[22] A. E. Faraggi and D. V. Nanopoulos, *Phys. Rev.* D **48**, 3288 (1993).

[23] T. Kobayashi and H. Nakano, *Nucl. Phys.* B **496**, 103 (1997) [arXiv:hep-th/9612066].

[24] G. B. Cleaver and A. E. Faraggi, *Int. J. Mod. Phys.* A **14**, 2335 (1999) [arXiv:hep-ph/9711339].

[25] M. Dine, N. Seiberg and E. Witten, *Nucl. Phys.* B **289**, 589 (1987).

[26] J. J. Atick, L. J. Dixon and A. Sen, *Nucl. Phys.* B **292**, 109 (1987).

[27] G. B. Cleaver, *Proceedings of Orbis Scientiae on Physics of Mass*, Miami, Florida, 12-15 Dec 1997.

[28] J. L. Lopez and D. V. Nanopoulos, *Nucl. Phys.* B **338**, 73 (1990).

[29] M. Cvetic, L. L. Everett and J. Wang, *Phys. Rev.* D **59**, 107901 (1999) [arXiv:hep-ph/9808321].

[30] A. E. Faraggi, *Nucl. Phys.* B **487**, 55 (1997) [arXiv:hep-ph/9601332].

[31] J. L. Lopez and D. V. Nanopoulos, *Phys. Rev. Lett.* **76**, 1566 (1996) [arXiv:hep-ph/9511426].

[32] A. B. Lahanas and D. V. Nanopoulos, *Phys. Rept.* **145**, 1 (1987).

[33] J. R. Ellis, K. Enqvist and D. V. Nanopoulos, *Phys. Lett.* B **151**, 357 (1985).

[34] G. B. Cleaver, D. V. Nanopoulos, J. T. Perkins and J. W. Walker, arXiv:hep-th/0512020.

[35] G. Dvali, R. Kallosh and A. Van Proeyen, *JHEP* **0401**, 035 (2004) [arXiv:hep-th/0312005].

[36] A. E. Faraggi, E. Manno and C. Timirgaziu, *Eur. Phys. J. C* **50**, 701 (2007) [arXiv:hep-th/0610118].

[37] W. Buchmuller, K. Hamaguchi, O. Lebedev and M. Ratz, *Phys. Rev. Lett.* **96**, 121602 (2006) [arXiv:hep-ph/0511035].

[38] W. Buchmuller, K. Hamaguchi, O. Lebedev and M. Ratz, *Nucl. Phys. B* **785**, 149 (2007) [arXiv:hep-th/0606187].

[39] J. Wess and J. Bagger, "Supersymmetry and supergravity," http://www.slac.stanford.edu/spires/find/hep/www?irn=5426545SPIRES entry

[40] S. Hamidi and C. Vafa, *Nucl. Phys. B* **279**, 465 (1987).

[41] A. Font, L. E. Ibanez, H. P. Nilles and F. Quevedo, *Phys. Lett.* **210B**, 101 (1988) [Erratum-ibid. B **213**, 564 (1988)].

[42] G. Cleaver, V. Desai, H. Hanson, J. Perkins, D. Robbins and S. Shields, *Phys. Rev. D* **67**, 026009 (2003) [arXiv:hep-ph/0209050].

[43] J. Perkins, B. Dundee, R. Obousy, E. Kasper, M. Robinson, K. Stone and G. Cleaver, arXiv:hep-ph/0310155.

[44] J. Perkins *et al.*, *Phys. Rev D* **75**, 026007 (2007) [arXiv:hep-ph/0510141].

[45] J. Giedt, *Mod. Phys. Lett. A* **18**, 1625 (2003) [arXiv:hep-ph/0205224].

[46] V. Braun, Y. H. He, B. A. Ovrut and T. Pantev, *JHEP* **0605**, 043 (2006) [arXiv:hep-th/0512177].

[47] V. Braun, B. A. Ovrut, T. Pantev and R. Reinbacher, *JHEP* **0412**, 062 (2004) [arXiv:hep-th/0410055].

[48] V. Braun, Y. H. He, B. A. Ovrut and T. Pantev, *Phys. Lett. B* **618**, 252 (2005) [arXiv:hep-th/0501070].

[49] V. Braun, Y. H. He, B. A. Ovrut and T. Pantev, *JHEP* **0506**, 039 (2005) [arXiv:hep-th/0502155].

[50] V. Braun, Y. H. He, B. A. Ovrut and T. Pantev, *Adv. Theor. Math. Phys.* **10**, 4 (2006) [arXiv:hep-th/0505041].

[51] V. Braun, Y. H. He and B. A. Ovrut, *JHEP* **0604**, 019 (2006) [arXiv:hep-th/0601204].

[52] V. Braun, Y. H. He, B. A. Ovrut and T. Pantev, *JHEP* **0603**, 006 (2006) [arXiv:hep-th/0510142].

[53] V. Braun, Y. H. He and B. A. Ovrut, *JHEP* **0606**, 032 (2006) [arXiv:hep-th/0602073].

[54] V. Braun and B. A. Ovrut, *JHEP* **0607**, 035 (2006) [arXiv:hep-th/0603088].

[55] T. L. Gomez, S. Lukic and I. Sols, *Commun. Math. Phys.* **276**, 1 (2007) [arXiv:hep-th/0512205].

[56] V. Bouchard and R. Donagi, *Phys. Lett. B* **633**, 783 (2006) [arXiv:hep-th/0512149].

[57] V. Braun, E. I. Buchbinder and B. A. Ovrut, *Phys. Lett. B* **639**, 566 (2006) [arXiv:hep-th/0606166].

[58] V. Braun, E. I. Buchbinder and B. A. Ovrut, *JHEP* **0610**, 041 (2006) [arXiv:hep-th/0606241].

[59] K. Intriligator, N. Seiberg and D. Shih, *JHEP* **0604**, 021 (2006) [arXiv:hep-th/0602239].

[60] V. Bouchard, M. Cvetic and R. Donagi, *Nucl. Phys. B* **745**, 62 (2006) [arXiv:hep-th/0602096].

[61] R. Blumenhagen, S. Moster and T. Weigand, *Nucl. Phys. B* **751**, 186 (2006) [arXiv:hep-th/0603015].

[62] D. Bailin and A. Love, *AIP Conf. Proc.* **881**, 1 (2007) [arXiv:hep-th/0607158].

[63] D. Bailin and A. Love, *Nucl. Phys. B* **755**, 79 (2006) [arXiv:hep-th/0603172].

[64] C. Kokorelis, arXiv:hep-th/0406258.

[65] C. Kokorelis, arXiv:hep-th/0410134.

[66] R. Blumenhagen, B. Kors, D. Lust and T. Ott, *Nucl. Phys. B* **616**, 3 (2001) [arXiv:hep-th/0107138].

[67] M. Berkooz and R. G. Leigh, *Nucl. Phys. B* **483**, 187 (1997) [arXiv:hep-th/9605049].

[68] M. Cvetic, G. Shiu and A. M. Uranga, *Phys. Rev. Lett.* **87**, 201801 (2001) [arXiv:hep-th/0107143].

[69] J. F. G. Cascales, M. P. Garcia del Moral, F. Quevedo and A. M. Uranga, *JHEP* **0402**, 031 (2004) [arXiv:hep-th/0312051].

[70] F. Marchesano and G. Shiu, *Phys. Rev. D* **71**, 011701 (2005) [arXiv:hep-th/0408059].

[71] E. Witten, *JHEP* **9812**, 019 (1998) [arXiv:hep-th/9810188].

[72] A. M. Uranga, *Nucl. Phys. B* **598**, 225 (2001) [arXiv:hep-th/0011048].

[73] C. M. Chen, T. Li and D. V. Nanopoulos, *Nucl. Phys. B* **732**, 224 (2006) [arXiv:hep-th/0509059].

[74] F. Marchesano and G. Shiu, *JHEP* **0411**, 041 (2004) [arXiv:hep-th/0409132].

[75] D. Cremades, L. E. Ibanez and F. Marchesano, *JHEP* **0307**, 038 (2003) [arXiv:hep-th/0302105].

[76] M. Grana, *Phys. Rev. D* **67**, 066006 (2003) [arXiv:hep-th/0209200].

[77] P. G. Camara, L. E. Ibanez and A. M. Uranga, *Nucl. Phys. B* **689**, 195 (2004) [arXiv:hep-th/0311241].

[78] M. Grana, T. W. Grimm, H. Jockers and J. Louis, *Nucl. Phys. B* **690**, 21 (2004) [arXiv:hep-th/0312232].

[79] D. Lust, S. Reffert and S. Stieberger, *Nucl. Phys. B* **706**, 3 (2005) [arXiv:hep-th/0406092].

[80] J. Kumar and J. D. Wells, *JHEP* **0509**, 067 (2005) [arXiv:hep-th/0506252].

[81] M. Cvetic, T. Li and T. Liu, *Phys. Rev. D* **71**, 106008 (2005) [arXiv:hep-th/0501041].

[82] R. Blumenhagen, L. Goerlich, B. Kors and D. Lust, *JHEP* **0010**, 006 (2000) [arXiv:hep-th/0007024].

[83] G. Aldazabal, S. Franco, L. E. Ibanez, R. Rabadan and A. M. Uranga, *JHEP* **0102**, 047 (2001) [arXiv:hep-ph/0011132].

[84] G. Aldazabal, S. Franco, L. E. Ibanez, R. Rabadan and A. M. Uranga, *J. Math. Phys.* **42**, 3103 (2001) [arXiv:hep-th/0011073].

[85] R. Blumenhagen, B. Kors and D. Lust, *JHEP* **0102**, 030 (2001) [arXiv:hep-th/0012156].

[86] C. Angelantonj, I. Antoniadis, E. Dudas and A. Sagnotti, *Phys. Lett. B* **489**, 223 (2000) [arXiv:hep-th/0007090].

[87] J. F. G. Cascales and A. M. Uranga, *JHEP* **0305**, 011 (2003) [arXiv:hep-th/0303024].

[88] R. Blumenhagen, D. Lust and T. R. Taylor, *Nucl. Phys. B* **663**, 319 (2003) [arXiv:hep-th/0303016].

[89] M. Cvetic, G. Shiu and A. M. Uranga, *Nucl. Phys. B* **615**, 3 (2001) [arXiv:hep-th/0107166].

[90] M. Cvetic, P. Langacker and J. Wang, *Phys. Rev. D* **68**, 046002 (2003) [arXiv:hep-th/0303208].

[91] M. Cvetic and I. Papadimitriou, *Phys. Rev. D* **67**, 126006 (2003) [arXiv:hep-th/0303197].

[92] M. Cvetic, I. Papadimitriou and G. Shiu, *Nucl. Phys. B* **659**, 193 (2003) [Erratum-ibid. B **696**, 298 (2004)] [arXiv:hep-th/0212177].

[93] M. Cvetic, T. Li and T. Liu, *Nucl. Phys. B* **698**, 163 (2004) [arXiv:hep-th/0403061].

[94] M. Cvetic, P. Langacker, T. Li and T. Liu, *Nucl. Phys. B* **709**, 241 (2005) [arXiv:hep-th/0407178].

[95] C. M. Chen, G. V. Kraniotis, V. E. Mayes, D. V. Nanopoulos and J. W. Walker, *Phys. Lett. B* **611**, 156 (2005) [arXiv:hep-th/0501182].

[96] M. Cvetic, P. Langacker and G. Shiu, *Phys. Rev. D* **66**, 066004 (2002) [arXiv:hep-ph/0205252].

[97] M. Cvetic, P. Langacker and G. Shiu, *Nucl. Phys. B* **642**, 139 (2002) [arXiv:hep-th/0206115].

In: String Theory Research Progress
Editor: Ferenc N. Balogh, pp. 133-154

ISBN 1-60456-075-6
© 2008 Nova Science Publishers, Inc.

Chapter 3

INTRODUCTION TO NON-CRITICAL STRING THEORY: CLASSICAL AND QUANTUM ASPECTS

Stanislav Klimenko and Igor Nikitin
Institute of Computing for Physics and Technology, Protvino, Russia

Since the creation of the special theory of relativity, relativistic models have been successfully used to describe complex physical phenomena both in micro- and macro-world. However, the development of such models revealed that all of them, starting from a certain level of complexity, possess a number of common problems, whose investigation continues to the present. These problems are, particularly, the presence of singularities on the classical solutions and anomalies appearing in the quantization.

The relativistic models are formulated in Minkowski space-time, generally d-dimensional, i.e. pseudo-Euclidean space with coordinates $(x_0, x_1, ..., x_{d-1})$, where x_0 is identified with physical time, and the other x_i are spatial coordinates. The space-time is endowed with the scalar product $(ab) = a_0 b_0 - a_1 b_1 ... - a_{d-1} b_{d-1}$. An important role is played by the transformations, preserving this scalar product, which form *the Poincare group*. This group consists of d-dimensional translations and *Lorentz subgroup*, which, in its turn, is decomposed to rotations of $(d-1)$-dimensional Euclidean space and so called hyperbolic rotations or *boosts*. As a whole, the Poincare group exactly coincides with the group of transformations, describing the transitions between all possible inertial reference frames in special relativity [1].

The relativistic models are based on the following paradigm. Space-time is filled by various geometrical objects (curves, surfaces, volumes, field distributions ...), for each of them a functional is introduced, called *action*. In classical mechanics, the system occupies only those states, on which its total action reaches the extremum under fixed boundary conditions. In the relativistic models the action functional depends on the shape and mutual position of the objects, and does not depend on the details of their representation, such as a choice of coordinate frames or parametrization of the curves and surfaces. One can say that the algebraic content of the theory plays an auxiliary role, providing the explicit representation of the objects, but the physical results are completely defined by the geometry of the objects and do not depend on their representation. For example, the dynamics of a material point, whose world line is specified parametrically as $x_\mu(\tau)$, can be described by the action proportional to the length of the world line:

$$A_p = m \int d\tau \sqrt{\dot{x}^2} = m \int d\tau \sqrt{g^{(1)}}, \qquad (1)$$

for 2-dimensional surfaces $x_\mu(\tau, \sigma)$, the simplest action is proportional to the area of the surface:

$$A_s = \gamma \int d\tau d\sigma \sqrt{(\dot{x}x')^2 - \dot{x}^2 x'^2} = \gamma \int d\tau d\sigma \sqrt{-g^{(2)}}. \qquad (2)$$

Such surfaces are considered in string theory and are called *world sheets of strings*, while the curves spanning these surfaces in the motion through the space-time are called *the relativistic strings*. The strings can be homeomorphic to a segment (open strings), a circle (closed strings). They can also have a more complex topological type. In the given formulae, m is the mass of the material point, γ is the parameter defining the tension of the string; $\dot{x} = \partial_\tau x = \partial_1 x$, $x' = \partial_\sigma x = \partial_2 x$; $g^{(k)} = det(\partial_i x \partial_j x)|_{1 \leq i,j \leq k}$ is the determinant of the metric induced from the Minkowski space-time to respectively the world line ($k = 1$) and the world sheet ($k = 2$); conditions $g^{(1)} > 0$ and $g^{(2)} \leq 0$ restrict the class of considered world lines and world sheets to so called *time-like* curves and surfaces. Analogously, k-dimensional surfaces in d-dimensional Minkowski space can be considered, with the action proportional to the spanned k-dimensional *world volume*. At $k = p + 1$ such surfaces are called p-branes (particularly, the $p = 1$ case is the string, the $p = 2$ case is a membrane).

The interaction of various objects is introduced by the addition of corresponding terms in the action of the system. For example, the material point located at the end of the string corresponds to the sum of (1) and (2). For the obtained action during the determination of its extremum one needs to take into account the fact, that the world lines of the material point and the end of the string coincide. It will produce corresponding changes in boundary conditions. The interaction of the material point with the electromagnetic field is described by the term

$$A_{p+f} = e \int d\tau \dot{x}_\mu A_\mu \qquad (3)$$

where e is the electric charge, $A_\mu(x)$ is vector-potential, while the action for the electromagnetic field itself has a form

$$A_f = -\tfrac{1}{4} \int dx F_{\mu\nu} F_{\mu\nu}, \qquad (4)$$

where $F_{\mu\nu} = \partial_\mu A_\nu - \partial_\nu A_\mu$ is the intensity tensor of the electromagnetic field, and the integration is performed over d-dimensional volume of the space-time. We emphasize once more that the above described action functionals are invariant under the Poincare group, which for d-dimensional Minkowski space contains $d(d+1)/2$ generators, and also under the groups of reparametrization of the world lines, world sheets and world volumes, which possess *an infinite* set of generators. Therefore, the relativistic models necessarily contain large groups of symmetries.

For the determination of the solutions in relativistic models, the Hamiltonian formalism is appropriate. In this formalism, the objects are foliated to slices by a set of constant time planes in a certain reference frame (instead of the planes one can also use surfaces of more general form [2]). The symmetries of the action reveal themselves in Hamiltonian mechanics as follows. Those symmetries, which correspond to the transformation of the object as a whole, are called *global* symmetries. Such symmetries produce conserving variables (the first integrals). E.g. the symmetries under Poincare transformations correspond to the conservation laws for energy, momentum and angular momentum. Reparametrizations of the world lines and world sheet belong to the other type of transformation, which can be localized to a small part of the object, and are called *local* or *gauge* symmetries. For such transformations the corresponding first integrals are not simply conserved, but are constant in the whole phase space. As a result, so called Hamiltonian constraints appear – the relations of the form $F(u) = 0$, which reduce the dimension of the phase space and change its topology. In Hamiltonian mechanics each variable plays two roles: it is real-valued function $H(u)$ in the phase space with coordinates u and it is a generator of canonical transformations, described by a differential equation $\dot{u} = \{u, H\}$, where $\{,\}$ are Poisson brackets. This equation describes a phase flow, for which the given variable is used as a Hamiltonian. The first integrals and constraints of Hamiltonian mechanics are the generators of corresponding symmetries. There is also a canonical Hamiltonian, found from the Lagrangian, i.e. the integrand of the action, by means of the standard Legendre transformation. An important property of the relativistic models (1-3) is that their canonical Hamiltonians identically vanish. It does not mean vanishing energy, which in the relativistic models is a component p_0 of energy-momentum vector p_μ. The Hamiltonian, reproducing the physical evolution, corresponding to the extremal action, has a form of an arbitrary linear combination of the constraints [2], which, we remind, generate reparametrizations. It is one more expression of the fact that the evolution is performed by the shift of slices, i.e. by a certain reparametrization of the objects. Using an appropriate choice of the coefficients in the linear combination of the constraints one can considerably simplify the Hamiltonian equations, sometimes even reduce them to analytically solvable form.

The Hamiltonian formalism is also a basis for canonical quantization of the system. In quantum mechanics each variable is represented as a linear operator in Hilbert space. The representation should satisfy the following correspondence principle: Poisson brackets of two variables in Hamiltonian mechanics correspond to the commutator of the respective operators in quantum mechanics. If one requires that this principle would be satisfied for all variables, described by arbitrary functions in the phase space, the given problem would not have solutions [3]. If the correspondence principle is satisfied for a certain set of canonical variables in the phase space, then already polynomial functions of them will have ambiguities, related with ordering of operators, and their commutators will have so called *anomalous terms*. The common practice is the implementation of the correspondence principle only for a special set of physically important variables, containing particularly all generators of symmetries. It is equivalent to the implementation of the symmetries on the quantum level. It is evident from this that quantization of the relativistic systems results in extra difficulties with the quantum implementation of large groups of symmetries they possess. For some systems, including particularly the relativistic strings, these difficulties have not been resolved up to now.

As we have mentioned already, the classical string theory considers 2-dimensional surfaces of extremal area in d-dimensional Minkowski space-time, called the world sheets of strings. These surfaces are spanned in motion through the space-time of a 1-dimensional object, called relativistic string. The string theory is used in high energy physics to model the inner structure of elementary particles. For this purpose the world sheets, having microscopic spatial sizes and infinitely extended in temporal direction, are considered as structured world lines of elementary particles. As a result, the internal characteristics of particles, such as mass and spin, are expressed in terms of string dynamics and can be derived from a small set of fundamental constants.

Historically string theory has appeared in theoretical high energy physics in relation with the study of the internal structure of strongly interacting particles, the hadrons. Nowadays for the description of strong interaction, a special field theory is used – quantum chromodynamics, QCD [11, 5]. This theory describes the interaction of massive spinor fields representing *quarks*, by means of massless vector fields, representing *gluons*. Both the quarks and gluons possess a special quantum number, which takes three values (r, g, b) and conventionally called *color*. The color is analogous to the electric charge of quantum electrodynamics. The major difference is that the local symmetry of electrodynamics is a multiplication of the wave functions by a phase factor $e^{i\varphi(x)}$, which corresponds to the group $U(1)$ of unitary rotations of the complex plane, while the transformations in the color space are described by a special unitary group $SU(3)$. In the field theory, the presence of gauge symmetries leads to the fact that the charged particles necessarily enter into the interaction with massless vector fields, the quants of the interactions. In the case if the gauge group is non-commutative (non-Abelian) the quants necessarily interact with themselves. Because the group $U(1)$ is Abelian, the quants of electromagnetic field – the photons are not electrically charged and do not interact one with the other, while the group $SU(3)$ is non-Abelian, the gluons possess a color and intensively interact among themselves. As a result of this interaction, the gluonic field concentrates along the line connecting the quarks, in contrast to the electromagnetic field, which spreads over the whole space. Such a configuration appears to be energetically preferable [35, 4]. The same result is produced by the computations on the lattice [5, 6], evaluation of vacuum correlators [7, 8], investigation of chromodynamics in the limit of a large number of colors [29], investigation of the stationary field configurations in simplified models [27, 28]. These investigations have formed the modern picture of hadronic constitution as systems of quarks, connected by gluonic tubes of size 1 fm $= 10^{-13}$ cm and tension 1 GeV/fm $\approx 10^5$ N. The energy of the gluonic field, connecting the quarks, is proportional to the length of the tube, while in electrodynamics, the energy of interaction is inversely proportional to the distance between the charges. As a result, the separation of the quarks to an infinite distance requires an infinite amount of energy. Moreover, starting from some distances [9, 10] the break of the gluonic tube becomes energetically preferable, with the creation of quark-antiquark pair out of the vacuum. Therefore, the separation of quarks to an infinite distance becomes impossible, like the separation of the poles of the magnet. In this way, QCD explains the experimentally confirmed property of *quark confinement* [11]. Due to this property, only colorless bound quark-gluonic states, which, by the definition, are *the hadrons*, are observed in accelerator experiments.

Quantum chromodynamics successfully describes hard processes, i.e. the particle reactions possessing a large value of transferred momentum. In the region of small transferred momentum, the contribution of so called non-perturbative effects becomes considerable.

The main computational apparatus in the quantum field theory is perturbation theory by the parameter, defining the strength of the interaction. In quantum electrodynamics, the role of such a parameter is played by a fine structure constant $\alpha_{QED} = e^2/\hbar c \approx 1/137$, while in chromodynamics the corresponding parameter is defined by the formula $\alpha_{QCD} \sim 2\pi/(7ln(q/\Lambda_{QCD}))$, where q is the transferred momentum, $\Lambda_{QCD} \sim 0.1$ GeV [11]. For small transferred momentum, the parameter α_{QCD} becomes large (particularly, according to the given formula this parameter is unbounded at $q \sim \Lambda_{QCD}$). In this regime, perturbation theory is not applicable. Therefore QCD at the present time is not a complete theory, and for the description of the processes at the distances $1/\Lambda_{QCD} \sim 1$ fm, the effective models are used. The hadrons consisting of heavy quarks are well described by non-relativistic potential models [16]. For the bound states of light quarks the relativistic models are necessary. The string model has been proposed by Nambu, Hara and Goto [17, 18, 19] at the beginning of 1970s for the explanation of spin-mass spectra of the hadrons, as well as certain experimentally established properties of their interaction amplitudes [20, 21, 22]. In this model, the action of the string (2) is used to describe the gluonic tubes at the limit of zero thickness. Analogous relativistic systems have been also considered in earlier works [23] in the context of non-linear field models of Born-Infeld type [35]. We note that up to now the mathematically strict derivation of string action from QCD does not exist, although works in this direction are present [24, 25, 26, 27]. The string model of light hadrons [32] provides a natural explanation for such property of the observable hadronic spectrum, as linearity of Regge trajectories [36]. On the plane of parameters (M^2, S), where M is mass, S is spin, the families of hadrons possessing the same quark constitution form straight lines $S = \alpha' M^2 + \alpha_0$, called Regge trajectories. Here the parameter α' describes the slope of Regge trajectory, α_0 is the so called intercept. The model [32] uses a simplest solution, for which the string has a form of straight segment, performing a planar rotation about its middle at constant angular velocity in the center-of-mass frame. For this solution, the orbital moment of the string and its mass are related by the formula given above, where the slope α' is defined by the tension of the string γ as $\alpha' = 1/(2\pi\gamma)$. Nowadays the construction of string-inspired models of the hadrons is continued [28, 29, 30, 31, 32, 33, 34]. In these papers the idealized theoretical model has been equipped by necessary physical details: quarks at the ends of the string, carrying masses, electric charges, spin and other quantum numbers; different mechanisms of the string breaking, responsible for the decays of the hadrons. The progress of this construction has been reported in the book [35].

Meanwhile the development of string theory has selected a different way. It has been early recognized [37, 38, 39, 40] that at physical dimension of the space-time $d = 4$ the quantization of Nambu-Goto model possesses anomalies destroying its main symmetries: invariance under reparametrizations and Poincare group (namely, the anomalies appear in its Lorentz component). The anomalies are cancelled at $d = 26$. It has been also noticed that inclusion of additional fermionic degrees of freedom to the theory [41, 42] cancels the anomaly at the lower value of dimension $d = 10$. Later this approach has been combined with another idea: some of dimensions were considered as coordinates on a compact manifold of physically small size. Further developments, reviewed in the modern textbooks [43, 44, 45], have added more complex mathematical structures to the theory, such as the above described p-branes, as well as their supersymmetric analogs, and formed a powerful direction of theoretical physics, providing a basis for the construction of *the Grand Unifi-*

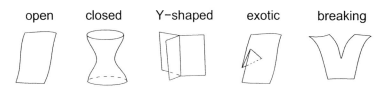

Figure 1. Main topological types of the world sheets.

cation Theory. The suggestion to use string theory for unification of fundamental interactions has been proposed by Sherk and Schwarz, based on the investigated in [58, 59, 57] properties of the scattering amplitudes of string states at the limit of zero slope of Regge trajectories (the infinite tension of the string). According to the results of these papers, the scattering matrix of the massless spin 1 states of open string coincides with S-matrix in Yang-Mills theory [5], while the massless spin 2 state of closed string interacts in a way identifying it with a graviton. The interesting results have been also obtained in the model of a heterotic string [60]. Topological effects appearing in the quantization of this model produce additional symmetries, not contained in the original formulation of model. With the aid of this mechanism (similar to the one used in Kaluza-Klein theory [61]), the string unification models describe the nature of internal quantum numbers and gauge symmetries of the elementary particles.

However, we emphasize, that the described approach cannot be immediately applied in the models of hadrons, where the subject of consideration is the bosonic 4-dimensional Nambu-Goto string, while the introduction of extra dimensions and additional degrees of freedom changes this system essentially. Hadronic models still need Nambu-Goto string theory quantized at the value of dimension $d = 4$.

The problems of Nambu-Goto model are analogous to those of modern theory of gravitation, *the general relativity*, which shares a lot of properties with the string theory. Particularly, their common features are the existence of singular classical solutions and quantum anomalies.

Singularities of classical solutions. String theory considers world sheets of various topological types, see Figure 1: open strings – surfaces, homeomorphic to bands $I \times \mathbf{R}^1$, closed strings – cylinders $S^1 \times \mathbf{R}^1$, Y-shaped strings – 3 bands, glued together along one edge, and also the surfaces of more complex topology, corresponding to transitions between the described types (decays and transmutations of particles).

The extremum of the action (2) for each topological type leads to Lagrange-Euler equations, satisfied at internal points of the world sheet, which have a form of local conservation of energy-momentum:

$$p_i^\mu = \delta A/\delta(\partial_i x_\mu), \quad \partial_i p_i^\mu = 0,$$

and boundary conditions, implying that total flow of momentum through the boundary vanishes. For example, an open string satisfies the equation $p_i^\mu \epsilon_{ij} d\sigma_j = 0$ on the boundary (here $d\sigma_j$ is tangent element of the world sheet boundary on the parameters plane, $\epsilon_{ij} d\sigma_j$ is the normal element); for Y-shaped string $\sum p_i^\mu \epsilon_{ij} d\sigma_j = 0$ on the world line of the node, where the sum is taken for the three surfaces attached to this line.

Further construction of string theory usually proceeds in the Hamiltonian approach. The coordinates on the world sheet are distinguished: σ – compact coordinate, $\sigma \in I$ for the open string, $\sigma \in S^1$ for the closed string; $\tau \in \mathbf{R}^1$ – non-compact coordinate, called evolution parameter. The following notation is introduced: $\dot{x} = \partial x/\partial \tau$, $x' = \partial x/\partial \sigma$, $p^\mu = p^\mu_\tau$, so that the string action can be written as $A = \int d\tau \int d\sigma\, \mathcal{L}$, where $\mathcal{L} = \sqrt{(\dot{x}x')^2 - \dot{x}^2 x'^2}$ is the Lagrangian density. Poisson brackets are introduced as $\{x_\mu(\sigma, \tau), p_\nu(\tilde{\sigma}, \tau)\} = g_{\mu\nu}\delta(\sigma - \tilde{\sigma})$, where $g_{\mu\nu} = \mathrm{diag}(+1, -1, ..., -1)$ is the metric tensor and $\delta()$ is Dirac's function. Cauchy problem is stated to find a solution $x(\sigma, \tau), p(\sigma, \tau)$ starting from the initial data $x(\sigma, 0), p(\sigma, 0)$. The evolution is governed by a system of autonomous differential equations, simultaneous by τ (so that τ-dependence is usually omitted, implying that all variables are estimated at the same value of evolution parameter).

String theory is Hamiltonian theory with the 1st class constraints [2]. This means the following. The canonical Hamiltonian density is defined by the Lagrangian one via Legendre transformation as $\mathcal{H}_c = \dot{x}p\, \mathcal{L}$. Substitution of p-definition in terms of $x', \dot{x}$ gives the vanishing canonical Hamiltonian $\mathcal{H}_c = 0$, and additionally creates the following identities: $\Phi_1 = x'p = 0$, $\Phi_2 = x'^2 + p^2 = 0$. As we have already mentioned, the appearance of these identities, called Dirac's constraints, is related with the symmetry of the action under the group of reparametrizations (right diffeomorphisms [1]) of the world sheet. Usually consideration of theories with constraints starts in extended phase space (x, p), where the Hamiltonian is defined as a linear combination of constraints, in our case $H = \int d\sigma (V_1\Phi_1 + V_2\Phi_2)$. Here the coefficients $V_{1,2}(\sigma)$ are arbitrary and called Lagrangian multipliers. On the surface $\Phi_i = 0$, the Hamiltonian vanishes, however its derivatives do not vanish and create the Hamiltonian vector field:

$$\dot{x}_\mu(\sigma) = \delta H/\delta p_\mu(\sigma),\ \dot{p}_\mu(\sigma) = -\delta H/\delta x_\mu(\sigma).$$

This field is tangent to the surface $\Phi_i = 0$ due to the fact that the Poisson brackets $\{\Phi_i(\sigma), \Phi_j(\tilde{\sigma})\}$ vanish on the the surface $\Phi_i = 0$. In this case the constraints are said to be of the 1st class. Phase trajectory, integrated from such a field, belongs to the surface $\Phi_i = 0$, and its projection to coordinate space $\{x\}$ gives a solution of the Lagrange-Euler equations. In string theory Φ_i-terms of Hamiltonian generate infinitesimal shifts of points in tangent directions to the world sheet: Φ_1 generates the shifts $\delta x \sim x'$, while Φ_2 generates $\delta x \sim p \perp x'$. Together the constraints generate all possible reparametrizations of the world sheet (a connected component of the right diffeomorphisms group). The coefficients $V_{1,2}$ influence only the parametrization of the world sheet. A choice $V_1 = 0$, $V_2 = 1$ corresponds to so called conformal parametrization $\dot{x}x' = 0$, $\dot{x}^2 + x'^2 = 0$ [44]. This choice linearizes the Hamiltonian equations. The equations were solved by different methods in [44, 70, 71, 72, 73, 84]. The resulting solutions allow simple geometric representation. E.g. the world sheets of open strings have a form $x_\mu(\sigma_1, \sigma_2) = (Q_\mu(\sigma_1) + Q_\mu(\sigma_2))/2$, where $\sigma_{1,2} = \tau \pm \sigma$ and $Q_\mu(\sigma)$ is a curve in Minkowski space-time, possessing the properties of light-likeness: $Q'^2 = 0$ and periodicity: $Q(\sigma + 2\pi) - Q(\sigma) = Const$. In a number of papers [30, 74, 65, 75, 76, 84, 77], the existence of singular points on such surfaces has

---
[1]For the mappings $x_\mu(\sigma, \tau)$ the transformations of image space x_μ are called *left*, while the transformations of pre-image space (σ, τ) are called *right*. A bijective continuous transformation is called *a homeomorphism*, while a bijective differentiable transformation, for which the inverse one is also differentiable, is called *a diffeomorphism*. E.g. the $\mathbf{R} \to \mathbf{R}$ mapping, given by the formula $t \to t^3$, is homeomorphism and is not diffeomorphism.

been mentioned. They appear to be topologically stable only in the space-time of dimension $d = 3, 4$ [83, 90]. The mathematical apparatus for the investigation of these phenomena is provided by *the theory of singularities on differentiable mappings* [97]. These methods have been applied for classification of singularities on the world sheets of strings in the paper [90].

From the physical point of view, the interest in singular points on strings is caused by several reasons. Singularities reveal themselves as stable formations inside the hadron, possessing a significant concentration of energy-momentum. This enables their physical interpretation as *valent gluons*, the elements entering in the structure of so called exotic hadrons [12, 13, 14, 15]. The investigation of singularities in 3-dimensional space-time shows many interesting phenomena. In this case, singularities behave as point-like particles, which propagate on the world sheet at light velocity, scatter or annihilate in collisions. The singularities on strings are closely related with singular solutions of the equations, appearing in frames of so called *geometrical approach* to string theory [35, 74]. The motion of singularities admits a Hamiltonian description. The obtained dynamical system is completely integrable [101, 102]. Local characteristics of singularities (*topological charges*) can be introduced, satisfying a global law of conservation [90].

The Nambu-Goto model also possesses a wide class of singular solutions, for which the corresponding world sheets are mapped to Minkowski space-time with a fold, see 1. Such solutions have been found explicitly in the paper [84]. Their presence has been also mentioned in earlier papers [65]. The analysis of these solutions shows that the density of energy is not everywhere positive for them. Solutions with negative energies and masses have been also found in the theory of gravitation, which uses for their denotation a term *"exotic matter"* [103]. In string theory, the typical processes for exotic world sheets are the creation of strings from vacuum, their recombination and annihilation. The necessity to consider such solutions is related with the fact that many standard approaches in string theory [44], particularly, *the covariant approach* described below, contain such solutions in unremovable way. On the quantum level due to uncertainty principle the exotic solutions are mixed with the "normal" ones, and the separation of these types of solutions makes the quantization procedure even more problematic.

The processes of string break are already present in classical mechanics. For example, the breaking of open string into two open strings corresponds to the diagram, shown on 1 right. This model of the break has been considered in the work [30], where the quasi-classical approximation of the corresponding first quantized theory has been also considered. The distinctive feature of this model is that the position of its break point is not defined by the initial data, but is a free parameter. Actually, this model states that the break of the string can happen at an arbitrary point. Further the decay products propagate according to the motion law of free strings, and as a whole the obtained world sheet is an extremal surface. In the application of this model to the description of the decays of hadrons, it should be supplied with a certain mechanism of string break [28, 31]. The hypothesis about the break of the string at a singular point looks physically reasonable. In this case the masses and spins of the decay products are completely defined by the dynamics of the string. The work [30] has also mentioned the relation between the break processes and singularities on the string: if the string breaks at a regular point, then singularities appear on the decay products due to an instantaneous change of boundary conditions. Also, it has been shown in the work [90], that a smooth world sheet can appear only as a result of a break of a singular world sheet at one of the singular points. This fact gives a possibility to construct the models of decay of elementary particles, where smooth world sheets describe the particles with long lifetime, while the singular world sheets decay finally to the smooth ones by a sequence of breaks at the singular points. The dimensions $d = 3, 4$ are naturally selected by these models as the only values of dimension where stable singularities of world sheets exist. Note that these special values of dimension are selected by the geometrical properties of solutions, in the contrast to the quantum critical dimensions, which, as we clarify further, are selected by specific properties of certain quantization schemes.

Problems of quantization in string theory are related to the fact that string theory, like the general relativity, contains reparametrization invariance. In such theories, all physically observable quantities are parametric invariants of geometric objects (length of the curve, area of the surface etc). As we have already mentioned, this symmetry leads to the infinite set of 1st class constraints in Dirac's terminology. The main problem, appearing in the quantization of strings, is quantum implementation of these constraints. The string possesses an infinite number of degrees of freedom. In quantum mechanics this leads to divergences. Elimination of divergences by standard methods developed in quantum field theory – introduction of normal ordering of non-commuting operators leaves a finite anomalous term (called also *central charge*) violating reparametrization invariance. The difficulties appear also in attempts to quantize the string theory in particular gauges. For example, in standard light cone gauge, Lorentz transformations are followed by reparametrizations of the world sheet, and quantum anomalies destroy the Lorentz-invariance of the theory.

During the last decades there were a number of investigations, in part contributed by the authors of this book, which show that *algebraic anomalies*, related with the violations of the quantum algebras, can be removed from the theory, opening a new layer of problems. They are, particularly, the presence of exotic solutions, complex topological structure of the phase spaces, and various violations of regularity in the quantum spectra which can be characterized by the term *spectral anomalies*. For instance, the quantities, which due to quasi-classical reasons should have integer-valued spectrum in the region of large quantum numbers, could have the deviations from this rule at small quantum numbers. In certain cases these violations can lead to critical consequences for the theory, e.g. the appearance of fractional spin; in other cases the spectral anomalies influence the quantities, for which the integer values are not so necessary, e.g. the square of mass. In any case, the spectral anomalies are softer than original algebraic ones, and they allow complete elimination with the proper choice of quantum definitions. The investigation of various approaches to self-consistent quantization of non-critical Nambu-Goto string theory is the subject of the second chapter of this book.

Before explaining the main idea of the construction, we remind that *canonical quantization* [2, 3] is a linear mapping $f \to \hat{f}$ of a pre-defined set of classical variables to a set of operators, satisfying the following correspondence principle: $i\hbar\{f, g\} \to [\hat{f}, \hat{g}]$. Here $\{,\}$ is Poisson bracket, $[,]$ is commutator, $\hbar$ is Planck's constant. Additionally it is required that real-valued dynamical variables should be represented by Hermitian operators, unity should be preserved: $1 \to \hat{1}$, the space of states should be positively defined and irreducible under the action of the constructed operators (does not have a smaller subspace invariant under their action).

It is shown in [3] that the correspondence principle cannot be satisfied for all dynamical variables in the theory. This property is related with ordering ambiguities for non-commuting operators. Indeed, let's consider two sets of classical variables, related by a certain non-linear transformation: $a_i \leftrightarrow b_i$. If the correspondence principle is satisfied for the set a_i: $\{a_i, a_j\} \to [\hat{a}_i, \hat{a}_j]$, then the commutator of operators $[\hat{b}_i(\hat{a}), \hat{b}_j(\hat{a})]$, represented as non-linear functions of $\hat{a}$, can contain ordering ambiguities, which create anomalous terms absent in Poisson bracket $\{b_i, b_j\}$. In spite of the fact that the contribution of such terms is suppressed by additional factor $\hbar$, their occurrence can lead to serious problems in quantum theory, especially if b_i represent the generators of symmetries of the system. On the other hand, if the correspondence principle is satisfied for the set b_i, the commutator $[\hat{b}_i, \hat{b}_j]$ is postulated directly from Poisson bracket $\{b_i, b_j\}$ and has no anomalies. The anomalies can appear in the commutator $[\hat{a}_i(\hat{b}), \hat{a}_j(\hat{b})]$. The common practice is to select a convenient basis of independent variables, where the quantization is performed, trying to keep all generators of symmetries being simple functions of independent variables and thus to avoid the occurrence of anomalies in their commutators.

To make this general consideration more concrete, let a_i represent the set of oscillator variables of string theory in the light cone gauge, and $b_i = (M_{\mu\nu}, \xi_i)$, where $M_{\mu\nu}$ are the generators of the Lorentz group and ξ_i is a set of variables complementing $M_{\mu\nu}$ to the full phase space. In selection of such a set, one needs to take care that ξ_i will have simple Poisson brackets with $M_{\mu\nu}$ and among themselves. Local existence of such variables is provided by Darboux theorem [106], while the determination of the global structure for their region of variation could be a complex task [70]. Further, performing the quantization of string theory in variables a_i, we obtain the anomaly in the commutator $[M_{\mu\nu}(a), M_{\rho\sigma}(a)]$. On the other hand, performing the quantization of string theory in variables b_i, we do not have the anomaly in $[M_{\mu\nu}, M_{\rho\sigma}]$. Anomalies can appear in the commutator $[a_i(b), a_j(b)]$. However, the point is that there is no more necessity to use the variables a_i in the theory, because the role of internal variables is now played by ξ_i, whose commutator does not have the anomalies. Also, if such a necessity would appear, the anomaly in $[a_i(b), a_j(b)]$ does not lead to further problems in the theory. In any case, it is not possible to remove the anomalies from all commutators, according to the theorem [3].

It becomes clear that the anomalies essentially depend on the choice of canonical variables in which the quantization is performed. This property has been previously discussed in [70]. Actually, this fact is well known and even presented in the textbooks on string theory, e.g. in [44] on p.157:

> "Should one not try to use a different representation of the string operators so as to avoid the central charge? Again, it might very well be possible to construct such a representation and, if so, it is very likely that the resulting quantum theory would be very different from the one explained here. It could be that this yet-to-be-constructed theory would possess an intrinsic interest of its own (e.g., through the occurrence of infinite-dimensional representations of the Lorentz algebra). Moreover, because this theory would not be based on the use of oscillator variables, it might be more easily extendable to higher-dimensional objects, such as the membrane."

Quantization of theories with gauge symmetries has own peculiarities. On the classical level such theories have Dirac's constraints L_n, which generate gauge symmetries via Poisson brackets. There are two main approaches for quantization of such theories. In the first one, called covariant quantization, the constraints are imposed on the state vectors $L_n\Psi = 0$, thus selecting gauge invariant wave functions in the space of states. If the algebra $[L_n, L_m]$ has the anomaly, it cannot be implemented by means of canonical formalism. But certain of its extensions can be used, such as an implementation only a half of constraints in the theory. In the second approach, the gauge fixing conditions are imposed in addition to the primary constraints, which select one representative from each gauge orbit. In this case the whole set of constraints belongs to the 2nd class in Dirac's terminology. The mechanics should be reduced to their surface by corresponding redefinition of Poisson brackets, and after that can be quantized. An approach closely related with gauge fixation is the reduced phase space formalism. In this formalism, the role of phase space is played by the factor-space with respect to the group of gauge symmetry. The factor-space contains gauge invariant variables only, that's why this formalism is often argued [48, 55, 56] to be most reasonable from the physical point of view, since the quantum effects are attributed only to the physically observable gauge invariants. The formalisms of gauge fixation and reduced phase space are completely equivalent, because the gauge fixation corresponds to a choice of basis in the space of gauge invariants. Indeed, in string theory the world sheet can be parameterized by a component of coordinate x_μ along some axis n_μ. The light-like vector n_μ corresponds to the light cone gauge, while $n_\mu = P_\mu/\sqrt{P^2}$, where P_μ is total momentum, is called Rohrlich's gauge. The oscillator variables are Fourier coefficients

$$a_k = \int d\sigma \, Q'(\sigma) e^{ik\sigma},$$

where $Q'(\sigma) = x'(\sigma) + p(\sigma)$ and σ is the light cone or Rohrlich's gauge parameter: $\sigma = \pi(nQ)/(nP)$. Obviously, the same variables can be written in the form of parametric invariants

$$a_k = \int dQ \, e^{ik\pi(nQ)/(nP)}.$$

For the light cone gauge, this formula gives the well known DDF variables [109]. In terms of these variables, the world sheet can be reconstructed (up to translations). Thus, the gauge fixation does not make the theory restricted or weak in some other sense, but provides the particular choice of basis in the space of gauge invariants. When n_μ is changed, the basis $a_k(n_\mu)$ is changed accordingly. As we already know, such a change influences the anomalies, which therefore become depending on the selected gauge.

Standard covariant quantization of string theory [44] is performed in the oscillator representation, similar to the representation of field theories by the creation and annihilation operators [107]. In this approach there is an infinite set of reparametrization generators L_n, $n \in \mathbf{Z}$, quadratic in the oscillator variables, which possess the property $L_n^* = L_{-n}$ and form, on the classical level, *the Virasoro algebra*: $\{L_k, L_n\} = i(k-n)L_{k+n}$. On the quantum level, this algebra acquires the anomaly term: $[L_k, L_n] = (n-k)L_{k+n} + \delta_{k,-n}c_k$, where $c_k = d/12 \cdot k(k^2 - 1)$. Note that the anomaly is present at any value of dimension d, therefore the imposition of constraints to the states: $L_n\Psi = 0$ for all n is never possible, because $[L_n, L_{-n}]\Psi = c_n\Psi \neq 0$, $|n| > 1$. On the other hand, the subset of constraints with $n \geq 0$ forms closed subalgebra and can be imposed on the states. Such a weak version of constraints imposition is taken from quantum electrodynamics [107], where it is used as a basis for *the Gupta-Bleuler's method*. In string theory, as well as in electrodynamics, the quantization is performed in pseudo-Hilbert space of states possessing *an indefinite metric*, i.e. the scalar product $\langle\Psi|\Psi\rangle$ in the space of states is not positively defined. As the explicit computation shows [44], the physical space of states, selected by the conditions $L_n\Psi = 0$, $n \geq 0$, is positively defined for $d < 26$, while for $d > 26$ this space contains vectors of negative norm. The value $d = 26$ corresponds to *critical* case, when the physical space is semi-definite and contains a subspace of so called *spurious* states, possessing zero norm and orthogonal to all vectors in the physical space. After a factorization of the physical space, identifying all spurious states with zero [109], the positively defined space of states is obtained, isomorphic to the result of string theory quantization in the light cone gauge [44] (which does not contain anomalies and is self-consistent only at $d = 26$). Similar factorizations have been also performed in quantum electrodynamics. An essential difference is that in quantum electrodynamics this procedure leads to a positively defined space for arbitrary d, and the system of constraints, which in quantum electrodynamics is free of anomalies, after factorization is satisfied in exact sense. Therefore, in quantum electrodynamics, the factorization of spurious states logically completes the quantization procedure and brings it in full agreement with Dirac's method. In string theory, the algebra of constraints L_n for all n is not closed even in the factor-space, and the theory is not quantized strictly by Dirac for all d.

Other constructions in quantum string theory have been also presented. The paper [32] demonstrates an experimentally observed spectrum of isospin 1 mesons superimposed with the prediction of string-inspired model of the hadrons constructed in this paper. The model is based on a single particular solution of the Nambu-Goto theory – the above described straight-line string. This solution is associated with a finite-dimensional Lorentz-invariant submanifold in the phase space of the full theory. The authors of [32] introduce convenient local coordinates in this submanifold, properly reduce full Hamiltonian mechanics to it and quantize the mechanics in these coordinates. The resulting quantum theory acts in the physical space-time of dimension $d = 4$ and does not have any anomalies. The spectrum of this mechanics contains a single Regge trajectory possessing the maximal slope among all possible motions of the string. Then the authors of [32] introduced the spins of quarks at the ends of the string, and obtained the splitting of the trajectories necessary to describe the experimentally observed mesonic spectrum. Further, in the paper [70] more general infinitely-dimensional submanifolds have been found, for which the same scheme works: the selected submanifolds are Lorentz-invariant. One can introduce in them the local coordinates possessing simple behavior under the action of the Lorentz group and quantize the theory in these coordinates without the anomalies, at the non-critical value of dimension of the space-time.

Is there any other evidence that anomaly-free quantization of string theory is possible beyond the critical dimension? There is a famous paper by Polyakov [111] which shows that in the frames of path integration formalism, the quantum Nambu-Goto string theory at $d = 26$ is equivalent to the collection of linear oscillators, while at $d \neq 26$ the theory exists as well, however it contains a non-linear field theory associated with so called *Liouville modes*, which makes explicit computations hard. The evidence of the second type are isolated examples scattered in the literature, e.g. one given in the textbook [44] pp.157-159, which explicitly presents quantum solutions of closed string theory in the class of non-oscillator representations possessing no anomaly in Virasoro algebra at the arbitrary even value of dimension. The third evidence has been provided in the paper by Rohrlich [104], which shows that one should use not the light cone gauge, but the above described timelike Lorentz-invariant gauge, related with total momentum P_μ, to remove gauge degrees of freedom from the theory. In this case the anomaly is also eliminated from the Lorentz algebra.

Are there any no-go theorems preventing anomaly-free quantization of non-critical string theory? If so, are there any ways to bypass them? Statements prohibiting quantization of non-critical string theory can be often found in the literature. Particularly, it is shown in [43] Vol.1 pp.162-165, that in a covariant formalism based on the Polyakov's string action using vacuum correlators estimated by Gaussian path integrals (essentially equivalent to the selection of Fock's vacuum), one cannot avoid the anomaly in Virasoro algebra. In [110] the anomaly in Virasoro algebra is reproduced using the technique of geometric quantization [3] at the choice of holomorphic polarization (again equivalent to Fock's representation). Further examples can be found, however, their pattern is the same: these examples prove the presence of an anomaly in *Virasoro algebra* and they are internally based on the usage of *Fock's representation* or the constructions equivalent to it. Thus there are basically two ways to bypass these theorems. The first one is to follow Rohrlich and consider the Lorentz-invariant gauges, which already at the classical level remove the reparametrization gauge degree of freedom from the theory, together with Virasoro algebra and anomaly contained in it. This is the class of theories we will consider in this book. The second way – usage of non-standard vacuums and non-oscillator representations is beyond of our scope, however, we give the references to the papers exploring this way: [44] (pp.154-159), [112, 80, 89, 113].

In this book we will not consider the interaction of quantum strings. In this area, one can find further no-go theorems [43] Vol.2, related particularly with the loss of unitarity and discrete symmetries in the theories of interacting strings at the non-critical values of dimension. These theorems are also bounded to Fock's representation, particular interaction scenarios and techniques of perturbation theory, and they can be bypassed as well. Positive examples of interacting non-critical strings were given in papers [32, 30].

Briefly speaking, our strategy is to consider the phase space of free open Nambu-Goto string theory, to factorize it explicitly by the action of the gauge group, i.e. the group of reparametrizations of the world sheets, to construct a convenient basis of local variables in the obtained reduced phase space, possessing good Lorentz-covariance properties, and to quantize the theory in these variables.

One should not wonder that non-critical string theory appears to be quite complex both algebraically and geometrically comparing with the critical case. That's why its investigation requires the usage of modern methods, and we apply for this purpose the techniques of computer modeling. Particularly, the methods of *computer visualization* acquire an increasing significance in such special scientific areas as theoretical and mathematical physics. The visualization of string dynamics has been earlier performed by A.Hanson et al [122, 123, 124]. Visualization methods have been used in [30] for the investigation of string breaking and in [33] for the description of Y-shaped string configurations. A combination of visualization and analytical methods has been used by the authors of this book for the classification of singularities on open strings [83], and also for the strings of other topological types [90]. Most of the lemmas, describing the structure of singularities, have been obtained in the form of hypotheses after the performed visualization of string dynamics, and then have been proven analytically. In works [92, 93, 94, 95, 96], the corresponding methods are described, as well as the software for visualization of relativistic string dynamics developed on their basis. The visualization becomes ultimately effective in virtual environment systems [96], based on personal computers and affordable projection equipment. Today many educational and scientific centers have become skilled in usage of such systems, both for research purposes and for the presentation of the results. These visualization methods have been also used for the preparation of an Internet course on string theory [88].

The most of the above considered problems in string theory require the investigation of polynomial systems, performed by the computation of a finite set of generators in the corresponding polynomial ideals, so called *Gröbner bases*. The technique of Gröbner bases is also used immediately for solution of polynomial systems, which e.g. in the case of overdetermined systems allows one to obtain the analytical solution. The necessity to solve such systems appears in the investigation of quantum string theory in certain of the above described approaches. For the determination of Gröbner bases effective algorithms exist [119], analogous to the reduction of the linear systems to triangular form. The fact that these algorithms are currently implemented in practically all computer algebra systems, such as *Mathematica* [120], allows one to accomplish the solution of the above mentioned complex problems.

In quantum mechanics, one often needs to investigate the matrix elements of linear operators in finite-dimensional subspaces of Hilbert space. It leads to a necessity of representation and algebraic operations with very large matrices. To give an idea of these sizes, we mention that in our typical problems the storage of one such matrix would require more than 500 Gb, if we would not apply special methods of compression described below. The operative memory of modern computers would be not sufficient to store such matrices. However, the presence of symmetries in quantum theories leads to the appearance of a multilevel block structure in such matrices, where on the lowest level the polynomial composition of the theory creates sparse fractal substructures. For such matrices, special compression methods are available [121], considerably reducing the usage of operative memory due to the storage of only non-zero elements of matrices and indexing structures for these elements. These methods also allow one to accelerate significantly the algebraic operations with sparse matrices. The description of these methods and the related algorithms for generation of quantum spectra and computation of matrix elements.

References

[1] B.A. Dubrovin, A.T. Fomenko, S.P. Novikov, *Modern Geometry: Methods and Applications, Vol.1-3*, Springer, (1995)

[2] P.A.M. Dirac, *Lectures on Quantum Mechanics*, New York, (1964)

[3] N.E. Hurt, *Geometric Quantization in Action*, D. Reidel, (1982)

[4] K.G.Wilson, Confinement of quarks, *Phys.Rev.D*, V.10, N8, pp.2445-2459 (1974)

[5] L.D. Faddeev, A. A. Slavnov, *Gauge Fields, Introduction to Quantum Theory*, Addison-Wesley, (1991)

[6] M. Bande, *Phys. Rep.*, V.75, N4, p.207 (1981)

[7] H.G.Dosch, Yu.A.Simonov, The area law of the Wilson loop and vacuum field correlators, *Phys.Lett.B*, V.205, N4, pp.339-349 (1988)

[8] Yu.A.Simonov, Vacuum background fields in QCD as a source of confinement, *Nucl.Phys.B*, V.307, N3, pp.512-528 (1988)

[9] K.S.Gupta, C.Rosenzweig, Semiclassical decay of excited string states on leading Regge trajectories, *Phys.Rev.D*, V.50, N5, pp.3368-3376 (1994)

[10] I.Yu.Kobzarev, B.V.Martemjanov, M.G.Shepkin, Decays of orbitally exited hadrons, *Sov.J.Nucl.Phys.*, V.48, N2, pp.344-355 (1988)

[11] L.B.Okun, *Physics of elementary particles*, Moscow, Nauka, (1988)

[12] L.G. Landsberg, *Exotic Hadrons*. Preprint IHEP 89-54, Serpukhov, (1989)

[13] L.G. Landsberg, Exotic baryons, *Uspekhi Fiz. Nauk*, V.164, N11, pp.1129-1165 (1994)

[14] V.V. Anisovich, Exotic mesons: search of glueballs, *Uspekhi Fiz. Nauk*, V.165, N11, pp.1225-1247 (1995)

[15] D.V.Bugg, M.Peardon, B.S.Zou, The glueball spectrum, *Phys.Lett.B*, V.486, N1, pp.49-53 (2000)

[16] A.I. Weinshtein et al, *Uspekhi Fiz. Nauk*, V.123, p.217 (1977)

[17] Y. Nambu, Quark model and factorization of the Veneziano amplitude, in *Lectures at the Copenhagen Symp. on Symmetries and Quark Models*, New York: Gordon and Breach Book Comp., p.269 (1970)

[18] O. Hara, *Prog. Theor. Phys.*, V.46, p.1549 (1971)

[19] T. Goto, *Prog. Theor. Phys.*, V.46, p.1560 (1971)

[20] R. Dolen, D. Horn, C. Schmidt, *Phys. Rev. Lett.*, V.19, N7, p.402 (1967)

[21] G. Veneziano, *Il Nuovo Cimento A*, V.57, N1, p.190 (1968)

[22] S. Mandelstam, *Phys. Rep.*, V.13, N6, p.259 (1974)

[23] B.M. Barbashov, N.A. Chernikov, *Comm. Math. Phys.*, V.5, p.313 (1966)

[24] Y.Nambu, QCD and the string model, *Phys.Lett.B*, V.80, N4-5, pp.372-376 (1979)

[25] A.Yu.Dubin, A.B.Kaidalov, Yu.A.Simonov, Dynamical regimes of the QCD string with quarks, *Phys.Lett.B,* V.323, N1, pp.41-45 (1994)

[26] M.Baker, R.Steinke, Effective string theory of vortices and Regge trajectories, *Phys.Rev.D,* V.63, N9, p.094013 (2001)

[27] Y. Nambu, *Phys. Rev. D,* V.10, N.12, p.4262 (1974)

[28] G.P. Pron'ko, *Nucl. Phys. B,* V.165, p.269 (1980)

[29] A.A. Migdal, *Nucl. Phys. B,* V.189, p.253 (1981)

[30] X. Artru, *Phys. Rep.,* V.97, p.147 (1983)

[31] E.V. Gedalin, E.G. Gurvich, *Sov.J.Nucl.Phys.* V.52, p.240 (1990)

[32] E.B. Berdnikov, G.G. Nanobashvili, G.P. Pron'ko, *Int. J. Mod. Phys. A*, V.8, N14, p.2447 (1993); V.8, N15, p.2551 (1993)

[33] M.S. Plyushchay, G.P. Pron'ko and A.V. Razumov, *Theor. Math. Phys.* **63**, 389 (1985); S.V. Klimenko, V.N. Kochin, M.S. Plyushchay, G.P. Pron'ko, A.V. Razumov and A.V. Samarin, *Theor. Math. Phys.* **64**, 810 (1986); M.S. Plyushchay, G.P. Pron'ko and A.V. Razumov, *Theor. Math. Phys.* **67**, 576 (1986).

[34] G.S. Sharov, *Phys. Rev. D,* V.62, N.9, p.094015 (2000)

[35] B.M. Barbashov, V.V. Nesterenko, *Introduction to the Relativistic String Theory*, Singapore, World Scientific, (1990)

[36] T.Regge, Introduction to complex orbital momenta, *Il Nuovo Cimento*, V.14, N5, pp.951-960 (1959)

[37] P.Goddard, J.Goldstone, C.Rebbi, C.B.Thorn, *Nucl.Phys.B*, V.56, p.109 (1973)

[38] R.C. Brouwer, *Phys. Rev. D*, V.6, p.1655 (1972)

[39] P. Goddard, C.B. Thorn, *Phys. Lett. B,* V.40, p.235 (1972)

[40] J. Scherk, *Rev. Mod. Phys.,* V.47, p.123 (1975)

[41] P. Ramond, *Phys. Rev. D,* V.3, N9, p.2415 (1971)

[42] A. Neveu, J.H. Schwarz, Nucl. Phys. B, V.312, N1, p.86 (1971); *Phys. Rev. D*, V.4, N4, p.1109 (1971)

[43] M. Green, J. Schwarz, E. Witten, *Superstring Theory Vol. 1,2*, Cambridge Univ. Press, (1987)

[44] L. Brink, M. Henneaux, *Principles of String Theory*, Plenum Press, New York and London, (1988)

[45] M. Kaku, *Introduction to superstrings*, Springer, (1988)

[46] M. Henneaux, BRST Symmetry in the Classical and Quantum Theories of Gauge Systems, pp.117-144, in: *Quantum mechanics of fundamental systems - 1*, (C.Teitelboim, ed.), Plenum Press, New York (1988)

[47] M. Henneaux, *Phys.Rev.Lett.* V.55, p.769 (1985)

[48] M. Henneaux, *Phys.Rep.* V.126, p.1 (1985)

[49] K. Fujikawa, *Phys.Rev. D,* V.25, p.2584 (1982)

[50] M. Kato, K. Ogawa, *Nucl.Phys. B*, V.212, p.223 (1983)

[51] S. Hwang, *Phys.Rev. D*, V.28 2614 (1983)

[52] E.S. Fradkin, G.A. Vilkovisky, *Phys.Lett. B*, V.55, p.224 (1975)

[53] I.A. Batalin, G.A. Vilkovisky, *Phys.Lett. B,* V.69, p.309 (1977)

[54] E.S. Fradkin, T.E. Fradkina, *Phys.Lett. B*, V.72, p.343 (1978)

[55] L.D. Faddeev, *Theor. Math. Phys*. V1, p.3 (1969)

[56] E.S. Fradkin, G.A. Vilkovisky, CERN Report TH-2332 (1977)

[57] Sherk J., Schwarz J.H. // *Nucl. Phys. B*, V.81, p.118 (1974)

[58] Neveu A., Scherk J. // *Nucl. Phys. B*, V.36, p.155 (1972)

[59] Yoneya T. // *Nuovo Cim. Lett.*, V.8, p.951 (1973); *Prog. Theor. Phys.*, V.51, p.1907 (1974)

[60] Gross D.J. et al // Nucl. Phys. B, V.256, p.253 (1985); *Nucl. Phys. B*, V.267, p.75 (1986)

[61] An introduction to Kaluza-Klein theories, *Proc. Chalk River workshop on Kaluza-Klein theories,* ed. H. C. Lee, World Scientic, Singapore, (1984)

[62] A. M. Polyakov, *Phys. Lett.,* **B59**, p.82 (1975)

[63] G. t'Hooft, *Phys. Rev. Lett.*, **37**, p.8 (1976)

[64] S.V. Ketov, Introduction to Quantum Theory of Strings and Superstrings, Novosibirsk, *Nauka*, (1990)

[65] A.A. Zheltuhin, *Sov.J.Nucl.Phys.* V.34, p.562, (1981)

[66] L.D. Landau, E.M. Lifshitz, *The Classical Theory of Fields,* Elsevier Science, (1976)

[67] L.D. Landau, E.M. Lifshitz, Quantum Mechanics: Non-Relativistic Theory, Butterworth-Heinemann, (1981)

[68] R.P.Feynman, *The Development of the Space-Time View of Quantum Electrodynamics*, Nobel Lecture, December 11, 1965. Preprint les Prix Nobel en 1965. The Nobel Foundation, Stockholm, (1966)

[69] G.S.Sharov, *Excitations of rotational states in string model of meson and in three-string baryon model*, hep-ph/0012334.

[70] G.P. Pron'ko, *Rev. Math. Phys.*, V.2, N.3, p.355 (1990)

[71] G.P. Pron'ko, *Theor. Math. Phys.*, V.59, N.2, p.240 (1984)

[72] E.B. Berdnikov, G.P. Pron'ko, *Theor. Math. Phys.*, V.81, N.1, 94 (1989)

[73] V.S. Vladimirov, *Equations of mathematical physics, ed.5*, Moscow, Nauka, (1988)

[74] G.P. Pron'ko et al, *Particles and Nuclei J.*, V.14, N3, p.558 (1983)

[75] S. Brundobler, V. Elser, *Am.J.Phys.*, V.60, N8, p.726 (1992)

[76] R. Dilão, R. Schiappa, *Phys.Lett. B*, V.404, p.57 (1997)

[77] G.S. Sharov, V.P. Petrov, *Theor. Math. Phys.*, V.109, p.187 (1996)

[78] I.N. Nikitin, *String configurations with anomaly-free quantization*, *Sov.J.Nucl.Phys.*, V.56, N9, p.1283, (1993)

[79] I.N. Nikitin, G.P. Pronko, *Electromagnetic interaction in the theory of straight line string*, *Sov.J.Nucl.Phys.*, V.58, N6, p.1123, (1995)

[80] I.N. Nikitin, *Quantum string theory in indefinite space of states*, *Theor.Math.Phys.*, V.107, N2, p.589, (1996)

[81] I.N. Nikitin, *Particular types of string motion with anomaly-free quantization*, *Theor.Math.Phys.*, V.109, N2, p.1400, (1996)

[82] I.N.Nikitin, Classical and quantum aspects of low dimensional relativistic string theory, *Proc. of 10th Int. Conf. on Problems of Quantum Field Theory*, pub. JINR, Dubna, Russia, p.355 (1996)

[83] S.V. Klimenko, I.N. Nikitin, Singularities on world sheets of open relativistic strings, *Theor.Math.Phys.*, V.114, num.3, pp.299-312, (1998)

[84] S.V. Klimenko, I.N. Nikitin, Exotic solutions in string theory, *Il Nuovo Cimento A*, V.111, pp. 1431-1456 (1998)

[85] I.N. Nikitin, *String theory in Lorentz-invariant light cone gauge*, hep-th/9906003.

[86] I.N. Nikitin, L.D. Nikitina, String theory in Lorentz-invariant light cone gauge - II, hep-th/0301204; I.Nikitin, Dirac quantization of Nambu-Goto open string theory in 4-dimensional Minkowski space-time, *Particles and Nuclei J.*, V.36, N1, p.51, (2005)

[87] I.N. Nikitin, L.D. Nikitina, String theory in Lorentz-invariant light cone gauge - III, hep-th/0306010; I.Nikitin, *Quantization of non-critical bosonic open string theory*, chapter in: New Developments in String Theory Research, Nova Science, (2005)

[88] I.N. Nikitin, *Introduction to string theory*, Internet course, <http://sim.ol.ru/~nikitin/course>

[89] I.N. Nikitin, *String theory in Lorentz-invariant time-like gauge*, hep-th/9907196.

[90] I.N. Nikitin, Structure of singularities on the world sheets of relativistic strings, *Particles and Nuclei J.*, V.34, N7, p.112, (2003); S.Klimenko and I.Nikitin, in: Concise Encyclopedia of Supersymmetry and noncommutative structures in mathematics and physics, Kluwer Academic Publishers, Dordrecht, (2003)

[91] I.N. Nikitin, L.D. Nikitina, S.V. Klimenko, *Classification of singularities on the world sheets of relativistic strings*, in Proc. of XXV Workshop on the Fundamental Problems of High Energy Physics and Quantum Field Theory, IHEP, Protvino (2002); hep-th/0110042.

[92] S.V. Klimenko, I.N. Nikitin, V.V. Talanov, Visualization of singularities on the world sheets of relativistic string, *Programming and Computer Software*, V.4, p.47, (1994)

[93] S.V. Klimenko, I.N. Nikitin, V.V. Talanov, Visualization of complex phenomena in string theory, in *Proc. of Artificial Intelligence in High Energy and Nuclear Physics conference*, Pisa, Italy, April, (1995)

[94] Stanislav Klimenko, Igor Nikitin, Valery Burkin, Vitaly Semenov, Oleg Tarlapan and Hans Hagen, Visualization in string theory, *Computers and Graphics*, V.24, N1, pp. 23-30 (2000)

[95] S.Klimenko, J.Luca, I.Nikitin, L.Nikitina, A.Straube, W.Urazmetov, Mathematical modeling and visualization of non-linear dynamical systems, *Proc. of 3d Int. Workshop VEonPC'2003*, Moscow, p.90, (2003)

[96] P.Brusentsev, M.Foursa, P.Frolov, S.Klimenko, S.Matveyev, I.Nikitin and L.Nikitina, Virtual Environment Laboratories Based on Personal Computers: Principles and Applications, *Proc. of 2nd Int. Workshop VEonPC'2002*, Protvino, pub. ICPT, pp.6-14 (2002)

[97] V.I. Arnold et al, *Singularities of differentiable maps - I*, Birkhäuser, Basel, (1985)

[98] M.A. Lavrentiev, B.V. Shabat, *The methods of theory of functions of complex variable*, Moscow, Nauka, (1973)

[99] R. Thom, G. Levin, *Singularities of differentiable maps*, in: Singularities of Smooth Maps, Gordon (1967).

[100] E. Noether, Göttinger Nachrichten, pp.235-257 (1918)

[101] G.P. Jorjadze et al, *Theor. Math. Phys.*, V.40, p.221, (1979)

[102] A.K. Pogrebkov, *Theor. Math. Phys.*, V.45, p.161, (1980)

[103] M. Alcubierre, *Classical and Quantum Gravity*, V.11, L73, (1994)

[104] F. Rohrlich, *Phys.Rev.Lett.*, V.34, p.842 (1975)

[105] M.S. Plyushchay and A.V. Razumov, *Int. J. Mod. Phys. A*, V.11, p.1427 (1996)

[106] V.I. Arnold, *Mathematical methods of classical mechanics*, Springer-Verlag, (1989)

[107] N.N. Bogoljubov, D.V. Shirkov, *Introduction to Theory of Quantized Fields*, Moscow, Nauka, (1984)

[108] L.D. Faddeev, S.L. Shatashvili, *Phys. Lett. B,* V.167, p.225, (1986)

[109] E. Del Giudice, P. Di Vecchia and A. Fubini, *Ann.Phys.*, V.70. p.378 (1972)

[110] M.J. Bowick, S.G. Rajeev, *The Complex Geometry of String Theory and Loop Space*, in Proc. of 11th John Hopkins Workshop on Current Problems in Particle Theory, Lanzhou, China, June 17-19, 1987, eds. G. Domokos, Duan Yi-Shi and S. Kovesi-Domokos, World Scientific, Singapore, (1987)

[111] A.M. Polyakov, *Phys. Lett. B*, V.103, p.207 (1981)

[112] E. Ramos, *The reduced covariant phase space quantization of the three dimensional Nambu-Goto string*, hep-th/9709131.

[113] S.N. Vergeles, *Two approaches to anomaly-free quantization of generally covaraint systems on an example of a two dimensional string*, hep-th/9906024.

[114] G. C. Wick, *Phys. Rev.*, V.96, p.1124 (1954)

[115] V. I. Borodulin, O. L. Zorin, G. P. Pron'ko, A. V. Razumov and L. D. Solov'ev, *Theor. Math. Phys.*, V.65, p.1050 (1986).

[116] G.P. Jorjadze, Constrained Quantization on Symplectic Manifolds and Quantum Distribution Functions, *J.Math.Phys.,* V.38, pp.2851-2879 (1997)

[117] V.N. Gribov, *Nucl. Phys.* **B139**, p.1 (1978)

[118] D. Cox, J. Little, D. O'Shea, *Ideals, Varieties, and Algorithms*, Springer, (1996)

[119] B.Buchberger, *Groebner bases: an algorithmic method in polynomial ideal theory*, in: Multidimensional Systems Theory, ed. by N.K.Bose, D.Reidel Pub., Dordrecht, pp.184-232 (1985)

[120] S. Wolfram, *The Mathematica Book*, Cambridge University Press, (1999)

[121] W.H. Press, S.A. Teukolsky, W.T. Vetterling, B.P. Flannery, *Numerical Recipes in C*, Cambridge University Press, (1988)

[122] I. Bars and A.J. Hanson. Quarks at the Ends of the String, *Physical Review* **D13**, pp.1744-1760 (1976)

[123] W.A. Bardeen, I. Bars, A.J. Hanson, and R.D. Peccei. Study of the Longitudinal Kink Modes of the String, *Physical Review* **D13**, pp.2364-2382 (1976)

[124] W.A. Bardeen, I. Bars, A.J. Hanson, and R.D. Peccei. Quantum Poincare Covariance of the Two-Dimensional String, *Physical Review* **D14**, pp.2193-2196 (1976)

[125] G.K. Francis, *A Topological Picturebook*, Springer-Verlag, (1988)

[126] U. Dierkes et al, *Minimal Surfaces I*, Springer-Verlag, (1991)

[127] M. Göbel et al, *Virtual Spaces: VR Projection System Technologies and Applications*. Tutorial Notes. Eurographics'97, Budapest, p.75 (1997)

[128] J.Rohlf and J.Helman. *IRIS Performer: A High Perfomance Multiprocessing Toolkit for Real Time 3D Graphic*. In A. Glassner, editor, Proceedings of SIGGRAPH '94, pp.381-395 (1994)

[129] R. Carey and G. Bell. *The VRML 2.0 Annotated Reference Manual*. Addison-Wesley, Reading, MA, USA, Jan, (1997)

[130] Josie Wernecke, *Open Inventor Architecture Group, The Inventor Mentor: Programming Object-Oriented 3D Graphics with Open Inventor, Release 2*, Addison-Wesley, Reading, Massachusetts, (1994);
Josie Wernecke, *Open Inventor Architecture Group, The Inventor Toolmaker: Extending Open Inventor, Release 2*, Addison-Wesley, Reading, Massachusetts, (1994)

[131] J.N. Mather, *Uspekhi Mat. Nauk* V.29, p.99, (1974)

[132] V.I. Arnold, *Uspekhi Mat. Nauk* V.27, p.119, (1972)

[133] A. Gray, *Monkey Saddle*. Modern Differential Geometry of Curves and Surfaces with Mathematica, 2nd ed. Boca Raton, FL: CRC Press.

[134] L.D. Kudryavcev, *Mathematical analysis course - III*, Moscow, (1988)

[135] G.N. Watson, *A Treatise on the Theory of Bessel Functions*, 2nd ed. Cambridge, England: Cambridge University Press, (1966)

[136] H. Weyl, *The theory of groups and quantum mechanics*, New York: Dover, (1950)

[137] H.B.G. Casimir, *Rotation of a Rigid Body in Quantum Mechanics*, J.B.Wolter's, Groningen, (1931)

[138] E. Wigner, *Group Theory*, Academic Press, New York, (1959)

[139] M. Najmark, *Theory of Group Representations*, Springer, (1982)

[140] N.Ya. Vilenkin, *Special Functions and the Theory of Group Representations*, AMS, Providence, (1968)

[141] L. Schulman, *Phys. Rev.*, V.176, N.5, pp.1558-1569 (1968)

[142] J.J. van Wijk, *Spot noise: texture synthesis for data visualization*, in: T.W. Sederberg, editor, Computer Graphics (Siggraph'91 Proceedings), V.25, pp.263-272 (1991)

[143] R.Davies, NewMat C++ Matrix Class,
<http://www.robertnz.net/nm_intro.htm>

[144] I. Gohberg, P. Lancaster, L. Rodman, *Matrix Polynomials*, Academic Press, (1982)

[145] J.H. Wilkinson, C. Reinsch, *Linear Algebra, vol.2 of Handbook for Automatic Computations*, New York: Springer Verlag (1971)

In: String Theory Research Progress
Editor: Ferenc N. Balogh, pp. 155-177

ISBN 1-60456-075-6
© 2008 Nova Science Publishers, Inc.

Chapter 4

STRING THEORY WITH A CONTINUOUS SPIN SPECTRUM

J. Mourad

Laboratoire de Physique Théorique,[*] Bât. 210, Université Paris XI,
91405 Orsay Cedex, France
and
Laboratoire Astro Particules et Cosmologie,[†] Université Paris VII,
2 place Jussieu - 75251 Paris Cedex 05, France

Abstract

A classical action is proposed which upon quantisation yields massless particles belonging to the continuous spin representation of the Poincaré group. The string generalisation of the action is shown to be given by a tensionless extrinsic curvature action. It shares with the classical string action the two-dimensional local reparametrization and Weyl symmetries as well as the global target space Poincaré invariance. It also introduces a mass scale similar to the string tension. It differs however in two important aspects: it is non-polynomial and it is a higher derivative action. We analyse the constraints that arise from the string action and show that they are are bilocal. The continuous spin string action is then quantized in the BRST formalism. We show that, in the critical dimension 28, the spectrum is ghost free: the vacuum carries a continuous spin representation of the Poincaré group and the higher level states are all of zero norm. Our results prove the consistency of the free theory opening the possibility to obtain continuous spin particles from a string theory. We also consider the generalisation to a string action with world-sheet supersymmetry and show that the resulting critical dimension is 12.

1. Introduction

Among the irreducible representations of the Poincaré group [1], the massless continuous spin representation [1]- [8] is very peculiar. On the one hand, contrary to the other representations, it does not seem to describe any physical system, but on the other hand it seems

---

[*]Unité Mixte de Recherche du CNRS (UMR 8627).
[†]Unité Mixte de Recherche du CNRS (UMR 7164).

to be realised by a sector of a string theory with zero tension. The latter is the subject of this chapter. It is described by the action

$$S[X^\mu, h_{mn}] = \mu \int d^2\sigma \sqrt{-h} \sqrt{\Box X^\mu \Box X^\nu \eta_{\mu\nu}}. \qquad (1)$$

It shares with the classical string action the two-dimensional local reparametrization and Weyl symmetries as well as the global target space Poincaré invariance. It also introduces a mass scale μ similar to the string tension. It differs however in two important aspects: it is non-polynomial and it is a higher derivative action. The action (1) was first considered by Savvidy in [11–13] where it was conjectured to play a role in the tensionless limit of the Nambu-Goto-Polyakov action [17–19], it can be also motivated as a generalization of the point particle action [24] describing a massless particle belonging to the continuous spin representations of the Poincaré group [1,2,4,5,7,8]. The quantization of the action (1), due to its non-polynomiality and higher derivative nature, is not straightforward. These two difficulties can be overcome if one introduces an auxiliary field, called Ξ^μ in the following and a Lagrange multiplier. The action becomes second order and polynomial. More explicitly the classically equivalent action is

$$S[X^\mu, \Xi^\mu, h_{mn}, \lambda] = -\mu \int d^2\sigma \sqrt{-h} \left[h^{mn} \partial_m X^\mu \partial_n \Xi^\nu \eta_{\mu\nu} + \lambda(\Xi^2 - 1) \right]. \qquad (2)$$

In turn, this action presents some difficulties: its kinetic term has signature (D, D) signalling potential ghosts and the constraints that arise from this action are bilocal constraints. It was shown in [24] that one can transform these bilocal constraints to local one at the expense of breaking manifest Lorentz covariance. The quantization can then be performed with the standard BRST techniques [14–16]. It was initiated in [24] where it was found that the critical dimension of the theory is 28. It this article, we pursue this quantization in more details and find that the ground state carries a continuous spin representation of the Poincaré group and that the BRST-closed higher level states are all of zero norm. This implies that the only physical state is the ground state. In this respect it has some similarities with the N=2 superstring [23] with a finite number of physical states. It differs however in that the signature of spacetime is $(1, D-1)$. Although the manifest Lorentz covariance was broken to transform the constraints, the resulting spectrum is relativistic. The absence of ghosts in the spectrum, in spite of the (D, D) signature of the sigma model (2), is due to the additional constraints contained in (2) and proves the consistency of the free theory.

This theory may prove to be useful to explore some properties of string theory. The states being all massless, it may be relevant to some very high energy regime of string theory. This remains to be explored. It is also of crucial importance to study interactions and to elucidate the issue of unitarity in the resulting amplitudes. In this respect let us note that consistent interactions of particles belonging to the massless continuous spin representations are not known. An interesting possibility would be that string theory is needed in order to get consistent interactions. The theory may also be considered as a simple toy model for the quantization of systems with higher derivatives or bilocal constraints.

The plan of this chapter is the following. Section 2 is devoted to the continuous spin representations of the Poincaré group in arbitrary spacetime dimensions. It contains the Wigner equations whose solutions span this irreducible representation. In Section 3, we show how to motivate the action 1 from the action of a point particle which upon quantisation leads to massless particles belonging to the continuous spin representation. In Section 4 we derive the resulting constraints. We show that the constraints are bilocal. They can be transformed to local ones for the case of the open string. The resulting spectrum is shown, in Section 5, to be given by massless standard helicity states. For the case of the closed string, the subject of Section 6, we show that the bilocal constraints can be transformed to local ones at the expense of breaking manifest Lorentz invariance. We then perform a BRST quantisation [14, 16] and show that the nilpotency of the BRST charge gives a critical dimension of 28. Section 7 is devoted to the BRST quantization where we calculate the critical dimension and the normal ordering constant. The latter is crucial for the ground state to belong to the continuous spin representation. In Section 8 we show that the higher level BRST-closed states are all of zero norm. The proof that the spectrum is ghost-free relies heavily on the requirement that the zero modes of the antighosts annihilate the physical states. In Section 9 we briefly sketch the generalisation to the world-sheet supersymmetric case and derive the corresponding critical dimension. Finally, Section 10 contains the conclusions.

2. The Continuous Spin Representation

Consider a massless particle in D-dimensional spacetime[1]. Let its momentum have zero components except for p^+ ($V^\pm = \frac{1}{\sqrt{2}}(V^0 \pm V^{D-1})$, $V^+ = -V_-$). The little group leaving the momentum invariant is generated by M_{ij} and $M_{+i} = \pi_i$, they verify the Lie algebra

$$[\pi_i, \pi_j] = 0, \quad [M_{ij}, M_{kl}] = i(\delta_{jk}M_{il} - \delta_{ik}M_{jl} - \delta_{jl}M_{ik} + \delta_{il}M_{jk})$$
$$[\pi_i, M_{kl}] = i(\delta_{ik}\pi_l - \delta_{il}\pi_k), \tag{3}$$

which is the Lie algebra of the $D - 2$ dimensional Euclidean group E_{D-2}. The Casimir $\pi^i\pi_i = \mu^2$ classifies the representations of E_{D-2}. If μ vanishes then the irreducible representations are given by those of $SO(D-2)$, these are the helicity states. When μ is nonvanishing we get the continuous spin representations. They are of infinite dimension and are determined by the UIR of the subgroup leaving a given π^i invariant, the short little group, $SO(D-3)$.

In fact, the second Casimir operator of the Poincaré group, is given by

$$W = -\frac{1}{2}p^2 M_{\mu\nu}M^{\mu\nu} + M_{\mu\alpha}p^\alpha M^{\mu\beta}p_\beta. \tag{4}$$

It reduces to μ^2 for the massless particle and to $m^2 s(s + D - 3)$ for the massive one. Here s corresponds to the rank of the traceless completely symmetric tensor in D spacetime dimensions.

---

[1] Our conventions are as follows: Greek indices such as $\mu, \nu, \ldots$ denote spacetime indices running from 0 to $D - 1$ while Latin indices such as $i, j, \ldots$ denote transverse indices running from 1 to $D - 2$. The Minkowski metric is mostly plus and reads in light-cone coordinates: $ds^2 = -2dx^+ dx^- + dx^i dx_i$. Dots denote contraction of (implicit) spacetime indices.

2.1. Wigner's Wave Equation

A wave equation whose physical content is the single valued continuous spin representation was proposed by Wigner [2]. The wave function depends on two vectors: the momentum p and the additional vector ξ which is dimensionless. There are two independent equations obeyed by the wave function $\overline{\Psi}(p,\xi)$ and two other which are consequences of these. They read

$$\mathcal{E}_1 \overline{\Psi} \equiv p \cdot \frac{\partial \overline{\Psi}}{\partial \xi} - i\mu \overline{\Psi} = 0, \tag{5}$$

$$\mathcal{E}_2 \overline{\Psi} \equiv (\xi^2 - 1)\overline{\Psi} = 0. \tag{6}$$

The first compatibility condition reads

$$[\mathcal{E}_1, \mathcal{E}_2]\overline{\Psi} \equiv 2\mathcal{E}_3 \overline{\Psi} = 2 p \cdot \xi \, \overline{\Psi} = 0, \tag{7}$$

and the second compatibility condition

$$[\mathcal{E}_1, \mathcal{E}_3]\overline{\Psi} \equiv \mathcal{E}_4 \overline{\Psi} = p^2 \overline{\Psi} = 0 \tag{8}$$

is the mass-shell constraint. There are no more compatibility conditions. These equations can be obtained as the first class constraints arising from a higher derivative classical action [7].

The equation (5) reflects the fact that the couples (p,ξ) and $(p,\xi + \alpha p)$ are physically equivalent for arbitrary $\alpha \in \mathbb{R}$. Indeed, one gets

$$\overline{\Psi}(p, \xi + \alpha p) = e^{i\alpha\mu} \overline{\Psi}(p, \xi) \tag{9}$$

from Equation (5). The equation (6) states that the internal vector ξ is a unit space-like vector while the mass-shell condition (8) states that the momentum is light-like. From the equation (7), one obtains that the internal vector is transverse to the momentum. All together, one finds that ξ lives on the unit hypersphere S^{D-3} of the transverse hyperplane $\mathbb{R}^{D-2}$. In brief, the "continuous spin" degrees of freedom essentially correspond to $D-3$ angular variables, whose Fourier conjugates are discrete variables analogous to the usual spin degrees of freedom.

2.2. The Fourier Transformed Wave Equation

In fact, it is useful for later purposes to write the equations (5) - (8) in terms of w, the Fourier conjugate to ξ. The equations now read

$$(p \cdot w + \mu) \Psi = 0, \tag{10}$$

$$\left(\frac{\partial}{\partial w} \cdot \frac{\partial}{\partial w} + 1 \right) \Psi = 0, \tag{11}$$

$$\left(p \cdot \frac{\partial}{\partial w} \right) \Psi = 0, \tag{12}$$

$$p^2 \Psi = 0. \tag{13}$$

In order to explicit the physical content of the equations, let us consider a plane wave,

$$\Psi(p, w) = \delta(p - p_0)\,\psi_{p_0}(w), \tag{14}$$

with $p_0^2 = 0$. Suppose that the only non-vanishing component of p_0^μ is p_0^+, then Equation (10) implies that

$$\psi_{p_0}(w) = \delta(w^- p_0^+ - \mu)\,\phi(w^+, w^i), \tag{15}$$

where w^i are the transverse coordinates. Equation (12) implies that ϕ does not depend on w^+ and, finally, Equation (11) becomes the Helmholtz equation

$$\frac{\partial^2 \phi}{\partial w^i \partial w_i} + \phi = 0. \tag{16}$$

There are several formal ways to write the solutions to the equation (16). A first way is to expand $\phi(w^i)$ in powers of w:

$$\phi(w^i) = \sum_{n=0}^{\infty} \frac{1}{n!} \phi_{i_1 \ldots i_r} w^{i_1} \ldots w^{i_n}, \tag{17}$$

with symmetric coefficients obeying

$$\phi_{i_1 \ldots i_n} = -\phi^j{}_{j i_1 \ldots i_n}. \tag{18}$$

A second way is to to expand in spherical harmonics as

$$\phi(w^i) = \sum_{n=0}^{\infty} \frac{f_n(r)}{n!} f_{i_1 \ldots i_n} \hat{w}^{i_1} \ldots \hat{w}^{i_n}, \tag{19}$$

where $r^2 = w_i w^i$, and $\hat{w}^i = w^i/r$ are the coordinates on the sphere S^{D-3}. In equation (19), the constant tensors $f_{i_1 \ldots i_n}$ are totally symmetric and traceless. The function f_n verifies the differential equation

$$f_n'' + \frac{D-3}{r} f_n' - \frac{n(n+D-4)}{r^2} f_n + f_n = 0, \tag{20}$$

which results from (16). The solution which is regular at $r = 0$ to equation (20) is given by

$$f_n(r) = r^{2 - \frac{D}{2}} J_{n + \frac{D}{2} - 2}(r), \tag{21}$$

where J_ν is the Bessel function of the first kind. Notice that each term in the expansion (19) is by itself a solution to the Helmoltz equation. This was not the case of the first expansion.

In both expansions, one gets totally symmetric tensors that one is tempted to compare with the fields appearing in the description of a massive higher-spin particle. The first expansion (17) turns out to be the one which will allow to make contact with the massive case. Rougly speaking, the point is that the physical components of a spin-s massive symmetric field correspond to a spin-s irreducible representation of the massive little group $SO(D-1)$, i.e. a rank-s traceless symmetric $D-1$ tensor $\phi_{I_1 \ldots I_s}$ ($I_k = 1, \ldots, D-1$), which decomposes as a tower of totally symmetric $D-2$ tensor $\phi_{i_1 \ldots i_r}$ of rank r running from zero to s and satisfying precisely (18). The second expansion (19) has the merit of exhibiting the physical content of the Wigner equations: the general solution is given by a sum of plane waves with functions over the (internal) hypersphere S^{D-3} as coefficients. We stress that the continuous spin wave function Ψ has a number of "components" which is infinite but *countable*. For a nonvaishing μ, a Lorentz transformation not belonging to the little group mixes the tensors of different ranks. The mixing disappears when μ vanishes and in this limit we get an infinite sum over all helicity states represented above by the tensors $f_{i_1...i_n}$.

The Hilbert space of functions on S^{D-3} carries the UIR of the massless little group E_{D-2} with a trivial representation of the short little group $SO(D-3)$. The cases with arbitrary irreducible representations of the latter are considered in [10].

3. Classical Action

We look for a classical action with dynamical variables x^μ and ξ^μ and conjugate momenta p_μ and q_μ which gives the constraints

$$\phi_1 = p.q - \mu = 0, \quad \phi_2 = \xi^2 - 1 = 0, \quad \phi_3 = \xi.p = 0, \quad \phi_4 = p^2 = 0. \tag{22}$$

Furthermore we require that the action be invariant under reparametrisation. In order to achieve that we introduce a one dimensional metric $ds^2 = e^2 d\tau^2$ and use it in the action

$$S[x^\mu, \xi, e, \lambda] = \int d\tau \frac{\dot{x}.\dot{\xi}}{e} + e\mu + \lambda(\xi^2 - 1), \tag{23}$$

here λ is a Lagrange multiplier which gives the constraint ϕ_2. The momenta associated to x and ξ are

$$p_\mu = \frac{\dot{\xi}_\mu}{e}, \quad q_\mu = \frac{\dot{x}_\mu}{e}. \tag{24}$$

Variation with respect to e gives

$$-\frac{\dot{x}.\dot{\xi}}{e^2} + \mu = 0, \tag{25}$$

this is equivalent to ϕ_1. The canonical Hamiltonian is

$$H_c = p.\dot{x} + q.\dot{\xi} + \pi_e \dot{e} + \pi_\lambda \dot{\lambda} - L = e(p.q - \mu) - \lambda(\xi^2 - 1), \tag{26}$$

The condition that the two constraints be preserved in time imply that the Poisson brackets of H_c with ϕ_1 and ϕ_2 must vanish. This gives a secondary constraint [20] $\xi.p = 0$ whose preservation in time produces the tertiary constraint $p^2 = 0$. The Poisson bracket of ϕ_1 and ϕ_2 gives ϕ_3 and the Poisson bracket of ϕ_3 with ϕ_1 gives ϕ_4. It is possible to add Lagrange multiplier to the action in order to get all the constraints as primary ones

$$S'[x, \xi, e, \lambda, v_1, v_2] = \int d\tau \frac{\dot{x}.\dot{\xi}}{e} + e\mu + \lambda(\xi^2 - 1) + v_1 \dot{\xi}^2 + v_2 \xi.\dot{\xi}. \tag{27}$$

An equivalent action can be obtained by eliminating e from the action (23) to get

$$S[x, \xi, \lambda] = 2\sqrt{\mu} \int d\tau \sqrt{\dot{x}.\dot{\xi}} + \lambda(\xi^2 - 1). \tag{28}$$

This action has however a singular $\mu \to 0$ limit. It is also possible to eliminate the Lagrange multiplier λ by solving for ξ^0 as $\sqrt{\vec{\xi}^2 - 1}$ and replacing it in the first part of the action. The action loses however its manifest covariance.

Notice that by fixing the gauge $\dot{x}.\dot{\xi} = 1$ the equations of motion obtained from the action, after elimination of the variable ξ give a fourth order equation for x which reads

$$\frac{d^2}{d\tau^2}\left[\frac{\ddot{x}^\mu}{\sqrt{\ddot{x}^\nu \ddot{x}_\nu}}\right] = 0, \tag{29}$$

this equation is also obtained from the higher derivative action proposed in [7] $S = \int d\tau \sqrt{\ddot{x}^\nu \ddot{x}_\nu}$, which is not invariant under reparametrisation[2].

4. String Action and Resulting Constraints

The world sheet action is now

$$S[X^\mu, \Xi^\mu, h_{mn}, \lambda] = -\tilde{\mu} \int d^2\sigma \sqrt{-h}\left[h^{mn}\partial_m X^\mu \partial_n \Xi^\nu \eta_{\mu\nu} + \lambda(\Xi^2 - 1)\right], \tag{30}$$

where the two-dimensional metric h_{mn} replaces e and the two coordinates X and Ξ depend on the two world-sheet coordinates σ^0 and σ^1. Notice that we did not add an analog of the second term in (23) to maintain conformal invariance. The equations of motion are

$$\Box \Xi^\mu = 0, \quad \Box X^\mu = 2\lambda \Xi^\mu, \tag{31}$$

where $\Box = \frac{1}{\sqrt{-h}}\partial_m \sqrt{-h} h^{mn} \partial_n$. The constraints obtained by varying with respect to λ and h_{mn} are

$$(\Xi^2 - 1) = 0, \tag{32}$$

$$\partial_m X^\mu \partial_n \Xi_\mu + \partial_n X^\mu \partial_m \Xi_\mu - h_{mn}\partial_l X^\mu \partial^l \Xi_\mu = 0. \tag{33}$$

Using (31) and (32) we get for λ

$$(2\lambda)^2 = (\Box X)^2, \tag{34}$$

The second equation in (31) together with (34) can be used to determine Ξ in terms of X as

$$\Xi = \frac{\Box X}{\sqrt{(\Box X)^2}}. \tag{35}$$

---

[2] Higher derivative actions with reparametrisation invariance describing fixed helicity massless states were proposed and studied in [9].

We can eliminate Ξ from the action to get the higher derivative action

$$\begin{aligned}S &= -\tilde{\mu}\int d^2\sigma\sqrt{-h}h^{mn}\partial_m X^\mu \partial_n \frac{\Box X_\mu}{\sqrt{(\Box X)^2}} \\ &= \tilde{\mu}\int d^2\sigma\sqrt{-h}\sqrt{\Box X^\mu \Box X_\mu},\end{aligned} \qquad (36)$$

which is the form proposed in [11].

Notice that the kinetic term in the action (30) is invariant under the $SO(D,D)$ inhomogeneous group. The (D,D) signature of the kinetic terms signals potential pathologies in the quantum theory like negative norm states and instabilities. We shall return to this point later. The constraints, as expected, will be crucial in solving this potential difficulty.

We now turn to determine the full set of constraints contained in the action. The canonical momenta P_μ and Q_μ associated respectively to X^μ and Ξ^μ are (we use the notation $\gamma^{mn} = \tilde{\mu}\sqrt{-h}h^{mn}$, we will set from now on $\tilde{\mu} = 1$)

$$P_\mu = -\gamma^{00}\partial_0 \Xi_\mu - \gamma^{01}\partial_1 \Xi_\mu, \quad Q_\mu = -\gamma^{00}\partial_0 X_\mu - \gamma^{01}\partial_1 X_\mu. \qquad (37)$$

They allow the determination of the canonical Hamiltonian density

$$H_c = N(P.Q + \partial_1 X.\partial_1 \Xi) + M(P.\partial_1 X + Q.\partial_1 \Xi) - \lambda(\Xi^2 - 1), \qquad (38)$$

where we have denoted $-(\gamma^{00})^{-1}$ by N and γ^{01} by M/N. The primary constraints are thus

$$\begin{aligned}\mathcal{H}_0 &= P.Q + \partial_1 X.\partial_1 \Xi = 0 \\ \mathcal{H}_1 &= P.\partial_1 X + Q.\partial_1 \Xi = 0, \\ \phi_1 &= \Xi^2 - 1 = 0.\end{aligned} \qquad (39)$$

The first two are the usual (Virasoro) constraints due to the reparametrisation invariance of the string world-sheet. They are first class. The Poisson bracket of $\mathcal{H}_1$ with ϕ_1 gives

$$\{\mathcal{H}_1(\sigma), \phi_1(\sigma')\} = \phi_1'(\sigma)\delta(\sigma - \sigma'), \qquad (40)$$

which is not a new constraint. The Poisson bracket of $\mathcal{H}_0$ with ϕ_1, on the other hand, gives the secondary constraint

$$\{\mathcal{H}_0(\sigma), \phi_1(\sigma')\} = P.\Xi(\sigma)\delta(\sigma - \sigma') = \phi_2\delta(\sigma - \sigma'). \qquad (41)$$

The new constraint ϕ_2 does not commute with $\mathcal{H}_0$, it generates a tertiary constraint ϕ_3 which is given by

$$\phi_3 = P^2 - (\partial_1 \Xi)^2. \qquad (42)$$

It is easy to convince onself that the procedure does not end. It is however possible to find all the constraints by solving explicitly the Ξ equations of motion. It will be convenient to work in the conformal gauge $M = 0, N = 1$, where the solution for Ξ is

$$\begin{aligned}\Xi(\sigma, \tau) &= \Xi_+(\tau + \sigma) + \Xi_-(\tau - \sigma) \\ &= \Xi(\sigma) + \sum_{n=1}^{\infty}\frac{\tau^n}{n!}\left(\frac{(1-(-1)^n)}{2}P^{(n-1)}(\sigma) + \frac{(1+(-1)^n)}{2}\Xi^{(n)}(\sigma)\right) \\ &= \frac{\Xi(\sigma + \tau) + \Xi(\sigma - \tau)}{2} + \frac{1}{2}\int_{\sigma - \tau}^{\sigma + \tau}d\tilde{\sigma}\, P(\tilde{\sigma}),\end{aligned} \qquad (43)$$

where $\Xi(\sigma) = \Xi(\sigma, 0)$. From the requirement that $\Xi^2 = 1$ be true for all time we get the two bilocal constraints

$$(\Xi(\sigma) + \Xi(\sigma')) \cdot \int_{\sigma'}^{\sigma} d\tilde{\sigma}\, P(\tilde{\sigma}) = 0, \tag{44}$$

$$\Xi(\sigma).\Xi(\sigma') + \frac{1}{2}\left(\int_{\sigma'}^{\sigma} d\tilde{\sigma}\, P(\tilde{\sigma})\right)^2 = 1, \tag{45}$$

where σ and σ' are arbitrary. Taking the derivative of (44) with respect to σ gives the equivalent constraint

$$\partial_1 \Xi(\sigma) \cdot \int_{\sigma'}^{\sigma} d\tilde{\sigma}\, P(\tilde{\sigma}) + (\Xi(\sigma) + \Xi(\sigma')).P(\sigma) = 0. \tag{46}$$

If we set $\sigma = \sigma'$ in the above equation we get

$$P(\sigma).\Xi(\sigma) = 0, \tag{47}$$

which is our previous secondary constraint ϕ_2. Taking the derivative of (46) with respect to σ' gives

$$-\partial_1 \Xi(\sigma).P(\sigma') + \partial_1 \Xi(\sigma').P(\sigma) = 0. \tag{48}$$

Finally we get that two constraints (47) and (48) are equivalent to (44).

We turn to the second bilocal constraint (45). Setting $\sigma = \sigma'$ gives the primary constraint ϕ_1. Its derivative with respect to σ gives

$$\partial_1 \Xi(\sigma).\Xi(\sigma') + P(\sigma) \cdot \int_{\sigma'}^{\sigma} d\tilde{\sigma}\, P(\tilde{\sigma}) = 0. \tag{49}$$

Setting $\sigma = \sigma'$ gives the derivative of ϕ_1 with respect to σ and taking the derivative with respect to σ' yields

$$\partial_1 \Xi(\sigma).\partial_1 \Xi(\sigma') - P(\sigma).P(\sigma') = 0. \tag{50}$$

The $\sigma = \sigma'$ part of (50) gives ϕ_3. In summary, all the constraints are contained in (39), (47), (48) and (50). If we expand (48) and (50) around $\sigma = \sigma'$ we get an infinite number of local constraints.

An equivalent way of writing the two bilocal constraints (48) and (50) is obtained by taking their sum and difference. The sum gives

$$\chi(\sigma, \sigma') = (P(\sigma) + \partial_1 \Xi(\sigma)).(P(\sigma') - \partial_1 \Xi(\sigma')) = 0. \tag{51}$$

The difference gives the equivalent constraint

$$(P(\sigma) - \partial_1 \Xi(\sigma)).(P(\sigma') + \partial_1 \Xi(\sigma')) = 0, \tag{52}$$

which can be obtained from (51) by permuting σ and σ'. So both (48) and (50) are contained in (51).

The Poisson brackets of $\mathcal{H}_0$ and $\mathcal{H}_1$ with χ can be readily calculated

$$\{\mathcal{H}_0(\sigma), \chi(\sigma', \sigma'')\} = -\delta'(\sigma - \sigma')\chi(\sigma, \sigma'') + \delta'(\sigma - \sigma'')\chi(\sigma', \sigma), \quad (53)$$
$$\{\mathcal{H}_1(\sigma), \chi(\sigma', \sigma'')\} = -\delta'(\sigma - \sigma')\chi(\sigma, \sigma'') - \delta'(\sigma - \sigma'')\chi(\sigma', \sigma). \quad (54)$$

If we define $\mathcal{H}_\pm$ as $\mathcal{H}_0 \pm \mathcal{H}_1$, $P_R(\sigma) = P + \partial_1 \Xi$, $P_L(\sigma) = P - \partial_1 \Xi$, then we get

$$\chi(\sigma, \sigma') = P_R(\sigma).P_L(\sigma'), \quad \mathcal{H}_+ = P_R(\sigma).Q_R(\sigma), \quad \mathcal{H}_- = P_L(\sigma).Q_L(\sigma), \quad (55)$$
$$\{\mathcal{H}_+(\sigma), \chi(\sigma', \sigma'')\} = -2\delta'(\sigma - \sigma')\chi(\sigma, \sigma''), \quad (56)$$
$$\{\mathcal{H}_-(\sigma), \chi(\sigma', \sigma'')\} = 2\delta'(\sigma - \sigma'')\chi(\sigma', \sigma). \quad (57)$$

5. Open Strings

For open strings with Neumann boundary conditions we have $\partial_1 \Xi(0) = \partial_1 \Xi(\pi) = \partial_1 X(0) = \partial_1 X(\pi) = 0$. As usual this implies that the left movers and right movers are not independent, rather one has $P_R(\sigma) = P_L(-\sigma)$ and $Q_R(\sigma) = Q_L(-\sigma)$. The constraints now, keeping the right movers, become

$$P_R(\sigma).Q_R(\sigma) = 0, \quad \Xi^2(\sigma) = 1, \quad (58)$$
$$P_R(\sigma).\Xi(\sigma) = P_R(-\sigma).\Xi(\sigma) = 0, \quad P_R(\sigma).P_R(\sigma') = 0. \quad (59)$$

All the variables are 2π periodic.

Let us first analyse the last bilocal constraint. Setting $\sigma = \sigma'$ we get

$$P_R^-(\sigma) = \frac{1}{2P_R^+(\sigma)} P_R^i(\sigma) P_{R\,i}(\sigma). \quad (60)$$

Plugging this in the constraint yields

$$\left(P_R^i(\sigma) - \frac{P_R^+(\sigma)}{P^+(\sigma')} P_R^i(\sigma')\right) \left(P_{R\,i}(\sigma) - \frac{P_R^+(\sigma)}{P^+(\sigma')} P_{R\,i}(\sigma')\right) = 0, \quad (61)$$

which is verified provided

$$\frac{P_R^i(\sigma)}{P_R^+(\sigma)} = \frac{P_R^i(\sigma')}{P_R^+(\sigma')} = \frac{p^i}{p^+}, \quad (62)$$

where p^μ does not depend on σ. This is equivalent to

$$P_R^\mu(\sigma) = p^\mu F_R(\sigma), \quad (63)$$

that is all the components of P_R are determined by a single function F_R and a constant momentum p^μ with $p^2 = 0$. The remaining constraints, supposing that F_R does not vanish, become

$$p.Q_R = 0, \quad \Xi^2 - 1 = 0, \quad p.\Xi = 0. \quad (64)$$

The function $F_R(\sigma)$ is arbitrary and by using the conformal invariance can be set to 1 so that

$$P^\mu(\sigma) = p^\mu, \quad \Xi^\mu(\sigma) = \xi^\mu, \quad (65)$$

where ξ^μ does not depend on σ. The first equation removes the oscillator modes of X as dynamical variables and the second removes the oscillator modes of Ξ. The problem has been greatly simplified since the dynamical variables now are just the zero modes x and p as well as ξ and its conjugate momentum q subject to the constraints (64) which become

$$p.q = 0, \ \xi^2 = 1, \ p.\xi = 0, \ p^2 = 0. \tag{66}$$

where q and p are canonically conjugate to ξ and x. The quantisation of the system is straightforward: the wave function $\psi(\xi, p)$ is subject to the constraints

$$(\xi^2 - 1)\psi = 0, \ p.\frac{\partial}{\partial \xi}\psi = 0, \ p^2\psi = 0. \tag{67}$$

The solution can be written as

$$\psi = \sum_n h_{\mu_1 \ldots \mu_n}(p)\xi^{\mu_1}\ldots\xi^{\mu_n}, \tag{68}$$

where $h_{\mu_1 \ldots \mu_n}$ is a traceless completely symmetric tensor which verifies the massless spin n equations of motion (see for instance [21] and references therein)

$$p^2 h_{\mu_1 \ldots \mu_n} = 0, \ p^{\mu_1} h_{\mu_1 \ldots \mu_n} = 0. \tag{69}$$

Notice that the first constraint is equivalent to the tracelessness of h. The open string does not possess physical modes with the continuous spin representation. It is important to note that the quantisation reduces to the point particle quantisation and so does not ask for a critical dimension.

6. Closed Strings

We start by analyzing the bilocal constraint (51) rewritten as $P_R(\sigma).P_L(\sigma') = 0$. Let V_R be the vector space spanned by $P_R(\sigma)$ when σ varies from 0 to 2π and similarly for V_L, then V_R and V_L are orthogonal. Let p be the common zero mode of P_R and P_L. By taking the integral on both σ and σ' of the constraint (51) we get that $p^2 = 0$. All the string modes are thus massless. Furthermore p is contained in both V_R and V_L. Suppose that the only nonvanishing component of p is p^+, and split the spacelike and transverse indices $i = 1, \ldots D - 2$ into a which belong to V_R and a' which belong to V_L. We thus have

$$P_L^a(\sigma) = 0, \ a = 1, \ldots N, \quad P_R^{a'}(\sigma) = 0, \ a' = N + 1, \ldots D - 2, \tag{70}$$

and since p^+ is in both V_R and V_L we also have

$$P_L^-(\sigma) = 0 = P_R^-(\sigma). \tag{71}$$

The latter two constraints are equivalent to $\Xi^- = \xi^-$ and $P^- = 0$, with ξ^- a constant zero mode. We have thus transformed the bilocal constraints into local ones at the expense of breaking the manifest Lorentz covariance.

The remaining constraints are

$$P.\Xi = 0, \quad \Xi^2 - 1 = 0. \tag{72}$$

We wish to transform them, using (51), into constraints involving only left movers or right movers. We shall be able to do so except for a zero mode which involves the sum of right moving and left moving variables. We have

$$P = \frac{P_R + P_L}{2}, \quad \Xi' = \frac{P_R - P_L}{2\mu}. \tag{73}$$

Let $\bar{P}_R$ and $\bar{P}_L$ be the nonzero mode parts of P_R and P_L and define $\hat{P}_R$ and $\hat{P}_L$ by $\hat{P}'_R = \bar{P}_R$ the integration constant being chosen so that $\hat{P}_R$ has no zero mode. Integrate the second equation in (73) as

$$\Xi(\sigma) = \frac{\hat{P}_R - \hat{P}_L}{2\mu} + \xi. \tag{74}$$

Here ξ is a constant D-vector. Notice that we have $\hat{P}_R.\hat{P}_L = 0$ and $P_R.\hat{P}_L = 0$. Using the derivative of ϕ_1, $\Xi'.\Xi = 0$, we get

$$P_R.\Xi = 0, \quad P_L.\Xi = 0, \tag{75}$$

which yield

$$G = \xi.P_R + \frac{1}{2\mu}\hat{P}_R.P_R(\sigma) = 0, \tag{76}$$

and

$$\tilde{G} = \xi.P_L - \frac{1}{2\mu}\hat{P}_L.P_L(\sigma) = 0, \tag{77}$$

Notice that G and $\tilde{G}$ have the same zero mode

$$G_0 = \frac{\xi.p}{2\pi} = -\frac{p^+\xi^-}{2\pi}. \tag{78}$$

We have thus obtained one left moving and one right moving constraints. It remains to take into account the zero mode constraint contained in ϕ_1. This is accomplished by

$$g = \xi^2 + \frac{1}{4\mu^2}\int \frac{d\sigma}{2\pi}(\hat{P}_R^2 + \hat{P}_L^2) - 1 = 0. \tag{79}$$

Notice that it is the sum of left moving variables and right moving ones.

7. BRST Quantization

Recall that for a system with first class constraints G_i which form a Lie algebra $[G_i, G_j] = C_{ij}{}^k G_k$ the first step in the BRST quantization is the enlargement of the Hilbert space by the introduction of the ghosts c^i and antighosts b_j which verify $\{b_i, c^j\} = \delta_i^j, \{c^i, c^j\} = \{b_i, b_j\} = 0$. The BRST charge is defined by $Q = c^i G_j - \frac{1}{2}C_{ij}{}^k c^i c^j b_k$, and verifies $Q^2 = 0$. The physical states are in the kernel of Q and two states are equivalent if they differ by an exact state, i.e. of the form $Q|\phi>$ for some $|\phi>$.

In string theory, the classical constraints form a closed Lie algebra. When one turns the dynamical variables into operators, an anomaly appears in the commutators of the energy-momentum tensor. This anomaly has two contributions: the first is dependent on the normal ordering constant in the energy-momentum tensor and the second is the central charge. More precisely, if one defines the normal ordered matter energy-momentum tensor by $T^{(m)} = \frac{\pi}{\mu} : P_R.Q_R : -\frac{2D}{24} + a^m$, with a^m a constant, then it obeys the commutation relations[3]

$$[T^{(m)}(\sigma), T^{(m)}(\sigma')] = 2\pi i (T^{(m)}(\sigma) + T^{(m)}(\sigma'))\delta'(\sigma - \sigma')$$
$$- 4\pi i a^m \delta'(\sigma - \sigma') + 2\pi i \frac{c^m}{12}\delta'''(\sigma - \sigma'), \quad (80)$$

where c^m is the central charge of the matter sector given by $2D$.

The other constraints, for the right moving part, are given by

$$P_R^{a'}(\sigma) = 0, \quad (81)$$
$$P_R^-(\sigma) = 0, \quad (82)$$
$$G = \xi.P_R + \frac{1}{2\mu}\hat{P}_R.P_R(\sigma) = 0. \quad (83)$$

One has also to add the zero mode constraints $g = 0$. The constraints (81-83) commute among each other and with g and they have conformal weights equal to one, that is if we denote generically one of the constraints (81-82) by K then we have

$$\left[T^{(m)}(\sigma'), K(\sigma)\right] = -2\pi i K'(\sigma)\delta(\sigma - \sigma') - 2\pi i K(\sigma)\delta'(\sigma - \sigma'). \quad (84)$$

We also have

$$[T(\sigma), g] = -\frac{i}{\mu}G(\sigma), \quad (85)$$

and

$$\left[T^{(m)}(\sigma'), G(\sigma)\right] = -2\pi i G'(\sigma)\delta(\sigma - \sigma') - 2\pi i G(\sigma)\delta'(\sigma - \sigma')$$
$$+ i\delta(\sigma - \sigma')\frac{p^+}{2\mu}P_R^-(\sigma). \quad (86)$$

Equation (86) means that G is weakly a field of conformal weight 1. Let the ghosts fields associated to the constraints (81-83) and $T^{(m)}$ be denoted respectively by $c_{a'}, c_-, d$ and c and the corresponding antighosts by $b^{a'}, b^-, e$ and b. All the ghosts except c have conformal weights 0 and c has the conformal weight -1. Let the ghost and antighost associated to g be denoted by γ and ω. These are not fields, they have only zero modes. The naive BRST charge Q which results when ignoring the anomalous terms in the commutation relations of the energy-momentum tensor reads

$$Q = \int \frac{d\sigma}{2\pi}\left[c(\sigma)T^{(m)} + c_{a'}P_R^{a'} + c_- P_R^- + dG\right] + \gamma g$$
$$+ \int \frac{d\sigma}{2\pi} : c(\sigma)\left[\frac{1}{2}T^{(c)} + T^{(c_{a'})} + T^{(c_-)} + T^{(d)} - \frac{i}{2\pi\mu}\gamma e - \frac{ip^+}{4\pi\mu}db^- + a\right] : \quad (87)$$

---
[3]We are using the canonical commutation relations $[X^\mu(\sigma), P^\nu(\sigma')] = i\eta^{\mu\nu}\delta(\sigma - \sigma')$ and so on.

$T^{(c)}$ is the energy-momentum tensor of the ghost system c and b and so on for the other terms in (87). The energy-momentum tensor of the weight h ghost [4] system is given by

$$T_h = -i : [h\partial(c_h b_h) - c_h \partial b_h] : -\frac{1}{12} + a^h, \quad (88)$$

which satisfies commutation relations analogous to (80) with a^m replaced by a^h and a central charge given by [22, 23] $c_h = 1 - 3(2h-1)^2$. In (87), we also allowed for a normal ordering constant a. The BRST charge depends on the normal ordering constants only through the combination $a^m + \sum a^h + a$. Without loss of generality it is thus possible to choose $a^m = a^h = 0$.

The left moving BRST charge is similarly defined with its corresponding left ghosts and antighosts except for γ and ω and the zero modes of d and e which are the same. This is due to the structure of g as a sum of left movers and right movers and the fact that the zero modes of G and $\tilde{G}$ coincide.

The crucial property of Q is its nilpotency. The calculation of Q^2, using the full commutation relations gives

$$Q^2 = -i\frac{c_T}{24}\int \frac{d\sigma}{2\pi} (\partial^3 c(\sigma)) c(\sigma) - ia\int \frac{d\sigma}{2\pi} c(\sigma)\partial c(\sigma), \quad (89)$$

where $c_T = c^m + \sum c_{h_i}$ is the total central charge of the matter and ghost system which is given by $2D - 26 - 2(D - N)$. Thus the nilpotency of Q requires that the total central charge c_T and the normal ordering constant a vanish. A similar conclusion is of course valid for the left moving sector whose total central charge is given by $2D - 26 - 2(N+2)$. The vanishing of both central charges gives $D = 28$ and $N = 13$.

8. Spectrum

The physical states are equivalence classes of states annihilated by Q, two states being equivalent if their difference is Q-exact. One has also to add some supplementary conditions originating from the ghosts zero modes. Notice that among the ghosts c, d and of course γ together with their antighosts have zero modes. This results in a degeneracy of the ground state. By analogy with the usual string theory [23] we impose that the zero modes of the antighosts annihilate the physical states

$$b_0|\Psi> = \tilde{b}_0|\Psi> = e_0|\Psi> = w|\Psi> = 0. \quad (90)$$

This allows to impose correctly the zero mode constraints on the physical states.
It will be convenient to use the following Fourier expansions

$$P_R(\sigma) = \frac{p}{2\pi} + 2\mu \sum_{n\neq 0} \beta_n e^{in\sigma}, \quad Q_R(\sigma) = \frac{q}{2\pi} + \frac{1}{2\pi}\sum_{n\neq 0} \alpha_n e^{in\sigma}, \quad (91)$$

$$c(\sigma) = \sum_{n=-\infty}^{+\infty} c_n e^{in\sigma}, \quad b(\sigma) = \sum_{n=-\infty}^{+\infty} b_n e^{in\sigma} \quad (92)$$

---

[4]Here h is the conformal weight of the antighosts b_h which is the same as that of the corresponding constraint; we use the conventions $\{c_h(\sigma), b_h(\sigma')\} = 2\pi\delta(\sigma-\sigma')$.

$$T(\sigma) = \sum_{n=-\infty}^{+\infty} L_n e^{in\sigma}, \quad G(\sigma) = \sum_n G_n e^{in\sigma} \tag{93}$$

$$\hat{P}_R(\sigma) = -2i\mu \sum_{n\neq 0} \frac{\beta_n}{n} e^{in\sigma}. \tag{94}$$

The Fourier coefficients of the fields satisfy the commutation relations

$$[\beta_n^\mu, \alpha_m^\nu] = m\eta^{\mu\nu}\delta_{n+m,0}, \quad \{c_n, b_m\} = \delta_{n+m,0}, \quad [q^\nu, \xi^\mu] = -i\eta^{\mu\nu}, \tag{95}$$

the others being zero. We have

$$\begin{aligned}
L_0^{(m)} &= \frac{p.q}{4\pi\mu} + \sum_{n\neq 0} :\beta_n.\alpha_{-n}: + \frac{7}{3}, \quad L_m^{(m)} = \frac{p.\alpha_m}{4\pi\mu} + q.\beta_m + \sum_{n\neq 0,m} \beta_n.\alpha_{m-n}, \\
L_0^{(c_h)} &= \sum_n n :c_n b_{-n}: - \frac{1}{12}, \quad L_m^{(c_h)} = \sum_n [(h-1)m + n] c_n b_{m-n}, \\
G_n &= 2\mu \left(\xi.\beta_n - i \sum_{m\neq 0} \frac{\beta_m}{m}.\beta_{n-m} \right).
\end{aligned} \tag{96}$$

In equation (96) we used $\beta_0 = p/(4\pi\mu)$.

Notice that the total normal ordering constant coming from the ghost sector is $(D+4)/24 = -2/3$ which when added to that of the matter sector gives 1.

If one defines similarly the left moving Fourier modes and denotes them with a tilde then the total BRST charge, including left and right contributions, reads

$$\begin{aligned}
Q &= \sum_n c_n(L_{-n}^{tot} - \frac{L_{-n}^{(c)}}{2}) + 2\mu \left(\sum_{n\neq 0} c_{a'n}\beta_{-n}^{a'} + \sum_{n\neq 0} c_{-n}\beta_{-n}^{-} \right) + \sum_n d_n G_{-n} \\
&+ \sum_n \tilde{c}_n(\tilde{L}_{-n}^{tot} - \frac{\tilde{L}_{-n}^{(c)}}{2}) + 2\mu \left(\sum_{n\neq 0} \tilde{c}_{an}\tilde{\beta}_{-n}^{a} + \sum_{n\neq 0} \tilde{c}_{-n}\tilde{\beta}_{-n}^{-} \right) + \sum_{n\neq 0} \tilde{d}_n \tilde{G}_{-n} \\
&- i\frac{p^+}{4\pi\mu}[\sum_{n,m\neq 0} c_n d_m b_{-(n+m)}^- + \tilde{c}_n \tilde{d}_m \tilde{b}_{-(n+m)}^-] \\
&+ \gamma \left[g + \frac{i}{2\pi\mu} \left(\sum_{n\neq 0}(c_{-n}e_n + \tilde{c}_{-n}\tilde{e}_n) + (c_0 + \tilde{c}_0)e_0 \right) \right] \\
&+ d_0 \left[G_0 + i\frac{p^+}{4\pi\mu} \sum_{n\neq 0} c_n b_{-n}^- + \tilde{c}_n \tilde{b}_{-n}^- \right],
\end{aligned} \tag{97}$$

where

$$g = \xi^2 - 1 + \sum_{n\neq 0} \left(\frac{\beta_n.\beta_{-n}}{n^2} + \frac{\tilde{\beta}_n.\tilde{\beta}_{-n}}{n^2} \right). \tag{98}$$

Notice that as anticipated γ and d_0 multiply sums of right movers and left movers. The vacuum state is defined by

$$\alpha_n^\mu |0> = \beta_n^\mu |0> = c_n^{(h)}|0> = b_n^{(h)}|0> = 0, \ \forall n > 0 \tag{99}$$

and for all the ghosts labelled by h here. The BRST operator acting on a vacuum state gives

$$Q|0> = \left[c_0\left(\frac{p.q}{4\pi\mu}+1\right) + d_0 G_0 + \gamma\left(g + \frac{i}{2\pi\mu}c_0 e_0\right)\right]|0>. \tag{100}$$

If we now use the supplementary conditions (90) then we get the following constraints on the zero-mode part of the state

$$(p.q + 4\pi\mu)|0> = 0, \quad \xi.p|0> = 0, \quad (\xi^2 - 1)|0> = 0. \tag{101}$$

These are precisely the conditions (5-8) defining a continuous spin state [2].

The additional zero mode constraints (90), when anticommuted with the BRST charge Q give the compatibility conditions:

$$L_0^{tot}|\Psi> = 0, \quad \tilde{L}_0^{tot}|\Psi> = 0 \tag{102}$$

$$\left[G_0 + i\frac{p^+}{4\pi\mu}\sum_{n\neq 0} c_n b_{-n}^- + \tilde{c}_n \tilde{b}_{-n}^-\right]|\Psi> = 0, \tag{103}$$

$$\left[g + \frac{i}{2\pi\mu}\sum_{n\neq 0}(c_{-n}e_n + \tilde{c}_{-n}\tilde{e}_n)\right]|\Psi> = 0. \tag{104}$$

Two states are equivalent if they differ by a Q exact state of the form $Q|\Phi>$. The constraints (90) imply that $|\Phi>$ is not arbitrary but has to verify

$$b_0|\Phi> = 0, \quad \tilde{b}_0|\Phi> = 0, \quad L_0^{tot}|\Phi> = 0, \quad \tilde{L}_0^{tot}|\Phi> = 0 \tag{105}$$

$$w|\Phi> = 0, \quad [g + \frac{i}{2\pi\mu}\sum_{n\neq 0}(c_{-n}e_n + \tilde{c}_{-n}\tilde{e}_n)]|\Phi> = 0 \tag{106}$$

$$e_0|\Phi> = 0, \quad [G_0 + i\frac{p^+}{4\pi\mu}\sum_{n\neq 0}c_n b_{-n}^- + \tilde{c}_n \tilde{b}_{-n}^-]|\Phi> = 0. \tag{107}$$

Let us first determine the physical states with only matter excitations, the ghosts being in the ground state. The condition $Q|\psi>\otimes|0> = 0$ gives for $|\psi>$, the state in the matter sector,

$$\left(p.q + 4\pi\mu(\sum_{n\neq 0}:\beta_n.\alpha_{-n}: +1)\right)|\psi> = 0, \quad L_n^{(m)}|\psi> = 0, \ n > 0 \tag{108}$$

$$\beta_n^{a'}|\psi> = 0, \ n > 0, \quad \beta_n^-|\psi> = 0, \ n > 0 \tag{109}$$

$$G_n|\psi> = 0, \ n \geq 0, \quad g|\psi> = 0. \tag{110}$$

There are of course similar conditions for the left moving sector. The first condition determines the continuous spin parameter as $p.q = 4\pi\mu(N-1)$, where N is the level of the state. Notice that for $N = 1$, we have a reducible representation of the Poincaré group containing an infinite number of fixed helicity states.

Consider in more details the first level states

$$|\psi> = (A_\mu \alpha^\mu_{-1} + B_\mu \beta^\mu_{-1})|0>, \tag{111}$$

The L_0 condition gives $p.q = 0$, this state belongs to standard helicity representations. The L_1 condition yields

$$\frac{p.B}{4\pi\mu} + q.A = 0. \tag{112}$$

The G_0 and G_1 conditions give

$$\xi.p = 0, \quad (\xi - \frac{i}{4\pi\mu}p)^\mu A_\mu = 0. \tag{113}$$

The β_1 conditions implies

$$A_+ = 0 = A_{a'}. \tag{114}$$

It remains to consider the g condition which gives

$$(\xi^2 - 1)A_\mu = 0, \quad (\xi^2 - 1)B_\mu = A_\mu. \tag{115}$$

The nonzero solution of equation (115) is

$$(\xi^2 - 1) = 0, \quad A_\mu = 0. \tag{116}$$

The first level physical state is thus

$$|\psi> = B_\mu \beta^\mu_{-1}|0>, \tag{117}$$

with $p.B = 0$. The norm of the state is proportional to $A_\mu B^\mu$ which is zero. The standard helicity states are thus of zero norm.

Notice that the most constraining condition was $g|\psi> = 0$ which implied alone the vanishing of A_μ. We shall prove that this is also true for the higher level states, that is the states annihilated simultaneously by L_0 and g do not contain α oscillators and are thus of zero norm. Define N_α by $-\sum_{m=1}^\infty \alpha_{-m}.\beta_m/m$ and $N_\beta = -\sum_{m=1}^\infty \beta_{-m}.\alpha_m/m$. N_α counts the number of α oscillators acting on the ground state and N_β counts the number of β oscillators, for example if

$$|\phi> = \alpha^{\mu_1}_{-n_1} \ldots \alpha^{\mu_n}_{-n_n}|0>, \tag{118}$$

with all the n_i strictly positive then

$$N_\alpha|\phi> = n|\phi>. \tag{119}$$

Let A be the part of g depending on the right oscillators $A = \sum_{m\neq 0} \beta_m.\beta_{-m}/m^2$. Then A decreases the number of α oscillators by one and increases the number of β oscillators by one:

$$[N_\alpha, A] = -A, \quad [N_\beta, A] = A. \tag{120}$$

We want to show that an eigenstate of A with a finite number of oscillators (an eigenstate of L_0) has necessarily eigenvalue zero and does not contain α oscillators. For this decompose the eigenstate as

$$|\psi> = \sum_{n_\alpha=0}^{N} |\psi_{n_\alpha}>, \qquad (121)$$

with $|\psi_{n_\alpha}>$ a state containing n_α α oscillators acting on the vacuum, and $A|\psi> = a|\psi>$, a being the eigenvalue. Since A decreases the number of α oscillators by one, we have

$$A|\psi> = \sum_{n_\alpha=1}^{N} |\chi_{n_\alpha-1}>, \qquad (122)$$

where $|\chi_{n_\alpha-1}> = A|\psi_{n_\alpha}>$. Equation (122) implies that $|\chi_{n_\alpha}> = a|\psi_{n_\alpha}>$ for $n_\alpha = 0, \ldots N-1$ and $a|\psi_N> = 0$. Suppose that a is not zero then $|\psi_N> = 0$ and $a|\psi_{N-1}> = |\chi_{N-1}> = A|\psi_N> = 0$ and so $|\psi_{N-1}>$ vanishes and all the other $|\psi_n>$ are zero. So a is necessarily 0 and $|\psi>$ does not contain any α oscillator. We conclude from the g condition that physical states are linear combinations of states of the form

$$\beta^{\mu_1}_{-n_1} \cdots \beta^{\mu_N}_{-n_N} |0>, \qquad (123)$$

and these are all of zero norm.

Consider now the general state in the matter and ghosts Hilbert space. The condition $g|\psi> = 0$ is now replaced by (104) which implies again that the state does not contain α oscillators and also that the state does not contain b nor d oscillators. The absence of α oscillators implies that a physical state of non-zero norm is necessarily of the form $|0> \otimes |\chi>$, where $|\chi>$ depends only on ghosts excitations and $|0>$ is the ground state of the matter sector. The physical state condition implies, in addition, that $|\chi>$ does not contain b, $b^{a'}$, b^- or e fermionic oscillators. In turn, this implies that $|\chi>$ has a positive norm. This completes the proof that there are no physical states with a strictly negative norm.

9. Supersymmetric Extension

In this Section, we consider the action with world-sheet supersymmetry. Since the analysis is analogous to the bosonic case we will be very brief. The main outcome will be a modified space-time critical dimension. World-sheet supersymmetry will give 12 as a critical space-time dimension.

We generalise the action and the constraints in the conformal gauge. We shall implement the (1,1) supersymmetry with two Majorana-Weyl supercharges Q_+ and Q_-. All the world-sheet fields become superfields depending on the supercoordinates $\sigma_\pm$ and $\theta^\pm$. A typical superfield ϕ will have the expansion

$$\tilde{\phi}(\sigma, \theta^+, \theta^-) = \phi(\sigma) + \theta^+ \psi_+(\sigma) + \theta^- \psi_-(\sigma) + \theta^+ \theta^- A(\sigma). \qquad (124)$$

The supercharges are given by

$$Q_+ = \frac{\partial}{\partial \theta^+} + \theta^+ \partial_+, \quad Q_- = \frac{\partial}{\partial \theta^-} + \theta^- \partial_-. \qquad (125)$$

They anticommute with the supersymmetric derivatives

$$\mathcal{D}_+ = \frac{\partial}{\partial \theta^+} - \theta^+ \partial_+, \quad \mathcal{D}_- = \frac{\partial}{\partial \theta^-} - \theta^- \partial_-. \tag{126}$$

The supersymmetric action in the conformal gauge reads

$$-\tilde{\mu} \int d^2\sigma d\theta^+ d\theta^- \; \mathcal{D}_-\tilde{X}\mathcal{D}_+\tilde{\Xi} + \tilde{\lambda}(\tilde{\Xi}^2 - 1), \tag{127}$$

the constraints associated to superconformal invariance are

$$\partial_+\tilde{X}.\mathcal{D}_+\tilde{\Xi} + \partial_+\tilde{\Xi}.\mathcal{D}_+\tilde{X} = 0, \quad \partial_-\tilde{X}.\mathcal{D}_-\tilde{\Xi} + \partial_-\tilde{\Xi}.\mathcal{D}_-\tilde{X} = 0. \tag{128}$$

The equations of motion read

$$\mathcal{D}_+\mathcal{D}_-\tilde{\Xi} = 0, \tag{129}$$
$$\mathcal{D}_+\mathcal{D}_-\tilde{X} - 2\tilde{\lambda}\tilde{\Xi} = 0, \tag{130}$$
$$\tilde{\Xi}^2 = 1. \tag{131}$$

The solution to $\mathcal{D}_+\tilde{\phi} = 0$ is given by

$$\tilde{\phi}_R = \phi(\sigma_-) + \theta^- \psi_-(\sigma_-), \tag{132}$$

this will be called a chiral superfield. An antichiral superfield is a solution to $\mathcal{D}_-\tilde{\phi} = 0$ and is given

$$\tilde{\phi}_L = \phi(\sigma_+) + \theta^+ \psi_+(\sigma_+). \tag{133}$$

The solution to the $\tilde{\Xi}$ equation of motion is thus given by

$$\tilde{\Xi} = \tilde{\Xi}_R + \tilde{\Xi}_L. \tag{134}$$

The constraint (131) on $\tilde{\Xi}$ implies

$$\mathcal{D}_-\tilde{\Xi}_R.\tilde{\Xi}_R = 0, \tag{135}$$
$$\mathcal{D}_+\tilde{\Xi}_L.\tilde{\Xi}_L = 0, \tag{136}$$
$$\mathcal{D}_-\tilde{\Xi}_R.\mathcal{D}_+\tilde{\Xi}_L = 0. \tag{137}$$

The solution to the $\tilde{X}$ equation of motion (130) may be written as

$$\mathcal{D}_+\tilde{X} = \mathcal{D}_+\tilde{X}_L - \theta^- \tilde{\lambda}\tilde{\Xi} + \int d\sigma^- d\theta^- \tilde{\lambda}\tilde{\Xi} \tag{138}$$

$$\mathcal{D}_-\tilde{X} = \mathcal{D}_+\tilde{X}_R + \theta^+ \tilde{\lambda}\tilde{\Xi} - \int d\sigma^+ d\theta^+ \tilde{\lambda}\tilde{\Xi} \tag{139}$$

The superconformal constraints are thus

$$\partial_+\tilde{X}_L.\mathcal{D}_+\tilde{\Xi}_L + \partial_+\tilde{\Xi}_L.\mathcal{D}_+\tilde{X}_L = 0, \tag{140}$$
$$\partial_-\tilde{X}_R.\mathcal{D}_-\tilde{\Xi}_R + \partial_-\tilde{\Xi}_R.\mathcal{D}_-\tilde{X}_R = 0. \tag{141}$$

To each superconstraint are now associated a superghost and a super anti-ghost. Let C_h be the superfield for a right moving ghost

$$C_h(z.\theta) = c_h(z) + \theta \gamma_h(z), \tag{142}$$

and B the corresponding antighost

$$B_h(z,\theta) = \beta_h(z) + \theta b_h(z), \tag{143}$$

with the operator product expansion

$$B(z_1, \theta_1) C(z_2, \theta_2) = \frac{1}{z_1 - z_2}(\theta_1 - \theta_2). \tag{144}$$

If c has a conformal weight h then γ has conformal weight $h + 1/2$. The ghost energy-momentum tensor is given by [22]

$$\tilde{T} = T_f + \theta T_b = -\frac{1}{2} :\partial BC: + (h - \frac{1}{2}) :\partial(BC): -2 :\mathcal{D}B\mathcal{D}C: . \tag{145}$$

and has central charge $3(3 - 4h)$. The OPE are

$$T_b(z)T_b(0) = \frac{c}{2z^4} + \frac{2}{z^2}T_b(0) + \frac{1}{z}\partial T_b(0), \tag{146}$$

$$T_f(z)T_f(0) = \frac{2c}{3z^3} + \frac{2}{z}T_b(0), \tag{147}$$

$$T_b(z)T_f(0) = \frac{3}{2z^2}T_f(0) + \frac{1}{z}\partial T_f(0). \tag{148}$$

To every superconstraint $\tilde{G} = G_f + \theta G_b$ we associate a superghost and its corresponding superantighost.

As for the bosonic case, neglecting zero modes which do not contribute to the total central charge, the BRST charge is defined by

$$Q = \int dz d\theta \, j_{BRST}, \tag{149}$$

with

$$j_{BRST} = C_2\tilde{T} + \frac{1}{2} :(C_2\tilde{T}^{(g)}): + \sum_i C^{(i)}\tilde{G}^{(i)} + \sum_i C_2\tilde{T}^{(i)}, \tag{150}$$

with the superghost energy momentum tensor is given in (145). The BRST charge is nilpotent if the total central charge is zero which now implies that the spacetime dimension is $D = 12$.

10. Conclusion

We have first analysed a classical action which gives rise to the constraints associated with the massless continuous spin representations. The generalisation to a two-dimensional world-sheet action leads to a string action with bilocal constraints. It is possible to transform these bilocal constraints to local one at the expense of breaking manifest Lorentz covariance. The quantization can then be performed with the standard BRST techniques. In this respect, notice that the quantization of the Polyakov string action can be done either in the BRST formalism or in the light cone gauge. In the first case, one has manifestly Lorentz covariant first class constraints and the critical dimension appears as the condition of nilpotency of the BRST charge. In the second quantisation scheme, one adds gauge fixing conditions and solves the constraints at the price of breaking manifest Lorentz invariance. The critical dimension appears as the condition of Lorentz invariance at the quantum level. In the present case, we have a mixture of the two above formalisms; the transformation of the bilocal constraints to local ones was done at the expense of breaking manifest Lorentz covariance. We ended with a Lie algebra of first class constraints and the nilpotency condition of the BRST charge gave us the critical dimension of 28 (12 in the SUSY case). This condition is equivalent to the vanishing of the total central charge, the anomaly of the Weyl symmetry. What we have thus shown is that a *necessary* condition for the consistency of the theory is that the space-time dimension be 28 (12 in the SUSY case).

We have proved the absence of ghosts in the spectrum and the presence of a physical state carrying the continuous spin representation. These results do not seem to be dependent on the particular way we used to handle the bilocal constraint, that is on its replacement by a number of equivalent local ones in a given frame. Notice in this respect that, although the intermediate steps were not manifestly covariant, the physical spectrum we obtained is Lorentz covariant. It would be interesting to treat the bilocal constraint in a manifestly covariant way. The results are however strongly dependent on the requirement that physical states are annihilated by the zero modes of the antighosts which seems to be the key point in this BRST quantization. Our results prove the consistency of the free theory. This is but the first step towards a consistent theory with interactions where the role of the critical dimension and the zero norm states should be very important.

References

[1] E. P. Wigner, *Annals Math.* **40** (1939) 149 [*Nucl. Phys. Proc. Suppl.* **6** (1989) 9].

[2] E.P. Wigner, *Z. Physik* **124** (1947) 665.

[3] V. Bargmann and E. P. Wigner, *Proc. Nat. Acad. Sci.* **34**, 211 (1948).

[4] J. Yngvason, *Commun. Math. Phys.* **18** (1970) 195; A. Chakrabarty, *J. Math. Phys.* **12**, 1813 (1971); G.J. Iverson and G. Mack, *Annals of Physics* **64** (1971) 211.

[5] L. F. Abbott, *Phys. Rev. D* **13**, 2291 (1976); K. Hirata, *Prog. Theor. Phys.* **58** (1977) 652.

[6] J. Mund, B. Schroer and J. Yngvason, *Phys. Lett. B* **596** (2004) 156 [arXiv:math-ph/0402043].

[7] D. Zoller, *Class. Quant. Grav.* **11**, 1423 (1994).

[8] L. Brink, A. M. Khan, P. Ramond and X. z. Xiong, *J. Math. Phys.* **43** (2002) 6279 [arXiv:hep-th/0205145].

[9] M. S. Plyushchay, *Mod. Phys. Lett. A* **4** (1989) 837; D. Zoller, *Phys. Rev. Lett.* **65**, 2236 (1990).

[10] X. Bekaert and J. Mourad, *JHEP* **0601** (2006) 115 [arXiv:hep-th/0509092].

[11] G. K. Savvidy, *Phys. Lett. B* **552** (2003) 72.

[12] G. K. Savvidy, *Int. J. Mod. Phys. A* **19** (2004) 3171 [arXiv:hep-th/0310085]. I. Antoniadis and G. Savvidy, arXiv:hep-th/0402077.

[13] G. Savvidy, arXiv:hep-th/0409047; G. Savvidy, arXiv:hep-th/0502114.

[14] C. Becchi, A. Rouet and R. Stora, *Annals Phys.* **98** (1976) 287; I.V. Tyutin, Lebedev Institute preprint N39 (1975). E. S. Fradkin and G. A. Vilkovisky, *Phys. Lett. B* **55** (1975) 224. I. A. Batalin and G. A. Vilkovisky, *Phys. Lett. B* **69** (1977) 309.

[15] M. Kato and K. Ogawa, *Nucl. Phys. B* **212** (1983) 443.

[16] M. Henneaux, *Phys. Rept.* **126** (1985) 1.

[17] A. Schild, *Phys. Rev. D* **16** (1977) 1722.

[18] A. Karlhede and U. Lindstrom, *Class. Quant. Grav.* **3**, L73 (1986); F. Lizzi, B. Rai, G. Sparano and A. Srivastava, *Phys. Lett. B* **182** (1986) 326; J. Isberg, U. Lindstrom and B. Sundborg, *Phys. Lett. B* **293**, 321 (1992) [arXiv:hep-th/9207005]; U. Lindstrom, B. Sundborg and G. Theodoridis, *Phys. Lett. B* **253**, 319 (1991); S. Hassani, U. Lindstrom and R. von Unge, *Class. Quant. Grav.* **11**, L79 (1994); *JHEP* **0201**, 034 (2002) [arXiv:hep-th/0112206].

[19] B. Sundborg, *Nucl. Phys. Proc. Suppl.* **102** (2001) 113 [arXiv:hep-th/0103247]; U. Lindstrom and M. Zabzine, *Phys. Lett. B* **584** (2004) 178 [arXiv:hep-th/0305098]; G. Bonelli, *Nucl. Phys. B* **669** (2003) 159 [arXiv:hep-th/0305155]; A. Sagnotti and M. Tsulaia, *Nucl. Phys. B* **682** (2004) 83 [arXiv:hep-th/0311257].

[20] P. A. M. Dirac, *Can. J. Math.* **2** (1950) 129; for a review see K. Sundermeyer, *Lect. Notes Phys.* **169** (1982) 1.

[21] N. Bouatta, G. Compere and A. Sagnotti, arXiv:hep-th/0409068.

[22] D. Friedan, E. J. Martinec and S. H. Shenker, *Nucl. Phys. B* **271** (1986) 93.

[23] M. Green, J. Schwarz and E. Witten, *Superstring theory*, Vol. 1, Cambridge university press (1987), chap. 3; J. Polchinski, String theory, Vol. 1, Cambridge university press (1998).

[24] J. Mourad, "Continuous spin and tensionless strings," arXiv:hep-th/0410009; J. Mourad, "Continuous spin particles from a string theory," arXiv:hep-th/0504118, to appear in the Proceedings of the Albert Einstein Century International Conference (Editors: J.-M. Alimi and A. Fzfa).

[25] L. Edgren, R. Marnelius and P. Salomonson, arXiv:hep-th/0503136, *JHEP* **0505** (2005) 002

In: String Theory Research Progress
Editor: Ferenc N. Balogh, pp. 179-194

ISBN 1-60456-075-6
charcoo∋ 2008 Nova Science Publishers, Inc.

Chapter 5

THE COULOMB GAS REALIZATION OF $SL(2,\mathbb{R})_k$ STRUCTURE CONSTANTS AND STRING INTERACTIONS IN AdS_3

Gaston Giribet
Department of Physics, Universidad de Buenos Aires
Ciudad Universitaria, 1428. Buenos Aires, Argentina

Abstract

By considering the integral representation of Liouville five-point function, and considering the four-point functions for $\hat{sl}(2)_k$ admissible representations as the staring point, we propose an analytic continuation of these correlators and rederive the generic structure contants of the $SL(2,\mathbb{R})_k$ WZNW theory. The analytic continuation we consider here employs standard tricks to adapt the Coulomb gas prescription to the case of non-rational conformal field theory, and it is shown to lead to the exact result. The structure constants we rederive here have direct application in string theory as these describe scattering amplitudes in AdS_3 space and in the 2D black hole background.

This note was written as a contribution to *String Theory Research Progress*, Nova Publishers, NY, 2007.

1. Introduction

The three-point correlation functions in the $SL(2,\mathbb{R})_k$ Wess-Zumino-Novikov-Witten theory (WZNW) describe string scattering amplitudes in AdS_3 space and in the 2D black hole. Because of this, these observables are important objects in the study of string theory formulated on non-compact curved backgrounds. This is actually the motivation we have for studying this subject here. We will present a rederivation of these WZNW correlators by using the often called Coulomb gas realization. The $SL(2,\mathbb{R})_k$ WZNW three-point functions were already computed in the literature by using several methods, see for instance [1, 2, 3, 4, 5, 6, 7, 8]. Unlike previous calculations, the way of deriving these correlation functions we propose here turns out to be remarkably simple. It makes use of simple techniques of the free field representation and the Coulomb gas integral representation.

Coulomb gas-like representation has shown to be a very powerful tool to work out the functional properties of correlation functions in conformal field theory (CFT). While developed for the case of rational conformal field theories, such representation was eventually employed to study non-compact conformal theories. More precisely, even though the Coulomb gas representation was first considered within the context of the minimal models [9, 10] and the $SU(2)_k$ WZNW theory [11], it was subsequently applied with appropriate modifications to the case of non-rational conformal field theories, *e.g.* the Liouville field theory [12] and the non-compact $SL(2,\mathbb{R})_k$ WZNW [2, 3]. The minimal models coupled to the 2D gravity, often called minimal gravity, were also studied in this framework with success [13]. It turns out that the adaptation of the Coulomb gas prescription to the case of non-rational CFT requires analytic continuation in order to make the integral representation of the free field theory to make sense. This aspect was particularly emphasized in Refs. [14, 13, 12] where some tricks were needed in order to make sense of the resulting formulas. Namely, divergent overall factors standing in correlators [14], a complex number of integral to be performed [12, 2], products whose indices run from zero to a negative integer, a fractional number of Wick contraction to be done [15], etc. are just some examples of the expressions we get when trying to naively adapt the formulae of the Coulomb gas representation to the case of non-rational CFT. In [14], it was explained that certain divergent factors $\sim \Gamma(-n)$ arising in the integral representation of Liouville correlation functions can be understood as due to the non-compactness of the target space in the stringy interpretation of correlator as representing bulk scattering processes, and one has to learn how to deal with such a divergent expression. In [16] it was pointed out that the results of the free field computation correspond to the residues on the pole of resonant correlators, and it is precisely what makes the analytic continuation possible. Besides the divergences arising in non-rational CFT computations, there is an additional problem: It turns out that the amount of Wick contractions and the amount of the integrations to be performed to compute a generic correlator are not necessarily a positive integer numbers. For instance, the non-real number of integrations required when realizing Liouville correlation functions by these means was discussed in [12]; and the usual trick in these cases is to assume that such amount of integrations was ruled by integer numbers through the computation and, then, the eventual expression has to be analytically extended to the complex plane [2]. In general, when a complex number of integrals are required, namely when a complex number of screening charges is considered, the expressions are highly formal and one has to deal with the problem of analytically continue such expressions. After such analytic continuation of the integral representation one usually gets expressions containing products running from a positive integer to a non-positive integer value in the upper limit. Dealing with these products is actually much simpler than performing the mentioned analytic continuation of the integral expressions. It was discussed in detail in Ref. [13] for the computation of three-point functions in minimal gravity. There, some tricks were employed through the Coulomb gas-like realization and formal expressions like

$$\prod_{r=1}^{-n} f(r) = \prod_{r=0}^{n-1} \frac{1}{f(-r)}$$

were considered to make sense of the products.

Here, we are interested in computing three-point functions in the $SL(2, R)_k$ WZNW model by similar techniques, and we will do this by using the relation existing between this model and the Liouville field theory. We will employ an integral realization such that it will not be necessary for us to deal with the case of complex amount of integrals, as we did in [3], but just with the case of product of quotient of Γ-functions whose indices run from positive values to negative values in the upper bound. So that, this is indeed analogous to the case of the computation in minimal gravity, and, consequently, we will be able to carefully describe each step in the analytic continuation through the calculations.

2. Coulomb Gas Realization

The starting point of our analysis is the following remark: The relation between the $SU(2)_k$ Knizhnik-Zamolodchikov equation (KZ) and the Belavin-Polyakov-Zamolodchikov equation (BPZ) of minimal models worked out in Ref. [17] can be straightforwardly generalized of the non-compact case of $SL(2, \mathbb{R})_k$. This implies that the BPZ-like equation satisfied by five-point functions in Liouville CFT coincides with the KZ equation satisfied by four-point functions in the WZNW model on $SL(2, \mathbb{R})$. Actually, this result is a powerful tool which enables us to infer properties of the WZNW model from what is known for the case of Liouville field theory. In particular, it permitted to prove the crossing symmetry of the $SL(2, \mathbb{C})/SU(2)$ WZNW four-point function in Ref. [18]; and also it has led to show how to write representations for the monodromy of the conformal blocks in the $SL(2, \mathbb{C})/SU(2)$ model by means of the corresponding quantities in Liouville theory [19]. This was also employed in Ref. [20, 5] to study properties of the $SL(2, \mathbb{R})_k$ four-point correlation functions, as its logarithmic divergences.

More precisely, the relation pointed out by Fateev and Zamolodchikov in Ref. [17] states that the BPZ equation satisfied by five-point correlators including degenerate fields $\psi_{2,1}$ (or $\psi_{1,2}$) in the minimal models coincides with the KZ equation for four-point functions in the $SU(2)_k$ WZNW model when the following relations between the conformal dimensions Δ of degenerate operators and the Kac-Moody level k hold

$$4\Delta_{(2,1)} + 2 = 3(k+2), \quad 4\Delta_{(1,2)} + 2 = 3(k+2)^{-1}$$

for the case $\psi_{2,1}$ (resp. $\psi_{1,2}$). On the other hand, we notice that the non-compact $SL(2, \mathbb{R})_k$ case (it is in what we are interested here) is related to the $SU(2)_k$ case by replacing $k \to -k$, while the degenerate states of minimal models $\psi_{2,1}$ (and $\psi_{1,2}$) are in correspondence with the state(s) $V_{\alpha=-\frac{1}{2b}}$ (and resp. $V_{\alpha=-\frac{b}{2}}$) of the Liouville field theory, being $b^{-2} = k - 2$. So that, the non-compact generalization of the correspondence pointed out in [17] leads to write down an explicit form for four-point functions in WZNW model on $SL(2, \mathbb{R})$ in terms of the five-point functions of Liouville field theory. These five-point correlators include degenerate states (so that contain null descendants) that are represented by Liouville vertex operators of the form $e^{-\sqrt{k-2}\varphi(x)}$. Besides, the Liouvilel correlation functions in the Coulomb-gas like prescription are realized by inserting a precise amount of screening operators of the form $e^{2(k-2)^{\mp 1/2}\varphi(w)}$; see (10) below.

Let us begin by writing down the WZNW four-point functions [1]

$$\mathcal{A}^{WZNW}_{j_1,j_2,j_3,j_4} = \left\langle \prod_{i=1}^{N=4} \Phi_{j_i}(z_i|x_i) \right\rangle_{WZNW} \tag{1}$$

in terms of the Liouville five-point functions

$$\mathcal{A}^{Liouville}_{\alpha_1,\alpha_2,-\frac{1}{2b},\alpha_3,\alpha_4} = \left\langle \prod_{i=1}^{N+1=5} V_{\alpha_i}(z_i) \right\rangle, \tag{2}$$

which include a fifth operator with momentum $\alpha_5 = -\frac{1}{2b}$. The key point is that (1) and (2) are related as follows

$$\mathcal{A}^{WZNW}_{j_1,j_2,j_3,j_4} = \mathcal{N}_{j_1,j_2,j_3,j_4} |x|^{-2\alpha_2/b} |1-x|^{-2\alpha_3/b} |x-z|^{-2\alpha_1/b} \times$$
$$\times |z|^{-4(b^2 j_1 j_2 - \alpha_1 \alpha_2)} |1-z|^{-4(b^2 j_3 j_1 - \alpha_3 \alpha_1)} \mathcal{A}^{Liouville}_{\alpha_1,\alpha_2,-\frac{1}{2b},\alpha_3,\alpha_4} \tag{3}$$

where $x_1 = x$, $x_2 = 0$, $x_3 = 1$, and $x_4 = \infty$ are auxiliary variables used to classify $SL(2,\mathbb{R})$ representations (see Ref. [4] for details, and see also (25) below), and where

$$2\alpha_1 = -b(j_1 + j_2 + j_3 + j_4 + 1) \, , \quad 2\alpha_i = -b(j_1 - j_2 - j_3 - j_4 + 2j_i - b^{-2} - 1),$$

with $i = \{2,3,4\}$ and, as mentioned, $b^{-2} = k - 2$. We also made use of projective invariance to fix the inserting points, $z_1 = z$, $z_2 = 0$, $z_3 = 1$, $z_4 = \infty$. In the expression above the symbol $\mathcal{N}_{j_1,j_2,j_3,j_4}$ represents an overall factor which is determined by the normalization of the reflection coefficients of the theory. Notice that it does depend on k as well.

Then, (3) allows to study the solutions to KZ equation indirectly by studying correlators on the *Liouville side* of the correspondence between both CFT's [5]. With this purpose, let us remind basic aspects of Liouville conformal field theory and its free field representation (see [21] for the details).

Consider the quantum Liouville action

$$S_L = \int d^2z \left((\partial \varphi)^2 + 4\pi Q R \varphi + \mu e^{2b\varphi} \right) \tag{4}$$

being

$$Q = (b + b^{-1})$$

and where R is the two-dimensional scalar curvature. The Liouville φ field propagator is given by

$$\langle \varphi(z,\bar{z}) \varphi(w,\bar{w}) \rangle = -\frac{1}{2} \log |z-w|^2 \tag{5}$$

and the stress-tensor takes the form

$$T = -\partial \varphi \partial \varphi + Q \partial^2 \varphi, \tag{6}$$

---

[1] This is the usual notation for $SL(2,\mathbb{R})_k$ WZNW correlation functions. The complex coordinates x_i here represent auxiliary variables which classify the $SL(2,\mathbb{R})$ representations; see [4] for details; see also (25) below.

which leads to the central charge

$$c = 1 + 6Q^2 > 1. \tag{7}$$

The exponential primary fields are important objects of the theory; namely

$$V_\alpha(z) =: e^{2\alpha\varphi(z)} :, \tag{8}$$

which have conformal dimension

$$\Delta_\alpha = \alpha(Q - \alpha)$$

with respect to the stress tensor (6).

Notice that the formula for the conformal dimension is invariant under the called Liouville reflection symmetry $\alpha \to Q - \alpha$; and corresponding states would be related by such conjugation through an operator valued relation that involves the reflection coefficient of the theory.

In Liouville CFT, there are two different screening operators that can be used to compute correlation functions; namely

$$S_\pm = \int d^2 z V_{b^{\pm 1}}(z) = \mu_\pm \int d^2 z e^{2b^{\pm 1}\varphi(z)} \tag{9}$$

being

$$\mu_- = \frac{\Gamma(1 - b^{-2})}{\pi \Gamma(b^{-2})} \left(\pi \frac{\Gamma(b^2)}{\Gamma(1 - b^2)} \mu_+\right)^{b^{-2}}.$$

One of these, S_+, corresponds to the Liouville cosmological term itself, and is the one we will consider here. So, the N-point correlation functions can be written as follows

$$A^{Liouville}_{\alpha_1,\ldots\alpha_N} = \langle V_{\alpha_1}(z_1) V_{\alpha_2}(z_2)\ldots V_{\alpha_N}(z_N) \rangle =$$

$$= \frac{\Gamma(-n)}{b} \left\langle \prod_{\mu=1}^{N} e^{2\alpha_\mu \varphi(z_\mu)} \prod_{r=1}^{n} \int d^2 w_r e^{2b\varphi(w_r)} \right\rangle \delta\left(\sum_{\mu=1}^{4} \alpha_\mu - \frac{1}{2b} + nb - b - \frac{1}{b}\right) \tag{10}$$

where n refers to the amount of screening operators of the type S_+ required to satisfy the charge symmetry condition coming from the integration over the zero-mode of φ, namely $\sum_{\mu=1}^{N} \alpha_\mu + nb = Q$. It is worth pointing out that in the first line of Eq. (10) the expectation value $\langle\ldots\rangle$ is defined by the self-interacting theory (4), while the expectation value in the second line is defined by considering the free theory $\mu = 0$. Taking this into account, the Coulomb gas like realization for the five-point correlators (3) is achieved by considering the propagator (5) and by considering

$$\sum_{\mu=1}^{4} \alpha_\mu - \frac{1}{2b} + nb = Q \tag{11}$$

where the contribution $-\frac{1}{2b}$ comes from the presence of the fifth operator representing the degenerate state with $\alpha_5 = -\frac{1}{2b}$. Condition (11) comes from the integration over the zero mode of φ, and it is necessary for the correlators to be non vanishing. The integration over the zero mode also gives rise the factor $\Gamma(-n)$ in (10). This factor is divergent for configurations obeying $n \in \mathbb{N}$, and can be interpreted as due to the fact that these configurations correspond to bulk scattering and such a divergence is due to the non-compactness of the space [14]. From this, it is clear that for the case $n \notin \mathbb{N}$ the integral expression above has to be understood just formally and a kind of analytic continuation should be required in order to make the non-integer number of integrals to make sense. Regarding this, when taking into account the relation existing between quantum numbers of the states of both models, namely α_μ and j_μ, one finds

$$n = 2j_1.$$

That is, the three-point functions of the $SL(2,\mathbb{R})_k$ model would correspond to a five-point Liouville correlators with no insertion of screening charges, namely $2j_1 = n = 0$. This implies that the $\widehat{sl}(2)_k$ structure constants are already encoded in the normalization factor $\mathcal{N}_{j_1,j_2,j_3,j_4}$ when considering $j_1 = 0$. This fact was used in the literature to rederive the Awata-Yamada fusion rules for the particular case of admissible representations of $\widehat{SL}(2)_k$ algebra [22], and it was done by emulating what had been previously done for the compact $\widehat{su}(2)_k$ case [17]. Instead, here we will focus our attention on the case of generic representations, and the analytic continuation of the products in $\mathcal{N}_{j_1,j_2,j_3,j_4}$ is required to that end. First, it is convenient to introduce an integral representation for four-point functions in $SL(2,\mathbb{R})_k$ WZNW model: The four-point function of correlation function for admissible representations of $\widehat{sl}(2)_k$ algebra admits an integral representation [22]; namely

$$A^{WZNW}_{j_1,j_2,j_3,j_4} = \mathcal{N}_{j_1,j_2,j_3,j_4} \Gamma(-2j_1) |z|^{-\frac{4j_1j_2}{k-2}} |1-z|^{-\frac{4j_1j_3}{k-2}}$$

$$\times \prod_{r=1}^{2j_1} \int d^2w_r |w_r|^{\frac{2}{k-2}(1-k+j_1+j_2-j_3-j_4)} |1-w_r|^{\frac{2}{k-2}(1-k+j_1-j_2+j_3-j_4)} \times$$

$$\times \prod_{r=1}^{2j_1} |z - w_r|^{\frac{2}{k-2}s} |x - w_r|^2 \prod_{r<t}^{2j_1-1,2j_1} |w_r - w_t|^{-\frac{4}{k-2}}. \qquad (12)$$

where we denoted $s = 1 + j_1 + j_2 + j_3 + j_4$ for short. Representation (12) follows from the free field representation of Liouville theory [12] by considering Eqs. (9), (8) and (5). In Ref. [17, 22], the normalization factor $\mathcal{N}_{j_1,j_2,j_3,j_4}$ was inferred from the normalization of reflection coefficients. Actually, the functional form of $\mathcal{N}_{j_1,j_2,j_3,j_4}$ that holds for the case of $\widehat{sl}(2)_k$ admissible representations is straightforwardly inferred from the one for the $\widehat{su}(2)_k$ case; see (13) below. On the other hand, the form of $\mathcal{N}_{j_1,j_2,j_3,j_4}$ for the case of generic representations admits to be written in terms of the often called Zamolodchikov's Υ-functions [18, 19, 20]. Nevertheless, we will not write it in such way here since our intention is precisely that of pointing out how one could analytically extend the formula holding for $\widehat{su}(2)_k$ to the non-compact model. In doing this, one is led to make some tricks to give a meaning to some formulas that, otherwise, should remain just as formal expressions: As in the case of the Coulomb gas-like representation of the minimal gravity [13], such tricks deal with the task of proposing a meaning for the products defined for negative indices and similar monsters.

3. The Structure Constants

In this section we will derive the structure constants of the $\widehat{sl(2)}_k$ theory form the normalization $\mathcal{N}_{j_1,j_2,j_3,j_4}$. As mentioned, in the case of admissible representations this was done in Ref. [22]. Here, we will analytically extend such expression in order to incorporate the whole set of representations and prove that

$$\mathcal{A}^{WZNW}_{j_1=0,j_2,j_3,j_4} = \prod_{a<b} |z_a - z_b|^{2(h_c - h_a - h_b)} |x_a - x_b|^{-2(j_c - j_a - j_b)} C(j_2, j_3, j_4)$$

being $C(j_2, j_3, j_4)$ the structure constants of the theory [1]. So, the main ingredient of our discussion will be the normalization factor $\mathcal{N}_{j_1,j_2,j_3,j_4}$ since we want to show that the case $j_1 = 0$ leads to the correct three-point function. Let us begin by simply considering $j_1 = 0$ in the formula for $\mathcal{N}_{j_1,j_2,j_3,j_4}$ given in [17, 22]. This yields

$$\mathcal{N}_{j_1=0,j_2,j_3,j_4} = \left(\pi \frac{\Gamma(1+\rho)}{\Gamma(1-\rho)}\right)^s \gamma(\rho) \prod_{a=2}^{4} \gamma(1 - (2j_a + 1)\rho) \times$$

$$\prod_{r=1}^{s} \gamma(r\rho) \frac{\prod_{r=1}^{j_2+j_3-j_4} \gamma(r\rho) \prod_{r=1}^{j_2-j_3+j_4} \gamma(r\rho) \prod_{r=1}^{-j_2+j_3+j_4} \gamma(r\rho)}{\prod_{r=1}^{2j_4} \gamma(r\rho) \prod_{r=1}^{2j_3} \gamma(r\rho) \prod_{r=1}^{2j_2} \gamma(r\rho)} \quad (13)$$

where we defined $\rho = -b^2$ for short, and we used the standard notation

$$\gamma(x) = \frac{\Gamma(x)}{\Gamma(1-x)}.$$

It is worth pointing out that, unlike the normalization considered in Ref. [22], where the two-point function was normalized as $<\Phi_{j_1}(z_1)\Phi_{j_1}(z_2)> = |z_1 - z_2|^{-4h}$, here we are using the standard normalization that yields $<\Phi_{j_1}(z_1)\Phi_{j_1}(z_2)> = B(j_1)|z_1 - z_2|^{-4h}$, where $B(j_1)$ is the reflection coefficient computed in [1]; see (24) below. This remark is important for comparing (13) with the expressions in the literature.

Besides, notice that a divergent factor $\Gamma(0)$ also arises in (13), and corresponds to the factor $\Gamma(-n) = \Gamma(-2j_1)$ in Eq. (10). This factor will be eventually cancelled out by another contribution $\Gamma^{-1}(0)$ arising when analytically extending this expression; see below.

The first step in rewriting (13) will be to consider the three factors of the form

$$\frac{\prod_{r=1}^{j_2+j_3+j_4-2j_a} \gamma(r\rho)}{\prod_{r=1}^{2j_a} \gamma(r\rho)};$$

Let us write them by splitting the product as follows

$$\prod_{r=1}^{j_2+j_3+j_4-2j_a} f(r) = \prod_{r=1}^{2j_a} f(r) \prod_{r=2j_a+1}^{j_2+j_3+j_4-2j_a} f(r)$$

which, of course, would be true in the case $j_2 + j_3 + j_4 - 2j_a > 2j_a$. As in [13], the trick here will be to assume some functional relations (arising from manipulating this kind of products) to be valid even in a range of indices for which the products are not necessarily well defined. So, at least formally, we can write

$$\frac{\prod_{r=1}^{j_2+j_3+j_4-2j_a}\gamma(r\rho)}{\prod_{r=1}^{2j_a}\gamma(r\rho)} = \prod_{r=1}^{2j_a}\gamma^{-1}(r\rho)\prod_{r=1}^{2j_a}\gamma(r\rho)\prod_{r=2j_a+1}^{j_2+j_3+j_4-2j_a}\gamma(r\rho) = \prod_{r=2j_a+1}^{j_2+j_3+j_4-2j_a}\gamma(r\rho).$$

Again, let us split the product, basically extending what would be valid for the case $2j_a+1 < -2j_a - 1 < j_2+j_3+j_4 - 2j_a$; namely

$$\prod_{r=2j_a+1}^{j_2+j_3+j_4-2j_a} f(r) = \prod_{r=2j_a+1}^{-2j_a-1} f(r) \prod_{r=-2j_a}^{j_2+j_3+j_4-2j_a} f(r).$$

Here, the analytic continuation becomes necessary when trying to make this splittig to be valid for the three indices j_a simultaneously within the same range of values. Then, such replacement in the formula has to be understood as a natural proposal for a sort of extension instead as an actual identity. The consistency of the results will give substance to these steps a posteriori. Continuing with this idea, we can write

$$\frac{\prod_{r=1}^{j_2+j_3+j_4-2j_a}\gamma(r\rho)}{\prod_{r=1}^{2j_a}\gamma(r\rho)} = \prod_{r=2j_a+1}^{-2j_a-1}\gamma(r\rho)\prod_{r=-2j_a}^{j_2+j_3+j_4-2j_a}\gamma(r\rho).$$

Now, a preliminary observation: We will replace the products of the form $\prod_{r=-x}^{x}\gamma(r\rho)$ appearing in the expression above by the quantity $\rho^{-2x-1}\gamma(-x)$. This can be heuristically motivated by the following expansion

$$\prod_{r=-x}^{x}\gamma(r\rho) = \frac{\Gamma(-x\rho)\Gamma((1-x)\rho)...\Gamma(-\rho)\Gamma(0)\Gamma(+\rho)...\Gamma(x\rho)}{\Gamma(1+x\rho)\Gamma(1+(x-1)\rho)...\Gamma(1+\rho)\Gamma(1)\Gamma(1-\rho)...\Gamma(1-x\rho)}$$

which, taking into account the properties of Γ-function, can be written as

$$\prod_{r=-x}^{x}\gamma(r\rho) = \frac{\Gamma(0)(-1)^x}{\rho^{2x+1}\Gamma(x+1)\Gamma(x+1)} = \frac{\Gamma(-x)}{\rho^{2x+1}\Gamma(x+1)},$$

where, in particular, we used that $\lim_{\varepsilon\to 0}\Gamma(\varepsilon)/\Gamma(\varepsilon-x) = (-1)^x\Gamma(x+1)$ and that $\Gamma(x+1) = x\Gamma(x)$. Consequently, we can write

$$\prod_{r=2j_a+1}^{-2j_a-1}\gamma(r\rho) = \rho^{4j_a+1}\gamma(2j_a+1)$$

and then (13) is proposed to take the form

$$\mathcal{N}_{j_1=0,j_2,j_3,j_4} = \left(\pi\frac{\Gamma(1+\rho)}{\Gamma(1-\rho)}\right)^s \gamma(\rho)\rho^{4(1-s)}\prod_{a=2}^{4}\frac{\gamma(2j_a+1)}{\gamma((2j_a+1)\rho)} \times$$

$$\times \prod_{r=1}^{s}\gamma(r\rho)\prod_{b=2}^{4}\prod_{r=-2j_b}^{j_2+j_3+j_4-j_b}\gamma(r\rho).$$

It is also convenient to rewrite the last three factors of the second line of this equation by redefining the index of the product as $l = r + 2j_b$, obtaining

$$\prod_{r=-2j_b}^{j_2+j_3+j_4-j_b} \gamma(r\rho) = \prod_{l=0}^{s-1} \gamma((l-2j_b)\rho).$$

Then, we write (13) as follows

$$\mathcal{N}_{j_1=0,j_2,j_3,j_4} = \left(\pi \frac{\Gamma(1+\rho)}{\Gamma(1-\rho)}\right)^s \gamma(\rho)\rho^{4(1-s)} \prod_{a=2}^{4} \frac{\gamma(2j_a+1)}{\gamma((2j_a+1)\rho)} \times$$

$$\times \prod_{r=1}^{s} \gamma(r\rho) \prod_{l=0}^{s-1} \gamma((l-2j_2)\rho) \prod_{l=0}^{s-1} \gamma((l-2j_3)\rho) \prod_{l=0}^{s-1} \gamma((l-2j_4)\rho). \quad (14)$$

Besides, it is also useful to write one of the last three factors of the second line of this equation by using that $\gamma(x) = \gamma^{-1}(1-x)$, namely

$$\prod_{l=0}^{s-1} \gamma(1+(l-2j_2)\rho) = \prod_{l=0}^{s-1} \gamma^{-1}((2j_2-l)\rho) =$$

$$= \frac{\Gamma(-2j_2\rho)\Gamma(-(2j_2-1)\rho)...\Gamma(-(2j_2-s+1)\rho)}{\Gamma(2j_2\rho)\Gamma((2j_2-1)\rho)...\Gamma((2j_2-s+1)\rho)} \times \frac{\Gamma(-2j_2+s)}{\Gamma(-2j_2)} \rho^{s+1}, \quad (15)$$

and we can also expand as

$$\prod_{l=0}^{s-1} \gamma((l-2j_2)\rho) =$$

$$= \frac{\Gamma(-2j_2\rho)\Gamma(-(2j_2-1)\rho)...\Gamma(-(2j_2-s+1)\rho)}{\Gamma(2j_2\rho)\Gamma((2j_2-1)\rho)...\Gamma((2j_2-s+1)\rho)} \times \frac{\Gamma(2j_2-s+1)\rho^{-s}}{\Gamma(2j_2-1)} \quad (16)$$

Then, by manipulating Γ-functions again, we get

$$\mathcal{N}_{j_1=0,j_2,j_3,j_4} = \left(\pi \frac{\Gamma(1+\rho)}{\Gamma(1-\rho)}\right)^s \gamma(\rho) \frac{\gamma(j_2-j_3-j_4)}{\gamma(2j_2+1)} \prod_{a=2}^{4} \frac{\gamma(2j_a+1)}{\gamma((2j_a+1)\rho)} \times$$

$$\times \rho^{2s} \prod_{r=1}^{s} \gamma(r\rho) \prod_{r=0}^{s-1} \left(\gamma(1+(r-2j_2)\rho)\gamma((r-2j_3)\rho)\gamma((r-2j_4)\rho)\right). \quad (17)$$

The reason why we preferred to write it in this way is that the second line of (17) can be identified as the contribution coming from a Dotsenko-Fateev (generalized Selberg) integral; namely

$$I_k = \Gamma(-s) \prod_{r=1}^{s} \int d^2 w_r \left(\prod_{r=1}^{s} \left(|w_r|^{-4j_2\rho} |1-w_r|^{-2-4j_3\rho}\right) \prod_{r<t}^{s-1,s} |w_r - w_t|^{4\rho}\right). \quad (18)$$

which can be performed and shown to be (see formula (B.9) of the Appendix of Ref. [10]) equal to

$$I_k = \Gamma(-s)\Gamma(s+1)\pi^s(-1)^s\rho^{2s}\left(\gamma(-\rho)\right)^s \prod_{r=1}^{s} \gamma(r\rho) \times$$

$$\times \prod_{r=0}^{s-1} \left(\gamma(1+(r-2j_2)\rho)\gamma((r-2j_3)\rho)\gamma((r-2j_4)\rho)\right).$$

It is worth noticing that integral (18) is precisely the one that arises when the Wakimoto free field representation is used to realize the three-point correlators [2]. For instance, the exponent of $|1-w_r|^{-2-4j_3\rho}$ in (18) can be thought of as coming from the Wick contraction between a $SL(2,\mathbb{R})_k$ vertex operator and the screening charges. On the other hand, the contributions $|w_r|^{-4j_2\rho}$ indicate the presence of highest weight states of discrete representations in the correlator.

Going back to expression (17), and by redefining index as $t = -r+s-1$, we find that the normalization factor can be written in terms of the Dotsenko-Fateev integral as follows

$$\mathcal{N}_{j_1=0,j_2,j_3,j_4} = \left(\frac{\Gamma(1+\rho)}{\Gamma(1-\rho)}\right)^s \frac{\gamma(j_2-j_3-j_4)}{\gamma(2j_2+1)} \times$$

$$\times \frac{(-1)^s\gamma(\rho)}{\gamma^s(-\rho)\Gamma(-s)\Gamma(1+s)} \prod_{a=2}^{4} \frac{\gamma(2j_a+1)}{\gamma((2j_a+1)\rho)} I_k. \qquad (19)$$

On the other hand, in Ref. [3] (see Eqs. (2.45) and (2.63) there) it was argued that the Dotsenko-Fateev integral referred above could be formally continued to be also expressed in terms of special functions as follows

$$I_k = -\rho\pi^s \left(\gamma(-\rho)\right)^s \gamma(-1-j_2-j_3-j_4)\gamma(2j_2+1)\gamma(-j_2-j_3+j_4)\gamma(-j_2+j_3-j_4) \times$$

$$\times \frac{G(-2-j_2-j_3-j_4)}{G(-1)} \prod_{a=2}^{4} \frac{G(-1-j_2-j_3-j_4+2j_a)}{G(-2j_a-1)}, \qquad (20)$$

where the functions $G(x)$ are k-dependent special functions that can be written in terms of Γ_2 Barnes functions as follows

$$G(x) \equiv (k-2)^{\frac{x(k-1-x)}{2k-4}} \Gamma_2(-x \mid 1, k-2)\Gamma_2(k-1+x \mid 1, k-2)$$

with

$$\log(\Gamma_2(x \mid 1, y)) \equiv \lim_{\varepsilon \to 0} \frac{\partial}{\partial \varepsilon} \left(\sum_{n=0}^{\infty}\sum_{m=0}^{\infty}(x+n+my)^{-\varepsilon} + \sum_{n=0}^{\infty}\sum_{m=0}^{\infty}(n+my)^{-\varepsilon}\right)$$

where the second sum $\sum_{n=0}^{\infty}\sum_{m=0}^{\infty}$ has to understood as excluding the step $m=n=0$. The way of proposing expression (20) makes use of the analytic extension of the functional properties that connect the function $G(x)$ with the Γ-function; see [3, 25] for the details. Formula (20) was also employed to compute correlation functions in a different free field realization [6], where the exact expression of the three-point functions was rederived. If we assume the same analytic continuation here, it enables to replace the piece

$$\prod_{r=1}^{s} \gamma(r\rho) \prod_{r=0}^{s-1} \left(\gamma(1+(r-2j_2)\rho)\gamma((r-2j_3)\rho)\gamma((r-2j_4)\rho)\right) =$$

$$= \frac{(-1)^s \rho^{-2s}}{\Gamma(-s)\Gamma(s+1)\pi^s \gamma^s(-\rho)} I_k$$

arising in our expression for $\mathcal{N}_{j_1=0,j_2,j_3,j_4}$, by the following contribution,

$$\frac{\rho^{-2s+1}(\gamma(-\rho))^s \gamma(-1-j_2-j_3-j_4)\gamma(-j_2-j_3+j_4)\gamma(-j_2+j_3-j_4)}{\Gamma(0)\gamma^s(-\rho)} \times$$

$$\times \frac{\gamma(2j_2+1)G(-2-j_2-j_3-j_4)}{G(-1)} \prod_{a=2}^{4} \frac{G(-1-j_2-j_3-j_4+2j_a)}{G(-2j_a-1)}. \quad (21)$$

where the factor $\Gamma^{-1}(0)$ arises from writing $\Gamma(-s)\Gamma(s+1) = (-1)^s \Gamma(0)$, and, as mentioned, precisely cancels the divergent factor $\Gamma(-2j_1) = \Gamma(0)$ standing from evaluating $j_1 = 0$ in (10). Next, by using functional properties of the G-function, namely

$$G(-1-x) = (-\rho)^{-2x-1}\gamma(1+x)G(x),$$

we can write

$$\gamma(2j_a-j_2-j_3-j_4)G(2j_a-j_2-j_3-j_4-1) = (-\rho)^{2(j_2+j_3+j_4-2j_a)+1}G(j_2+j_3+j_4-2j_a)$$

and

$$\gamma(-1-j_2-j_3-j_4)G(-2-j_2-j_3-j_4) = (-\rho)^{-2s-1}G(1+j_2+j_3+j_4).$$

Besides, we can also use that

$$\frac{G(x+1)}{\gamma((x+1)\rho)} = (-\rho)^{2x+1}\frac{G(-1-x)}{\gamma(1+x)}$$

to write

$$\frac{\gamma(2j_a+1)\gamma(1-(2j_a+1)\rho)}{G(-2j_a-1)} = (-\rho)^{4j_a+1}\frac{1}{G(2j_a+1)}.$$

Finally, the expression for the three-point correlation function reads

$$\mathcal{A}^{WZNW}_{j_1=0,j_2,j_3,j_4} = \widehat{c}_k \prod_{a<b} |x_{ab}|^{2J_{ab}} |z_{ab}|^{2h_{ab}} \left(\pi \frac{\Gamma(1+\rho)}{\Gamma(1-\rho)}\right)^{s+1} \times$$

$$\times \frac{G(1+j_2+j_3+j_4)}{G(-1)} \prod_{a=2}^{4} \frac{G(j_2+j_3+j_4-2j_a)}{G(2j_a+1)}, \quad (22)$$

where we have reinserted the dependence on z_a and x_a, and where we considered the following notation: $h_{34} = -h_2 + h_3 - h_4$, $h_{24} = h_2 - h_3 - h_4$, $h_{23} = h_4 - h_2 - h_3$, and $J_{34} = -j_2 + j_3 - j_4$, $J_{24} = j_2 - j_3 - j_4$, $J_{23} = -j_2 - j_3 + j_4$, and where $\widehat{c}_k$ represents a j-independent factor. Expression (22) corresponds exactly to the three-point function of the $SL(2,\mathbb{R})_k$ WZNW model [1]. Hence, through the analytic continuation whose steps we exhaustively described here, we were able to show that one can actually reconstruct the generic three-point function by means of this Coulomb gas-like prescription. By construction, when replacing $j_1 = j_2 = 0$ in expression (12), we also get the two-point function

$$\mathcal{A}^{WZNW}_{j_1=0,j_2=0,j_3,j_4} = |z_{34}|^{-4j_3(1+j_3)\rho}\left(|x_{34}|^{4j_3}B(j_3)\delta(j_3-j_4) + \delta^{(2)}(x_3-x_4)\delta(j_3+j_4+1)\right) \quad (23)$$

with

$$B(j_3) = -\frac{1}{\pi\rho}\left(\pi\frac{\Gamma(1+\rho)}{\Gamma(1-\rho)}\right)^s \frac{\Gamma(1-(1+2j_3)\rho)}{\Gamma((2j_3+1)\rho)}. \tag{24}$$

The three-point functions (22) describe string scattering amplitudes in AdS_3 space, and this is perhaps the most direct application of this computation within the context of string theory. Let us briefly describe the construction of the string theory on AdS_3 in terms of the $SL(2,\mathbb{R})_k$ WZNW model in the next section.

4. String Amplitudes in AdS_3 Space

The string theory on Euclidean AdS_3 is described by the $SL(2,\mathbb{C})_k/SU(2)$ WZNW model, while its Lorentzian version is constructed in terms of the $SL(2,\mathbb{R})_k$ WZNW model. In this framework, N-string scattering amplitudes in AdS_3 spaces turn out to be given by integrating over the inserting points z_a of the WZNW N-point correlation functions.

Within the context of the string theory interpretation, the Kac-Moody level k of the WZNW theory corresponds to the quotient between the typical string length scale l_s and the AdS radius $l = |\Lambda|^{-1/2}$ through the relation $k \sim l^2/l_s^2$. Then, k controls the coupling of the theory leading, in the large k limit, to both classical and flat limit. The Hilbert space of the theory is then given in terms of certain representations of $SL(2,\mathbb{R})_k \times SL(2,\mathbb{R})_k$ which is the symmetry of the model. Then, the string states are described by vectors $|j,m\rangle \otimes |j,\bar{m}\rangle$, which are defined by acting with the (Fourier transform of the) vertex operators

$$\Phi_{j,m,\bar{m}}(z) = \int d^2x\, \Phi_j(z|x)\, x^{j-m} \bar{x}^{j-\bar{m}} \tag{25}$$

on the $SL(2,\mathbb{R})_k$-invariant vacuum $|0\rangle$; namely

$$\lim_{z\to 0} \Phi_{j,m,\bar{m}}(z)|0\rangle = |j,m\rangle \otimes |j,\bar{m}\rangle. \tag{26}$$

In order to precisely define the string theory, it is also necessary to identify which is the subset of representations $|j,m\rangle$ that have to be taken into account to construct the Hilbert space. Such a subset has to satisfy several requirements: In the case of the free theory these requirements are associated to the normalizability and unitarity of the string states. However, it is worth pointing out that, even in the case of the free string theory, the fact of considering non-compact curved backgrounds is not trivial at all. The main obstacle in constructing the Hilbert space is the fact that, unlike what happens in flat space, the Virasoro constraints are not enough to completely decouple the negative-norm string states of the theory, which are often called ghosts. In early attempts for constructing a consistent string theory in AdS_3, additional *ad hoc* constraints were imposed on the vectors of the $SL(2,\mathbb{R})_k$ representations in order to decouple those ghost states. Usually, the vectors of $SL(2,\mathbb{R})$ representations are labeled by a pair of indices j and m, and the first proposals to achieve unitarity took the form of an upper bound for the index j of the discrete representations of the group (namely, $1-k < 2j < -1$ it was demanded). In modern approaches, the way of dealing with the "negative norm states problem" also implies such a kind of constraint, although it does not necessarily implies an upper bound on the mass spectrum as the old fashion ways do; see [4] for details. In the case of Euclidean AdS_3, the spectrum of string theory is given by the continuous C_λ^α series of $SL(2,\mathbb{C})$ and is parameterized by the values $j = -\frac{1}{2} + i\lambda$ with $\lambda \in \mathbb{R}$ and by real m. Besides, the case of string theory in Lorentzian AdS_3 is richer and its spectrum is composed by states belonging to both continuous C_λ^α and discrete $\mathcal{D}_j^\pm$ series. The continuous series C_λ^α have states with $j = -\frac{1}{2} + i\lambda$ with $\lambda \in \mathbb{R}$ and $m - \alpha \in \mathbb{Z}$, with $\alpha \in [0, 1) \in \mathbb{R}$. The vectors of these representations correspond to "long strings" in AdS_3. On the other hand, the states of discrete representations $\mathcal{D}_j^\pm$ satisfy $j = \pm m - n$ with $n \in \mathbb{Z}_{\geq 0}$. The states belonging to discrete representations have a discrete energy spectrum and represent the quantum version of those string states that are confined in the center of AdS_3 space; these are called "short strings". The long strings states describe massive strings that can escape to the infinity due to the coupling with the NS-NS field.

Moreover, in order to fully parameterize the spectrum in AdS_3 one has to introduce additional representations described by the often called "flowed" states, which are defined through the spectral flow automorphism of the $\hat{sl}(2)_k$ algebra and parameterized by an additional integer number ω. In other words, new states $|j, m, \omega\rangle \otimes |j, \bar{m}, \omega\rangle$ belonging to spectral flowed representations $\mathcal{D}_j^{\pm,\omega}$ and $C_\lambda^{\alpha,\omega}$ have to be included in the Hilbert space of the theory [4].

For the long strings, the quantum number ω can be actually thought of as a string winding number asymptotically. This winding number as a degree of freedom is due to the fact that strings in AdS_3 can couple to the NS-NS field, which makes the strings to expand to the boundary of the space. Then, this is not a topological winding and so it can be in principle violated in the scattering processes. In what we discussed in the previous section we did not take into account the winding number. Instead, we focused our attention on the $\omega_i = 0$ sector. However, this can be generalized, and we will return to this point below.

There exists evidence that suggests that the whole spectrum of the theory in AdS_3 is constructed just by the continuous and discrete representations $\mathcal{D}_j^{\pm,\omega}$ and $C_\lambda^{\alpha,\omega}$; even though these are not the only unitary representations of the group. This is particularly suggested by the modular invariance of the one-loop partition function in thermal AdS_3 and the fact that only long strings and short strings do appear in the spectrum. Nevertheless, other states, having momentum $j = -((n+1)(k-2)+1)/2$ with $n \in Z_{\geq 0}$, may also seem to appear in the theory. These are certainly non-perturbative states since, for heavy states, they have masses of the order j and, hence, of the order of the string tension. Since k controls the semi-classical limit, these states would decouple from the perturbative spectrum when k goes to infinity. These $j \sim -k$ states are associate to worldsheet instantons and a semi-classical interpretation of them relates their presence to the emergence of non-local effects in the dual CFT on the boundary, according to the AdS/CFT correspondence.

When string interactions are considered, additional restrictions to those imposed for unitarity on the free spectrum may appear. This is because, besides the requirement of normalizability and unitarity, it is necessary to guarantee that the fusion rules result closed among the unitary spectrum. Otherwise, it could be possible to produce negative norm states by scattering processes involving unitary incoming states. After integrating the N-point correlators over the worldsheet insertions z_a ($a \in \{1, 2, ...N\}$), the pole structure of scattering amplitudes and the factorization properties of them could also result in additional constraints on the external (incoming and outgoing) states, and in some cases such constraints can be even more restrictive than those required for unitarity. To be more precise, the N-point scattering amplitudes in AdS_3 are known to be well defined only if the external momenta of the states involved in the process satisfy the bound $\sum_{a=1}^{N} j_a > 3 - N - k$. This restriction is stronger than the unitarity bound $1 - k < j_a$ imposed on each individual state. In Ref. [4], such bound was analyzed from the viewpoint of the AdS/CFT correspondence and it was concluded that only those correlators satisfying such constraint would have a clear interpretation in terms of a local CFT on the boundary.

Here, we were concerned with the Coulomb gas-like computation of such correlators. Within this context, an interesting question one can rise is how to understand the unitarity constraints on WZNW correlators in terms of the Liouville five-point function. We will address this question in the future. Another interesting question is whether it is possible to describe also the three-point functions that violate the winding number conservation as we did for the conservative case (22). To answer this question it is useful to point out that such non-conservative three-point functions are obtained by a limiting procedure from certain particular four-point functions that involve a fourth operator with momentum $j_1 = -k/2$. Indeed, the violating winding three-point function was computed in Ref. [5] by employing the same relation between Liouville five-point function and WZNW four-point function we discussed here, and the result was shown to agree with the result previously obtained by Maldacena and Ooguri in Ref. [4]; see also [6] for a recent derivation, and see Refs. [3, 8, 4, 5] for further details[2].

5. Conclusion

Here, we discussed three-point functions in the $SL(2,\mathbb{R})_k$ WZNW model. These quantities describe string scattering amplitudes in AdS_3 space and are also related to the analogous quantities in the 2D black hole background. By means of the relation between Liouville field theory and WZNW theory we were able to present a rederivation of the three-point correlation functions (22) for this model. More precisely, we explicitly obtained the structure constants by analytically continuing an expression holding for the case of admissible representations of $\hat{SL}(2,\mathbb{R})_k$. First, we considered the Liouville five-point function as the staring point, and then, by taking into account the relation between this CFT and the WZNW model, we carefully described the steps involved in the analytic continuation of the normalization factor $\mathcal{N}_{j_1,j_2,j_3,j_4}$ that yielded the exact expression of $SL(2,\mathbb{R})_k$ WZNW structure constants. The Coulomb gas like realization and the free field approach had already shown to be a fruitful method to work out the functional form of correlators in this class of conformal models [12, 13, 2, 3, 7, 8, 24, 6]. Our aim here was to emphasize this aspect by presenting a concise example which shows how the Coulomb gas inspired computation works.

The three-point functions we rederived here were recently employed to perform highly non-trivial consistency checks of the AdS_3/CFT_2 Maldacena conjecture [26]. Our hope is that the relation between WZNW theory and Liouville theory will help in further understanding this correspondence beyond the semiclassical limit.

---

[2]The violation of the winding number was first suggested in a unpublished work by Fateev, Zamolodchikov and Zamolodchikov for the case of the 2D black hole background [23].

This work is inspired in my previous collaborations with Daniel López-Fogliani, Carmen Núñez and Claudio Simeone. This work was partially supported by CONICET, ANPCyT and University of Buenos Aires through grants PIP6160, PICT34557, UBACyT X861. The financial support of Fulbright Fellowship is also acknowledged. I wish also express my gratitude to the members of CCPP at New York University for the hospitality during my stay.

References

[1] J. Teschner, *Nucl. Phys.* **B546** (1999) pp. 390; *Nucl. Phys.* **B571** (2000) pp. 555.

[2] K. Becker, *Strings, Black Holes and Conformal Field Theory*, [arXiv:hep-th/9404157]; K. Becker and M. Becker, *Nucl.Phys.* **B418** (1994) pp. 206-230.

[3] G. Giribet and C. Núñez, *JHEP* **0006** (2000) pp. 033. G. Giribet and C. Núñez, *JHEP* **0106** (2001) pp. 010. G. Giribet and D. López-Fogliani, *JHEP* **0406** (2004) pp. 026.

[4] J. Maldacena and H. Ooguri, *Phys.Rev.* **D65** (2002) pp. 106006; *J.Math.Phys.* **42** (2001) pp. 2929.

[5] G. Giribet, *Phys. Lett.* **B628** (2005) pp. 148.

[6] G. Giribet, *Phys. Lett.* **B637** (2006) pp. 192.

[7] N. Ishibashi, K. Okuyama and Y. Satoh, *Nucl. Phys.* **B 588** (2000) 149. K. Hosomichi, K. Okuyama and Y. Satoh, *Nucl. Phys.* **B598** (2001) 451.

[8] T. Fukuda and K. Hosomichi, *JHEP* **0109** (2001) pp. 003.

[9] A.A. Belavin, A.M. Polyakov and A.B. Zamolodchikov, *Nucl.Phys.* **B241** (1984) pp. 333.

[10] V. S. Dotsenko and V. A. Fateev, *Nucl. Phys.* **B 251** (1985) PP. 691; *Nucl. Phys.* **B240** (1984) pp. 312.

[11] V. Dotsenko, *Nucl. Phys.* **B338** (1990) pp. 747; V. Dotsenko, *Nucl. Phys.* **B358** (1991) pp. 541.

[12] M. Goulian and M. Li, *Phys.Rev.Lett.* **66** (1991) pp. 2051-2055.

[13] V. S. Dotsenko, *Mod. Phys. Lett.* **A6** (1991) pp. 3601.

[14] P. Di Francesco and D. Kutasov, *Nucl. Phys.* **B375** (1992) 119; *Phys. Lett.* **B261** (1991) 385.

[15] J.L. Peterssen, J. Rasmussen and M. Yu, *Nucl.Phys.* **B481** (1996) 577. P. Furlan, A. Ganchev and V. Petkova, *Nucl. Phys.* **B491** (1997) 635.

[16] Al. Zamolodchikov and A.B. Zamolodchikov, *Nucl. Phys.* **B477** (1996) pp. 577.

[17] V. Fateev and A.B. Zamolodchikov, *Sov.J.Nucl.Phys.* **43** (4) (1987) pp. 657.

[18] J. Teschner, *Phys. Lett.* **B521** (2001) pp. 127.

[19] B. Ponsot, *Nucl. Phys.* **B642** (2002) pp.114.

[20] G. Giribet and C. Simeone, *Int. J. Mod. Phys.* **A20** A20 (2005) pp. 4821.

[21] Yu Nakayama, *Int. J. Mod. Phys.* **A19** (2004) pp. 2771.

[22] O. Andreev, *Phys. Lett.* **B363** (1995) 166.

[23] V. Fateev, Al.B. Zamolodchikov and A.B. Zamolodchikov, unpublished.

[24] G. Giribet, *Nucl. Phys.* **B739** (2006) pp 209.

[25] V. Fateev and A. Litvinov, *Multipoint correlation functions in Liouville field theory and minimal Liouville gravity*, [arXiv:0707.1664].

[26] A. Dabholkar and A. Pakman, *Exact chiral ring of AdS(3)/CFT(2)*, [arXiv:hep-th/0703022]. A. Pakman and A. Sever, *Exact N=4 correlators of AdS(3)/CFT(2)*, [arXiv:0704.3040]. M. Gaberdiel and I. Kirsch, *Worldsheet correlators in AdS(3)/CFT(2)*, [arXiv:hep-th/0703001].

SPECIAL BIBLIOGRAPHY ON STRING THEORY

John Harrington

1989 Summer School in High Energy Physics and Cosmology, Trieste, Italy, 26 June-18 August, 1989 / editors, J.C. Pati ... [et al.]. Published/Created: Singapore; Teaneck, N.J.: World Scientific, c1990. Related Names: Pati, J. C. Description: x, 696 p.: ill.; 22 cm. ISBN: 9810201745 Notes: Includes bibliographical references. Subjects: Superstring theories--Congresses. String models--Congresses. Supersymmetry--Congresses. Cosmology--Congresses. Series: The ICTP series in theoretical physics, 0218-0243; v. 6 LC Classification: QC794.6.S85 S85 1989 Dewey Class No.: 539.7/2 20

2D-gravity in non-critical strings: discrete and continuum approaches / E. Abdalla ... [et al.]. Published/Created: Berlin; New York: Springer-Verlag, c1994. Related Names: Abdalla, Elcio. Description: ix, 319 p.; 24 cm. ISBN: 0387578056 (New York: alk. paper): 3540578056 (Berlin: alk. paper): Notes: Includes bibliographical references (p. 309-316) and index. Subjects: Quantum gravity. Supergravity. String models. Sturm-Liouville equation--Numerical solutions. Mathematical physics. Series: Lecture notes in physics. New series m, Monographs; m20 LC Classification: QC178 .A12 1994 Dewey Class No.: 539.7/54 20

A first course in string theory / Barton Zwiebach. Published/Created: New York: Cambridge University Press, 2004. Description: xx, 558 p.: ill.; 26 cm. ISBN: 0521831431 Notes: Includes bibliographical references and index. Subjects: String models. LC Classification: QC794.6.S85 Z95 2004 Dewey Class No.: 530.14 22

A mathematical introduction to string theory: variational problems, geometric and probabilistic methods / Sergio Albeverio ... [et al.]. Published/Created: Cambridge; New York: Cambridge University Press, c1997. Related Names: Albeverio, Sergio. London Mathematical Society. Description: viii, 135 p.; 23 cm. ISBN: 0521556104 (pb) Notes: Includes bibliographical references (p. 126-132) and index. Subjects: String models--Mathematics. Mathematical physics. Series: London Mathematical Society lecture note series; 225 LC Classification: QC794.6.S85 M38 1997 Dewey Class No.: 539.7/2 20 Electronic File Information: Publisher description http://www.loc.gov/catdir/description/

cam027/95049052.ht ml Table of contents http://www.loc.gov/catdir/toc/cam027/95049052.html

A mechanical string model of adiabatic chemical reactions / W. Kliesch. Published/Created: Berlin; New York: Springer, c1998. Description: 128 p.: ill.; 24 cm. ISBN: 3540649786 (softcover: alk. paper) Notes: Includes bibliographical references (p. [123]-126) and index. Subjects: Chemical reactions--Mathematical models. String models. Series: Lecture notes in chemistry, 0342-4901; 69 LC Classification: QD501 .K7549 1998 Dewey Class No.: 541.3/9/015118 21

An introduction to black holes, information and the string theory revolution: the holographic universe / Leonard Susskind, James Lindesay. Portion of Title: Black holes, information and the string theory revolution Published/Created: Hackensack, NJ: World Scientific, c2005. Related Names: Lindesay, James. Description: xv, 183 p.: ill.; 24 cm. ISBN: 9812560831 9812561315 (pbk.) Notes: Includes bibliographical references (p. 179) and index. Subjects: Black holes (Astronomy) Holography. String models. LC Classification: QB843.B55 S97 2005 Dewey Class No.: 523.8875 22

Analytische Renormierung von String-Funktionalen / Thomas Filk. Published/Created: Bonn: Universität Bonn, Physikalisches Institut, 1982. Related Names: Universität Bonn. Physikalisches Institut. Description: 77 p.: ill.; 30 cm. Notes: "ISSN 0172-8741." Thesis (doctoral)--Rheinische Friedrich-Wilhelms-Universität zu Bonn, 1982. Includes bibliographical references. Subjects: String models. LC Classification: QC794.6.S85 F55 1982

Architecture des interactions fondamentales à courte distance = Architecture of fundamental interactions at short distances / édité par Pierre Ramond et Raymond Stora. Published/Created: Amsterdam; New York: North-Holland; New York, N.Y.: Sole distributors for the U.S.A. and Canada, Elsevier Science Pub. Co., 1987. Related Names: Ramond, Pierre, 1943- Stora, Raymond, 1930- Ecole d'été de physique théorique (Les Houches, Haute-Savoie, France) (44th: 1985) Related Titles: Architecture of fundamental interactions at short distances. Description: 2 v. (xxxvii, 1059 p.): ill.; 23 cm. Notes: At head of title: USMG, NATO ASI, Les Houches, session XLIV, 1er juillet-8 aout 1985. Lectures delivered at Les Houches Summer School of Theoretical Physics. Includes bibliographies. Subjects: Electroweak interactions--Congresses. Quantum chromodynamics--Congresses. String models--Congresses. Astrophysics--Congresses. LC Classification: QC794.8.E44 A73 1987 Dewey Class No.: 539.7/54 19 Language Code: engfre

Beyond extreme physics. Edition Information: 1st ed. Published/Created: New York: Rosen Pub. Group, 2008. Projected Publication Date: 0801 Related Titles: Scientific American. Description: p. cm. ISBN: 9781404214026 (library binding) Contents: "The first few microseconds" / by Michael Riordan and William A. Zajc -- "An echo of black holes" / by Theodore A. Jacobson and Renaud Parentani -- "The illusion of gravity" / by Juan Maldacena -- "The mysteries of mass" / by Gordon Kane -- "Inconstant constants" / by John D. Barrow and John K. Webb -- "Quantum black holes" / by Bernard J. Carr and Steven B. Giddings -- "The string theory landscape" / by Raphael Bousso and Joseph Polchinski. Notes: Includes index. Subjects: Relativity (Physics) Particles (Nuclear physics) Quantum theory. Gravity. String models. Series: Scientific American cutting-edge science LC

Classification: QC173.55 .B497 2008
Dewey Class No.: 539 22

Bosonic strings: a mathematical treatment / Jürgen Jost. Published/Created: Providence, R.I.: American Mathematical Society/International Press, c2001. Description: xi, 95 p.; 27 cm. ISBN: 0821826441 (alk. paper) Notes: Includes bibliographical references (p. 91-92) and index. Subjects: String models. Superstring theories. Series: AMS/IP studies in advanced mathematics, 1089-3288; v. 21 LC Classification: QC794.6.S5 J67 2001 Dewey Class No.: 530.14 21

Braided cord animals you can make / by Claude Morin; drawings by Wiegeist; photographs by Jean-Pierre Tesson; [translated by Maxine Hobson]. Published/Created: New York: Sterling Pub. Co., c1976. Related Names: Wiegeist. Tesson, Jean-Pierre. Related Titles: Animals you can make. Description: 32 p.: ill. (some col.); 22 cm. ISBN: 0806954000 0806954019 (lib. bdg.) Summary: Step-by-step directions for basic braiding techniques and several braided cord projects. Notes: Translation of La ficelle tressée. Subjects: String craft--Juvenile literature. Cordage--Juvenile literature. Zoological models--Juvenile literature. String craft. Braid. Handicraft. Series: Easy craft series LC Classification: TT880 .M6613 1976 Dewey Class No.: 746/.04/71 Language Code: engfre

Champs, cordes, et phénomènes critiques = Fields, strings, and critical phenomena / édité par E. Brézin et J. Zinn-Justin. Published/Created: Amsterdam; New York: North-Holland; New York, N.Y., U.S.A.: Sole distributors for the U.S.A. and Canada, Elsevier Science Pub. Co., 1990. Related Names: Brézin, E. Zinn-Justin, Jean. Université scientifique et médicale de Grenoble. NATO Advanced Study Institute. Related Titles: Fields, strings, and critical phenomena. Description: xxix, 640 p.: ill.; 23 cm. Notes: Text in English. At head of title: USMG, NATO ASI. Les Houches, session XLIX, 28 juin-5 aout 1988. Lectures presented at the Ecole d'été de physique théorique. Includes bibliographical references. Subjects: Field theory (Physics)--Congresses. String models--Congresses. Critical phenomena (Physics)--Congresses. Statistical mechanics--Congresses. LC Classification: QC173.68 .E26 1988 Dewey Class No.: 530.1/4 20

Charged stringlike solutions in low-energy heterotic string theory [microform] / by Daniel John Waldram. Published/Created: 1993. Description: p. 2528-2535; 28 cm. Notes: Offprint from: Physical review D, vol. 47, no. 6, 1993. Thesis (Ph. D.)--University of Chicago, 1993. Additional Formats: Microfilm. Chicago, Ill.: University of Chicago, Joseph Regenstein Library, Dept. of Photoduplication, 1993. 1 microfilm reel; 35 mm. Subjects: String models. LC Classification: Microfilm 94/2699 (Q)

Chern-Simons theory, matrix models, and topological strings / Marcos Mariño. Published/Created: Oxford; New York: Clarendon Press, 2005. Description: xii, 197 p.: ill.; 24 cm. ISBN: 0198568495 9780198568490 Notes: Includes bibliographical references (p. 187-196) and index. Subjects: String models. Gauge fields (Physics) Three-manifolds (Topology) Series: International series of monographs on physics; 131 LC Classification: QC794.6.S85 M37 2005 Dewey Class No.: 539.72 22 National Bibliography No.: GBA552101 bnb National Bibliographic Agency No.: 013225299 Uk

Complex differential geometry and supermanifolds in strings and fields: proceedings of the seventh Scheveningen conference, Scheveningen, The Netherlands, August 23-28, 1987 / P.J.M. Bongaarts, R. Martini, eds.

Published/Created: Berlin; New York: Springer-Verlag, c1988. Description: iv, 252 p.: ill.; 25 cm. ISBN: 0387503242 (U.S.) Cancelled ISBN: 3540503242 (U.S.) Notes: Includes bibliographies. Subjects: Global differential geometry--Congresses. Supermanifolds (Mathematics)--Congresses. String models--Congresses. Field theory (Physics)--Congresses. Series: Lecture notes in physics; 311 LC Classification: QA641 .C614 1988 Dewey Class No.: 516.3/62 19

Confluence of cosmology, massive neutrinos, elementary particles, and gravitation / edited by Behram N. Kurşunoğlu, Stephan L. Mintz, and Arnold Perlmutter. Published/Created: New York: Kluwer Academic/Plenum Publishers, 1999. Related Names: Kurşunoğlu, Behram, 1922- Mintz, Stephan L. Perlmutter, Arnold, 1928- Description: xvi, 246 p.: ill.; 26 cm. ISBN: 0306462087 Notes: "Proceedings of an international conference on confluence of cosmology, massive neutrinos, elementary particles, and gravitation, held December 17-21, 1998, in Fort Lauderdale, Florida"--T.p. verso. Includes bibliographical references and index. Subjects: Neutrinos--Congresses. String models--Congresses. Cosmology--Congresses. LC Classification: QC793.5.N42 C66 1999 Dewey Class No.: 539.7/215 21

Conformal field theory / Yavuz Nutku, Cihan Saçlioglu, and Teoman Turgut, editors. Variant Title: Other title information on cover: New non-perturbative methods in string and field theory Published/Created: Cambridge, Mass.: Perseus Pub., c2000. Related Names: Nutku, Yavuz. Saçlioĭlu, Cihan, 1948- Turgut, Teoman. Description: 1 v. (various pagings): ill.; 25 cm. ISBN: 0738202045 Notes: Includes bibliographical references and index. Subjects: String models. Conformal invariants. Quantum field theory. Series: Frontiers in physics; v. 102 LC Classification: QC794.6.S85 C66 2000 Dewey Class No.: 530.14 21 National Bibliography No.: GBA0-70120

Conformal field theory and its application to strings = Conforme veldentheorie en haar toepassing op strings / door Erik Peter Verlinde. Published/Created: [Netherlands: E.P. Verlinde, 1988] Related Titles: Conforme veldentheorie en haar toepassing op strings. Description: 100 p.: ill.; 24 cm. Notes: Summary in Dutch. Thesis (doctoral)--Rijksuniversiteit te Utrecht, 1988. Includes bibliographical references (p. 85-87). Subjects: String models. Conformal invariants. Quantum field theory. LC Classification: QC794.6.S85 V47 1988 Dewey Class No.: 530.1/43 20 Language Code: eng dut

Conformal invariance and string theory / edited by Petre Dita and Vladimir Georgescu. Published/Created: New York: Academic Press, c1989. Related Names: Dita, P. (Petre), 1942- Georgescu, V. (Vladimir), 1947- Institutul Central de Fizică (Romania) Summer School on Conformal Invariance and String Theory (1987: Poiana Brasov, Romania) Description: xiii, 557 p.: ill.; 24 cm. ISBN: 012218100X (alk. paper) Notes: Lectures delivered at the Summer School on Conformal Invariance and String Theory, held at Poiana Brasov, Romania, Sept. 1-12, 1987, sponsored by the Central Institute of Physics-Bucharest. Includes bibliographical references. Subjects: Statistical mechanics--Congresses. Conformal invariants--Congresses. String models--Congresses. Series: Perspectives in physics LC Classification: QC174.7 .C66 1989 Dewey Class No.: 530.1/3 20

Cosmic landscape: string theory and the illusion of intelligent design / Leonard Susskind. Edition Information: 1st ed. Published/Created: New York: Little, Brown and Co., 2005. Description: xii, 403 p.: ill.; 25 cm. ISBN: 0316155799 (hardcover: alk. paper) Contents: The world according to Feynman -- The mother of all physics problems -- The lay of the land -- The myth of uniqueness and elegance -- Thunderbolt from heaven -- On frozen fish and boiled fish -- A rubber band-powered world -- Reincarnation -- On our own? -- The branes behind Rube Goldberg's greatest machine -- A bubble bath universe -- The black hole war -- Summing up. Notes: Includes bibliographical references and index. Subjects: Cosmogony. Astrophysics. Intelligent design (Teleology) String models. LC Classification: QB981 .S886 2005 Dewey Class No.: 523.1/2 22

Cosmic string wakes and large-scale structure [microform] / by Jane C. Charlton. Published/Created: 1988. Description: p. 521-530: ill. Notes: Offprint from: Astrophysical journal, 325 (Feb. 1988). Thesis (Ph. D.)--University of Chicago, 1988. Additional Formats: Microfilm. Chicago, Ill.: University of Chicago, Joseph Regenstein Library, Dept. of Photoduplication, 1988. 1 microfilm reel; 35 mm. Subjects: Galaxies--Clusters. String models. Cosmology. LC Classification: Microfilm 94/2993 (Q)

Cosmological crossroads: an advanced course in mathematical, physical, and string cosmology / S. Cotsakis, E. Papantonopoulos (eds.). Published/Created: Berlin; New York: Springer, c2002. Related Names: Cotsakis, Spiros, 1963- Papantonopoulos, E. (Eleftherios) Description: xvi, 477 p.: ill.; 24 cm. ISBN: 3540437789 (acid-free paper) Notes: Includes bibliographical references. Subjects: Cosmology. String models. Series: Lecture notes in physics, 0075-8450; 592 LC Classification: QB985 .A44 2001 Dewey Class No.: 523.1 21

Cosmology in gauge field theory and string theory / David Bailin, Alexander Love. Published/Created: Bristol; Philadelphia: Institute of Physics Pub., c2004. Related Names: Love, Alexander. Description: xii, 313 p.: ill.; 24 cm. ISBN: 0750304928 (pbk.) Notes: Includes bibliographical references and index. Subjects: Gauge fields (Physics) String models. Cosmology. Series: Graduate student series in physics LC Classification: QC793.3.G38 B35 2004 Dewey Class No.: 530.14/35 22 National Bibliography No.: GBA469260 bnb National Bibliographic Agency No.: 012994456 Uk

Differential geometric methods in theoretical physics: physics and geometry / edited by Ling-Lie Chau and Werner Nahm. Published/Created: New York: Plenum Press, c1990. Related Names: Chau, Ling-Lie. Nahm, Werner. North Atlantic Treaty Organization. Scientific Affairs Division. Description: xvi, 830 p.: ill.; 26 cm. ISBN: 0306438070 Notes: "Proceedings of the NATO Advanced Research Workshop and the 18th International Conference on Differential Geometric Methods in Theoretical Physics: Physics and Geometry, held July 2-8, 1988, at the University of California, Davis, Davis, California"--T.p. verso. "Published in cooperation with NATO Scientific Affairs Division." Includes bibliographical references and indexes. Subjects: Geometry, Differential--Congresses. Mathematical physics--Congresses. Quantum field theory--Congresses. String models--Congresses. Conformal invariants--Congresses. Series: NATO ASI series. Series B, Physics; vol. 245 NATO advanced science institutes series. Series B, Physics; v. 245. LC Classification: QC20.7.D52 I58 1988 Dewey Class No.: 530.1/5636 20

Differential geometrical methods in theoretical physics / edited by K. Bleuler and M. Werner. Published/Created: Dordrecht; Boston: Kluwer Academic Publishers, c1988. Related Names: Bleuler, Konrad, 1912- Werner, M. Description: xvii, 471 p.; 25 cm. ISBN: 9027728208 Notes: "Proceedings of the NATO Advanced Research Workshop and the 16th International Conference on Differential Geometrical Methods in Theoretical Physics, Como, Italy, 24-29 August 1987"--T.p. verso. Includes bibliographies. Subjects: Geometry, Differential--Congresses. Gauge fields (Physics)--Congresses. String models--Congresses. Mathematical physics--Congresses. Series: NATO ASI series. Series C, Mathematical and physical sciences; vol. 250 NATO ASI series. Series C, Mathematical and physical sciences; no. 250. LC Classification: QC20.7.D52 I58 1987 Dewey Class No.: 530.1/5636 19

Dual resonance models and superstrings / by Paul H. Frampton. Published/Created: Singapore; [Philadelphia]: World Scientific, c1986. Description: xvi, 539 p.: ill.; 23 cm. ISBN: 9971500809: 9971500817 (pbk.): Notes: Includes bibliographies and index. Subjects: Resonance--Mathematical models. Duality (Nuclear physics)--Mathematical models. String models. LC Classification: QC794.6.R4 F73 1986 Dewey Class No.: 530.1 19

Dualities in gauge and string theories: proceedings of APCTP Winter School, Sorak Mountain Resort, Korea, 17-28 February 1997 / edited by Y.M. Cho, S. Nam. Published/Created: New Jersey: World Scientific Publishing, c1998. Related Names: Cho, Y. M. Nam, S. Description: viii, 399 p.: ill.; 26 cm. ISBN: 9810235860 (alk. paper) Subjects: Gauge fields (Physics)--Congresses. String models--Congresses. Duality (Nuclear physics)--Congresses. LC Classification: QC793.3.F5 A63 1997 Dewey Class No.: 530.14/35 21

Duality - strings & fields: proceedings of the 33rd Karpacz Winter School of Theoretical Physics, Karpacz, Poland, 13-22 February, 1997 / edited by Z. Hasiewicz, Z. Jaskólski, J. Sobczyk. Variant Title: Duality - strings and fields Published/Created: [Amsterdam]: North-Holland, c1998. Related Names: Hasiewicz, Z. (Zbigniew) Jaskólski, Z. (Zbigniew) Sobczyk, Jan, 1955- Description: xii, 187 p.: ill.; 27 cm. Contents: Duality in string theory / S. Förste and J. Louis - Duality and global symmetries / F. Quevedo -- Aspects of N=1 string dynamics / S. Kachru -- Lectures on heterotic-type I duality / I. Antoniadis, H. Partouche and T.R. Taylor -- Properties of p-branes, D-branes and M-branes / E.A. Bergshoeff - Introduction to D-branes / L. Thorlacius -- Solitins, black holes and duality in string theory / R.R. Khuri -- Black holes and D-branes / J.M. Maldacena -- Classical and quantum composite p-branes / I. Ya. Aref'eva -- Massive and massless supersymmetric black holes / T. Ortin -- Duality and supersymmetric monopoles / J.P. Gauntlett -- Dualities in supersymmetric field theories / P.C. Argyres -- Novel field theory phenomena from F theory and D-branes / J. Sonnenschein -- M-theory, torons and confinement / C. Gómez and R. Hernández - WDVV equations in Seiberg-Witten theory and associative algebras / A. Mironov. Notes: Includes bibliographical references and author index. Subjects: Duality (Nuclear physics)--Congresses. String models--Congresses. Field theory (Physics). Series: Nuclear Physics B (Proc. suppl.), 0920-5632; 61A Nuclear physics. B, Proceedings, supplements; v. 61A. LC Classification: QC770 .N772 vol. 61A Dewey Class No.: 539.7 s 530.14 21

Enumerative geometry and string theory / Sheldon Katz. Published/Created: Providence, RI: American Mathematical Society, c2006. Description: xiii, 206 p.;

22 cm. ISBN: 0821836870 (alk. paper) Notes: Includes bibliographical references (p. 197-200) and index. Subjects: Geometry, Enumerative. String models. Series: Student mathematical library v. 32 LC Classification: QA607 .K38 2006 Dewey Class No.: 516.3/5 22 Electronic File Information: Table of contents only http://www.loc.gov/catdir/toc/fy0702/2006040710.html

Extreme physics. Edition Information: 1st ed. Published/Created: New York: Rosen Pub., 2008. Projected Publication Date: 0801 Related Titles: Scientific American. Description: p. cm. ISBN: 9781404214064 (lib. bdg.) Contents: Negative energy, wormholes, and warp drive -- Quantum teleportation -- Parallel universes -- Information in the holographic universe -- The future of string theory: a conversation with Brian Greene -- Atoms of space and time -- The dawn of physics beyond the standard model. Notes: Includes index. Subjects: Cosmology. Quantum theory. String models. Particle physics. Series: Scientific American cutting-edge science LC Classification: QB981 .E98 2008 Dewey Class No.: 539 22

Field theory, quantum gravity, and strings II: proceedings of a seminar series held at DAPHE, Observatoire de Meudon, and LPTHE, Université Pierre et Marie Curie, Paris, between October 1985 and October 1986 / edited by H.J. de Vega and N. Sánchez. Published/Created: Berlin; New York: Springer-Verlag, c1987. Related Names: Vega, H. J. de (Héctor J.), 1949- Sanchez, N. (Norma), 1952- Description: 245 p.; 25 cm. ISBN: 0387179259 (U.S.): Subjects: Quantum field theory--Congresses. Quantum gravity--Congresses. String models--Congresses. Series: Lecture notes in physics; 280 LC Classification: QC174.45.A1 F55 1987 Dewey Class No.: 530.1/43 19

Field theory, quantum gravity, and strings: proceedings of a seminar series held at DAPHE, Observatoire de Meudon, and LPTHE, Université Pierre et Marie Curie, Paris, between October 1984 and October 1985 / edited by H.J. de Vega and N. Sánchez. Published/Created: Berlin; New York: Springer-Verlag, c1986. Related Names: Vega, H. J. de (Héctor J.), 1949- Sanchez, N. (Norma), 1952- Description: vi, 381 p.: ill.; 25 cm. ISBN: 0387164529 (U.S.: pbk.): Notes: Includes bibliographies. Subjects: Quantum field theory--Congresses. Quantum gravity--Congresses. String models--Congresses. Series: Lecture notes in physics; 246 LC Classification: QC174.45.A1 F54 1986 Dewey Class No.: 530.1/43 19

Fields, strings, and quantum gravity: proceedings of the CCAST (World Laboratory) Symposium Workshop held at the Temple of the Sleeping Buddha, Beijing, Peoples Republic of China, May 29-June 10, 1989 (cancelled June 4, 1989) / edited by Hanying Guo, Zhaoming Qiu, and Henry Tye. Published/Created: New York: Gordon and Breach Science Publishers, c1990. Related Names: Guo, Hanying. Qiu, Zhaoming, 1946- Tye, Henry. Description: xiii, 435 p.: ill.; 23 cm. ISBN: 2881247415 Notes: Includes bibliographical references and index. Subjects: Field theory (Physics)--Congresses. String models--Congresses. Quantum gravity--Congresses. Series: China Center of Advanced Science and Technology (World Laboratory) Symposium/Workshop proceedings, 0894-2536; v. 6 CCAST (World Laboratory) Symposium/Workshop. China Center of Advanced Science and Technology (World Laboratory) Symposium/Workshop proceedings; v. 6. LC Classification: QC173.68 .C33 1989 Dewey Class No.: 530.1/4 20 Electronic File Information: Publisher description http://www.loc.gov/catdir/enhancements/fy0653/90002999-d.html

Fourth Paris cosmology colloquium: phase transitions in cosmology EuroConference: within the framework of the International School of Astrophysics 'Daniel Chalonge' and sponsored by the Commission of European Communities: Observatoire de Paris, 4-9 June 1997 / editors, H.J. de Vega, N. Sánchez. Published/Created: Singapore; River Edge, NJ: World Scientific, c1998. Related Names: Vega, H. J. de (Héctor J.), 1949- Sanchez, N. (Norma), 1952- Description: xii, 319 p.: ill.; 23 cm. ISBN: 9810234384 (alk. paper) Notes: Includes bibliographical references. Subjects: Cosmology--Congresses. Astrophysics--Congresses. String models--Congresses. LC Classification: QB980 .P37 1998 Dewey Class No.: 523.1 21

Frontiers in particle theory: proceedings of the Johns Hopkins Workshop on Current Problems in Particle Theory 11, Lanzhou, 1987 June 17-19 / edited by Yi-shi Duan, G. Domokos, S. Kovesi-Domokos. Published/Created: Singapore; Teaneck, N.J.: World Scientific Pub. Co., c1988. Related Names: Domokos, G. Duan, Yi-shi. Kövesi-Domokos, S. Description: xi, 324 p.: ill.; 23 cm. ISBN: 9971503840 Notes: Includes bibliographies. Subjects: Particles (Nuclear physics)--Congresses. String models--Congresses. LC Classification: QC793 .J65 1987 Dewey Class No.: 530.1/4 19

Frontiers in quantum field theory: Toyonaka, Osaka, Japan, 14-17 December, 1995 / editors, H. Itoyama ... [et al.]. Published/Created: Singapore; River Edge, NJ: World Scientific, c1996. Related Names: Itoyama, H. Description: xxiii, 422 p.: ill.; 23 cm. ISBN: 9810228171 Notes: In honor of Keiji Kikkawa's 60th birthday--Jacket cover. "This volume contains the proceedings of an international physics conference"--Pref. Includes bibliographical references. Subjects: Kikkawa, K. (Keiji), 1935- Quantum field theory--Congresses. String models--Congresses. Quantum gravity--Congresses. Supersymmetry--Congresses. LC Classification: QC174.45.A1 F76 1996 Dewey Class No.: 530.14/3 21

Frontiers of high energy physics: lectures given at the 7th UK Institute for Theoretical High Energy Physics, Imperial College, London, 17 August-6 September 1986 / edited by I.G. Halliday. Published/Created: Bristol; Philadelphia: A. Hilger, c1987. Related Names: Halliday, I. G. (Ian Gibson), 1940- Description: x, 232 p.: ill.; 24 cm. ISBN: 0852743688 Notes: Includes bibliography and index. Subjects: Particles (Nuclear physics) String models. LC Classification: QC793.3.H5 U38 1986 Dewey Class No.: 539.7/2 19

Functional integration, geometry, and strings: proceedings of the XXV Karpacz Winter School of Theoretical Physics, Karpacz, Poland, 20 February-5 March, 1989 / edited by Zbigniew Haba, Jan Sobczyk. Published/Created: Basel; Boston: Birkhäuser Verlag, 1989. Related Names: Haba, Zbigniew, 1951- Sobczyk, Jan, 1955- Description: vii, 459 p.: ill.; 24 cm. ISBN: 3764323876 (Basel: alk. paper) 0817623876 (Boston: alk. paper): Notes: Includes bibliographical references. Subjects: String models--Congresses. Quantum field theory--Congresses. Mathematical physics--Congresses. Series: Progress in physics; v. 13 Progress in physics (Boston, Mass.); v. 13. LC Classification: QC794.6.S85 W56 1989 Dewey Class No.: 539.7/2 20

Future perspectives in string theory: Strings '95: University of Southern California, USA, 13-18 March 1995 / editors, I. Bars ... [et al.]. Portion of Title: Strings '95 Published/Created: River Edge, N.J.: World Scientific, c1996. Related Names: Bars, Itzhak. Description: xiii, 523 p.: ill.

(some col.); 23 cm. ISBN: 9810224729 (alk. paper) Subjects: String models--Congresses. LC Classification: QC794.6.S85 F88 1996 Dewey Class No.: 530.1 20

Gauge fields and strings / by A.M. Polyakov. Published/Created: Chur, Switzerland; New York: Harwood Academic Publishers, c1987. Description: x, 301 p.; 24 cm. ISBN: 3718603934 3718603926 (pbk.) Notes: Includes index. Bibliography: p. ix-x. Subjects: Gauge fields (Physics) String models. Series: Contemporary concepts in physics; v. 3 LC Classification: QC793.3.F5 P66 1987 Dewey Class No.: 530.1/43 19

Geometries of nature, living systems and human cognition: new interactions of mathematics with natural sciences and humanities / edited by Luciano Boi. Published/Created: New Jersey: World Scientific, c2005. Related Names: Boi, L. (Luciano), 1957- Description: xxiii, 419 p.: ill. (some col.); 24 cm. ISBN: 9812564748 UPC/EAN: 9789812564749 Contents: Pt. I. Mathematical ideas and techniques in classical and quantum physics: 1. T-duality, functional equation, and noncommutative string spacetime / M. Lapidus -- 2. On the soliton-particle dualities / F. Hélein and J. Kouneiher -- Pt. II. Knots in mathematics and in nature: 3. Knots / L.H. Kauffman -- 4. Topological knots models in physics and biology / L. Boi -- Pt. III. Mathematical and logical modeling in the natural sciences and living systems: 5. Beyond modeling: a challenge for applied mathematics / P.T. Saunders -- 6. Fondements cognitifs de la géométrie et expérience de l'espace / A. Berthoz -- 7. The reasonable effectiveness of mathematics and it cognitive roots / G. Longo -- 8. Pathways of deductions / A. Carbone -- 9. Phénoménologie et théories des catégories / F. Patras. Notes: Includes bibliographical references. Subjects: Geometry. Mathematical physics. Cognition--Mathematical models. LC Classification: QC20.7.G44 G464 2005 Dewey Class No.: 516 22

Gravity and strings / Tomás Ortín. Published/Created: Cambridge, UK; New York: Cambridge University Press, 2004. Description: xx, 684 p.: ill.; 26 cm. ISBN: 0521824753 Notes: Includes bibliographical references (p. 650-670) and index. Subjects: Quantum gravity. String models. Series: Cambridge monographs on mathematical physics LC Classification: QC178 .O78 2004 Dewey Class No.: 530.14 22

Gurdjieff, string theory, music / Mitzi DeWhitt. Published/Created: [Philadelphia, Pa.]: Xlibris Corp., c2006. Description: 134 p.: ill.; 22 cm. ISBN: 1425700241 (hardcover) 1425700233 (softcover) UPC/EAN: 9781425700249 9781425700232 Notes: Includes bibliographical references (p. 129-130) and index. Subjects: Gurdjieff, Georges Ivanovitch, 1872-1949. String models. Musicology. Music theory--Research. LC Classification: BP605.G94 D49 2006 Dewey Class No.: 110 22

High energy physics, 1985: proceedings of the Yale Theoretical Advanced Study Institute / editors, Mark J. Bowick, Feza Gürsey. Published/Created: Singapore: World Scientific, c1986. Related Names: Bowick, Mark J. Gürsey, Feza. Description: 2 v. (vii, 975 p.): ill.; 23 cm. ISBN: 997150006X (v. 1) 9971500078 (pbk.: v. 1) Notes: Includes bibliographies. Subjects: Particles (Nuclear physics)--Congresses. String models--Congresses. Cosmology--Congresses. LC Classification: QC793 .Y35 1985 Dewey Class No.: 539.7 19

Housecleaning / Dan Boyle. Published/Created: New York: Southern Tier Editions, Harrington Park Press, c2007. Projected Publication Date: 0710 Description: p. cm.

ISBN: 9781560237013 (pbk.: alk. paper) 1560237015 (pbk.: alk. paper) Subjects: Gay men--Fiction. Physicists--Fiction. String models--Fiction. Form/Genre: Love stories. gsafd Domestic fiction. gsafd LC Classification: PS3602.O97 H68 2007 Dewey Class No.: 813/.6 22

International Workshop on Superstrings, Cosmology, Composite Structures, March 11-18, 1987, Center for Theoretical Physics, University of Maryland at College Park / editors, S.J. Gates, Jr., R.N. Mohapatra. Published/Created: Singapore; Teaneck, NJ: World Scientific, c1987. Related Names: Gates, S. James. Mohapatra, R. N. (Rabindra Nath) Description: x, 592 p.: ill.; 23 cm. ISBN: 9971503735 9971503840 (pbk.) Notes: Includes bibliographies. Subjects: Superstring theories--Congresses. String models--Congresses. Cosmology--Congresses. LC Classification: QC794.6.S85 I58 1987 Dewey Class No.: 530.1/42 19

Introduction to the relativistic string theory / B.M. Barbashov, V.V. Nesterenko; translator, T. Yu. Dumbrajs. Published/Created: Singapore; Teaneck, N.J.: World Scientific, c1990. Related Names: Nesterenko, V. V. (Vladimir VitalÊ¹evich) Description: xi, 249 p.: ill.; 23 cm. ISBN: 9971506874 Notes: Includes bibliographical references (p. 233-244) and index. Subjects: String models. Relativity (Physics) Particles (Nuclear physics) Quantum theory. LC Classification: QC794.6.S85 B3713 1990 Dewey Class No.: 539.7/2 20 Language Code: eng rus

Last Workshop on Grand Unification: University of North Carolina at Chapel Hill, April 20-22, 1989 / editor, Paul H. Frampton. Published/Created: Singapore; Teaneck, NJ: World Scientific, c1989. Related Names: Frampton, Paul H., 1943- Description: xi, 362 p.: ill.; 23 cm. ISBN: 9810200145 Notes: Includes bibliographical references. Subjects: Grand unified theories (Nuclear physics)--Congresses. String models--Congresses. Astrophysics--Congresses. LC Classification: QC794.6.G7 W67 1989 Dewey Class No.: 539.7/2 20

Lectures on string theory / D. Lüst, S. Theisen. Published/Created: Berlin; New York: Springer-Verlag, c1989. Related Names: Theisen, S. (Stefan), 1957- Description: vii, 346 p.: ill.; 25 cm. ISBN: 0387518827 (U.S.: alk. paper): Notes: Includes bibliographical references. Subjects: String models. Series: Lecture notes in physics; 346 LC Classification: QC794.6.S85 L87 1989 Dewey Class No.: 530.1 20

L'universo prima del big bang: cosmologia e teoria delle stringhe / Maurizio Gasperini. Published/Created: Roma: F. Muzzio, c2002. Description: 197 p.: ill.; 21 cm. ISBN: 887413052X Notes: Includes bibliographical references (p. 196-197). Subjects: Big bang theory. Expanding universe. String models. Cosmology. Series: I Piaceri della scienza;2 LC Classification: QB991.B54 G37 2002 Dewey Class No.: 523.1/8 22

Mathematical, theoretical and phenomenological challenges beyond the standard model: perspectives of the Balkan collaborations / editors, G. Djordjević, L. Nešić, J. Wess. Published/Created: Hackensack, NJ: World Scientific, c2005. Related Names: Djordjević, G. Nešić, L. Wess, Julius. Description: xvii, 279 p.: ill.; 24 cm. ISBN: 9812561307 Notes: "The Balkan Workshop (BW2003): Mathematical, Theoretical and Phenomenological Challenges Beyond the Standard Model - Perspectives of Balkan collaborations - was held from 29 August to 2 September 2003, in Vrnjačka Banja, Serbia ... "--Pref. Includes bibliographical references. Subjects: Particles (Nuclear physics)--Congresses. String models--Congresses. Superstring theories--

Congresses. Supersymmetry--Congresses. LC Classification: QC793 .B85 2003 Dewey Class No.: 539.7/2 22

Mirror symmetry IV: proceedings of the Conference on Strings, Duality, and Geometry, Centre de recherches mathématiques of Université de Montréal (CRM), March 2000 / Eric D'Hoker, Duong Phong, and Shing-Tung Yau, editors. Published/Created: Providence, RI: American Mathematical Society; [Montréal]: Centre de recherches mathémathiques, CRM; [Somerville, MA]: International Press, c2002. Related Names: D'Hoker, Eric, 1956- Phong, Duong H., 1953- Yau, Shing-Tung, 1949- Description: ix, 381 p.: ill.; 27 cm. ISBN: 0821833359 (alk. paper) Notes: Includes bibliographical references. Subjects: Mirror symmetry--Congresses. Geometry, Differential--Congresses. String models--Congresses. Duality (Nuclear physics)--Congresses. Series: AMS/IP studies in advanced mathematics; v. 33 LC Classification: QC174.17.S9 C65 2000 Dewey Class No.: 516.3/62 21

ModelÊ¹ reliï, aï, ¡tivistskoĭ struny v fizike adronov / B. Barbashov, V.V. Nesterenko. Edition Information: Nauch. izd. Published/Created: Moskva: Energoatomizdat, 1987. Related Names: Nesterenko, V. V. (Vladimir VitalÊ¹evich) Description: 175 p.: ill.; 21 cm. Notes: Bibliography: p. 168-[174] Subjects: Hadrons. String models. Relativity (Physics) Quantum field theory. LC Classification: QC793.5.H32 B37 1987

Nelineinye sigma-modeli v kvantovoi teorii poliï, aï, ¡ i teorii strun / S.V. Ketov; otvetstvennyi redaktor V.G. Bagrov. Published/Created: Novosibirsk: Nauka, Sibirskoe otd-nie, 1992. Related Names: Bagrov, V. G. (Vladislav Gavrilovich) Description: 235, [1] p.: ill.; 22 cm. ISBN: 5020299359. Includes bibliographical references (p. 208-[236]). Subjects: String models. Quantum field theory--Mathematical models. Sigma particles. LC Classification: QC794.6.S85 K46 1992

New developments in fundamental interaction theories: 37th Karpacz Winter School of Theoretical Physics, Karpacz, Poland, 6-15 February 2001 / editors, Jerzy Lukierski, Jakub Rembieliński. Published/Created: Melville, N.Y.: American Institute of Physics, 2001. Related Names: Lukierski, Jerzy. Rembieliński, Jakub. Description: ix, 450 p.: ill.; 24 cm. ISBN: 0735400296 Notes: Includes bibliographical references and index. Subjects: Field theory (Physics)--Congresses. Gauge fields (Physics)--Congresses. Particles (Nuclear physics)--Mathematics--Congresses. Mathematical physics--Congresses. Series: AIP conference proceedings, 0094-243X; v. 589 AIP conference proceedings; no. 589. LC Classification: QC793.3.F5 W56 2001 Dewey Class No.: 539.7/54 21

New developments in quantum field theory / edited by Poul Henrik Damgaard and Jerzy Jurkiewicz. Published/Created: New York: Plenum Press, c1998. Related Names: Damgaard, P. H. (Poul Henrilk) Jurkiewicz, Jerzy. North Atlantic Treaty Organization. Scientific Affairs Division. NATO Advanced Research Workshop on New Developments in Quantum Field Theory (1997: Zakopane, Poland) Description: ix, 364 p.: ill.; 26 cm. ISBN: 0306458160 Notes: "Published in cooperation with NATO Scientific Affairs Division." "Proceedings of a NATO Advanced Research Workshop on New Developments in Quantum Field Theory, held July 14-20, 1997, in Zakopane, Poland"--T.p. verso. Includes bibliographical references and index. Subjects: Quantum field theory--Congresses. Matrices--Congresses. String models--Congresses. Mathematical physics--Congresses. Series: NATO ASI series. Series B, Physics; v. 366 LC

Classification: QC174.45.A1 N45 1998 Dewey Class No.: 530.14/3 21

New developments in string theory research / Susan A. Grece, editor. Published/Created: Hauppauge, N.Y.: Nova Science Publishers, c2006. Related Names: Grece, Susan A. Description: xi, 313 p.: ill.; 26 cm. ISBN: 1594544883 Notes: Includes bibliographical references and index. Subjects: String models--Research. LC Classification: QC794.6.S85 N49 2006 Dewey Class No.: 530.14 22 Electronic File Information: Table of contents http://www.loc.gov/catdir/toc/ecip0513/2005014612.html

New fields and strings in subnuclear physics: proceedings of the International School of Subnuclear Physics / edited by Antonino Zichichi. Published/Created: River Edge, NJ: World Scientific, c2002. Related Names: Zichichi, Antonino. Description: viii, 387 p.: ill.; 26 cm. ISBN: 9812381864 Notes: Includes bibliographical references. Subjects: String models--Congresses. Gauge fields (Physics)--Congresses. Particles (Nuclear physics)--Congresses. Series: The subnuclear series; v. 39 LC Classification: QC794.6.S85 I57 2001 Dewey Class No.: 539.7/2 21

Non-critical string theory: classical and quantum aspects / Stanislav Klimenko and Igor Nikitin. Published/Created: New York: Nova Science, c2007 Related Names: Nikitin, Igor. Description: xxvi, 194 p.: ill. (some col.); 27 cm. ISBN: 1594542678 9781594542671 Contents: Structure of classical solutions of Nambu-Goto model -- Structure of quantum solutions of Nambu-Goto model -- Computational methods and algorithms used in the investigation of Nambu Goto model. Notes: Includes bibliographical references (p. [181]-187) and index. Subjects: String models.

Nonperturbative aspects of strings, branes, and supersymmetry: proceedings of the Spring School on nonperturbative aspects of string theory and supersymmetric gauge theories, ICTP, Trieste, Italy, 23-31 March, 1998: proceedings of the Trieste Conference on Super-Five-Branes and Physics in 5 + 1 Dimensions, ICTP, Trieste, Italy, 1-3 April, 1998 / editors, M. Duff ... [et al.]. Published/Created: River Edge, N.J.: World Scientific, c1999. Related Names: Duff, M. J. Abdus Salam International Centre for Theoretical Physics. Unesco. International Atomic Energy Agency. Trieste Conference on Super-Five-Branes and Physics in 5 + 1 Dimensions (1998: Trieste, Italy) Description: xi, 454 p.: ill.; 22 cm. ISBN: 9810237855 (alk. paper) Notes: "1998 Trieste Spring School"--Pref. At head of title: The Abdus Salam International Centre for Theoretical Physics; United Nations Educational, Scientific and Cultural Organization; International Atomic Energy Agency. Subjects: String models--Congresses. Supersymmetry--Congresses. Branes--Congresses. LC Classification: QC794.6.S85 T75 1998 Dewey Class No.: 530.14 21

Not even wrong: the failure of string theory and the search for unity in physical law / Peter Woit. Published/Created: New York: Basic Books, c2006. Projected Publication Date: 0608 Description: p. cm. ISBN: 9780465092758 (alk. paper) 0465092756 (alk. paper) Contents: Particle physics at the turn of the millennium -- The instruments of production -- Quantum theory -- Quantum field theory -- Gauge symmetry and Gauge theories -- The standard model -- Triumph of the standard model -- Problems of the standard model -- Beyond the standard model -- New insights in quantum field theory and mathematics -- String theory: history -- String theory and supersymmetry: an evaluation -- On beauty and difficulty -- Is superstring theory science? -- The Bogdanov affair -- The only game in town: the power and the glory of string theory -- The landscape of string theory -- Other points of view. Notes: Includes bibliographical references and index. Subjects: String models. Supersymmetry. Quantum theory. Superstring theories. Physical laws. LC Classification: QC794.6.S85 W65 2006 Dewey Class No.: 539.7/258 22

Novelties in string theory: proceedings of the Johns Hopkins Workshop on Current Problems in Particle Theory 22, Göteborg, 1998 (August 20-22) / edited by L. Brink, R. Marnelius. Portion of Title: Proceedings of the Johns Hopkins Workshop on Current Problems in Particle Theory Published/Created: Singapore; River Edge, N.J.: World Scientific, 1999. Related Names: Brink, Lars, 1945- Marnelius, R. (Robert) Description: x, 372 p.: ill.; 23 cm. ISBN: 9810240848 (alk. paper) Notes: Includes bibliographical references. Subjects: String models--Congresses. Particles (Nuclear physics)--Congresses. LC Classification: QC794.6.S85 J64 1998 Dewey Class No.: 539.7/2 21

Our improbable universe: a physicist considers how we got here / Michael Mallary. Published/Created: New York: Thunder's Mouth Press, c2004. Description: vii, 227 p.: ill.; 21 cm. ISBN: 156858301X (pbk.) Notes: Includes bibliographical references (p. 225-227) Subjects: Cosmology. Astrophysics. Beginning. Big bang theory. String models. LC Classification: QB981 .M2555 2004 Dewey Class No.: 523.1 22

Particles, strings, and cosmology: 11th International Symposium on Particles, Strings, and Cosmology, PASCOS 2005, Gyeongju, Korea, 30 May-4 June 2005 / editors, Kiwoon Choi, Jihn E. Kim, Dongchul Son. Published/Created: Melville, N.Y.: American Institute of Physics, 2005. Related Names: Choi, Kiwoon. Kim, Jihn E. Son, Dongchul, 1952- Description: xi, 490 p.: ill.; 29 cm. ISBN: 0735402957 Notes: Includes bibliographical references and index. Subjects: Particles (Nuclear physics)--Congresses. String models--Congresses. Relativistic astrophysics--Congresses. Supersymmetry--Congresses. Cosmology--Congresses. Series: AIP conference proceedings, 0094-243X; v. 805 AIP conference proceedings; no. 805. LC Classification: QC793 .I575 2005 Dewey Class No.: 539.7/2 22

PASCOS 94: proceedings of the Fourth International Symposium on Particles, Strings and Cosmology, Syracuse University, Syracuse, New York, USA, 19-24 May 1994 / editor, Kameshwar C. Wali. Published/Created: Singapore; River Edge, NJ: World Scientific, c1995. Related Names: Wali, K. C. (Kameshwar C.) Description: xvii, 480 p.: ill.; 23 cm. ISBN: 9810221525 Notes: Includes bibliographical references. Subjects: Particles (Nuclear physics)--Congresses. Cosmology--Congresses. Nuclear astrophysics--Congresses. String models--Congresses. LC Classification: QC793 .I5617 1994 Dewey Class No.: 539.7/2 20

Perspectives in mathematical physics / edited by Robert Penner, Shing-Tung Yau. Published/Created: Cambridge, MA: International Press, c1994. Related Names: Penner, R. C., 1956- Yau, Shing-Tung, 1949- Description: 307 p.: ill.; 24 cm. ISBN: 1571460098 Contents: Preface -- Geometrical origin of integrability for Liouville and Toda theory / K. Aoki & E. D'Hoker -- Chern-Simons perturbation theory II / S. Axelrod & I.M. Singer -- Curved spacetime geometry for strings and affine non-compact algebras / I. Bars -- On the BRST structure of [omega]â, f gravity coupled to c=2 matter / P. Bouwknegt, J. McCarthy & K. Pilch -- Billiards, Poncelet's theorem, and integrable field theories / S.J. Chang -- Canonical quantization of Yang-Mills theories / H. Cheng -- Yang-Lee-Fisher zeros and Julia sets / B. Hu -- A construction of generalized spin models / V.G. Kac & M. Wakimoto -- Algebraic theory of the KP equations / M. Mulase -- On some algebraic structures arising in string theory / M. Penkava & A. Schwarz -- The Poincaré dual of Weil-Petersson Kähler two-form / R.C. Penner -- Topological properties of Calabi-Yau mirror manifold / S.S. Roan -- Closed string field theory, strong homotopy Lie algebras and the operad actions of moduli spaces / J. Stasheff. -- QCD: from low energy to high energy / T.M. Yan -- Fullerenes and carbon 60 / C.N. Yang. Notes: Combined proceedings of four conferences on mathematical physics, all held in 1992. Cf. pref. and bibliographical footnotes. Includes bibliographical references. Subjects: Mathematical physics--Congresses. Series: Conference proceedings and lecture notes in mathematical physics; v. 3 LC Classification: QC19.2 .P44 1994 Dewey Class No.: 530.15 21

Physics and mathematics of strings: proceedings of a Royal Society Discussion Meeting, held on 8 and 9 December 1988 / organized and edited by Sir Michael Atiyah ... [et al.]. Published/Created: London: Royal Society; Port Washington, N.Y.: Distributed by Scholium International, 1989. Related Names: Atiyah, Michael Francis 1929- Royal Society (Great Britain) Description: 95 p.: ill.; 31 cm. ISBN: 0854033939 Notes: "First published in Philosophical transactions of the Royal Society of London, series A, volume 329 (no. 1605), pages 317-413"--T.p. verso. Distributor from label on t.p. Pages also numbered 319-413. Includes bibliographical references. Subjects: String models--Congresses. Superstring theories--Congresses. Particles (Nuclear physics)--Congresses. LC Classification: QC794.6.S85 R69 1989 Dewey Class No.: 539.7/2 20

Physics from Planck scale to electroweak scale: proceedings of the US-Polish workshop, Warsaw, Poland, 21-24 September 1994 / editors, Pran Nath, Tomasz Taylor, Stefan Pokorski. Published/Created: Singapore; River Edge, NJ: World Scientific, c1995. Related Names: Pran Nath, 1939-. Taylor, Tomasz. Pokorski, Stefan, 1942- Description: x, 495 p.: ill.; 23 cm. ISBN: 9810221843 Notes: Includes bibliographical references. Subjects: Supersymmetry--Congresses. Grand unified theories (Nuclear physics)--Congresses. String models--Congresses. LC Classification: QC174.17.S9 P49 1995

Principles of string theory / Lars Brink and Marc Henneaux. Published/Created: New York: Plenum Press, c1988. Related Names: Henneaux, Marc. Description: xiii, 297 p.: ill.; 24 cm. ISBN: 0306426579 Notes: Includes index. Bibliography: p. 289-292. Subjects: String models. Series: Series of the Centro de Estudios Científicos de Santiago LC Classification: QC794.6.S85 B75 1988 Dewey Class No.: 530.1 19

Proceedings of the 10th International Symposium on Particles, Strings and Cosmology: PASCOS 2004, Northeastern University, Boston, 16-22 August 2004 / editors, George Alverson ... [et al.]. Portion of Title: Particles, strings and cosmology PASCOS 2004 Published/Created: Hackensack, N.J.: World Scientific, c2005. Related Names: Alverson, George Oscar, 1951- Description: xv, 573 p.: ill.; 24 cm. ISBN: 9812563911 9812564799 (set) Notes: Two days of the Symposium were devoted to a Fest in honor of Professor Pran Nath and the papers are published in a separate companion volume titled: Themes in unification: the Pran Nath festschrift. Includes bibliographical references.

Subjects: Particles (Nuclear physics)--Congresses. String models--Congresses. Relativistic astrophysics--Congresses. Supersymmetry--Congresses. Cosmology--Congresses. LC Classification: QC793 .I5617 2005

Proceedings of the 7th International Symposium on Particles, Strings and Cosmology: PASCOS 99, Lake Tahoe, California, 10-16 December 1999 / editors, Kingman Cheung, John F. Gunion, Stephen Mrenna. Variant Title: Proceedings of the seventh international symposium on particles, strings and cosmology Portion of Title: Particles, strings, and cosmology PASCOS 99 Published/Created: Singapore; River Edge, N.J.: World Scientific, c2000. Related Names: Cheung, Kingman. Gunion, J. F. (John Francis), 1943- Mrenna, Stephen. Description: xv, 563 p.: ill.; 23 cm. ISBN: 981024388X Notes: "The 7th International Symposium on Particles, Strings and Cosmology (PASCOS 99) was hosted by the University of California at Davis Physics Department at Granlibakken, North Lake Tahoe from December 10 to December 16"--Pref. Includes bibliographical references. Subjects: Particles (Nuclear physics)--Congresses. String models--Congresses. Relativistic astrophysics--Congresses. Supersymmetry--Congresses. Cosmology--Congresses. LC Classification: QC793 .I575 1999

Proceedings of the eighth international conference, particles, strings, and Cosmology: University of North Carolina, Chapel Hill, April 10-15, 2001 / editors, Paul Frampton and Jack Ng. Portion of Title: Particles, strings, and cosmology Published/Created: Princeton, N.J.: Rinton Press, c2001. Related Names: Frampton, Paul H., 1943- Ng, Y. Jack (Yeo Jack) Description: xiv, 427 p.: ill.; 27 cm. ISBN: 1589490150 Notes: Includes bibliographical references. Subjects: Particles (Nuclear physics)--Congresses. String models--Congresses. Relativistic astrophysics--Congresses. Supersymmetry--Congresses. Cosmology--Congresses. LC Classification: QC793 .I575 2001 Dewey Class No.: 539.7/2 21

Proceedings of the First International Symposium on Particles, Strings, and Cosmology, Northeastern University, Boston, USA, 27-31 March 1990 / editors, Pran Nath & Stephen Reucroft. Published/Created: Singapore; Teaneck, N.J.: World Scientific, c1991. Related Names: Pran Nath, 1939- Reucroft, Stephen. Description: xiv, 688, [12] p. of plates: ill.; 22 cm. ISBN: 9810203926 9810203934 (pbk.) Notes: Includes bibliographical references. Subjects: Particles (Nuclear physics)--Congresses. String models--Congresses. Cosmology--Congresses. LC Classification: QC793 .I5617 1990 Dewey Class No.: 539.7/2 20

Proceedings of the II Mexican School of Particles and Fields, Cuernavaca-Morelos, 1986 / editors, J.L. Lucio and A. Zepeda. Published/Created: Singapore; Teaneck, NJ: World Scientific, c1987. Related Names: Lucio, J. L. (José Luis) Zepeda, A. (Arnulfo) Related Titles: Second Mexican School of Particles and Fields. Description: ix, 503 p.: ill.; 22 cm. ISBN: 9971504340 Contents: Fermilab fixed-target program / R.L. Dixon--Physics goals of future colliders / G.L. Kane--Physics at [square root symbol](s) = 2 TeV / R. Raja--Introduction to string field theory / P. Ramond and M. Ruiz-Altaba--String and superstring theories / J. Govaerts--Topics in elementary particle physics / D.B. Cline. Notes: Proceedings of a conference held 4-12 December, 1986. Subjects: Particles (Nuclear physics)--Congresses. String models--Congresses. Series: Lecture notes on particles and fields LC Classification: QC793 .M49 1986 Dewey Class No.: 530.1/4 19

Proceedings of the Second International Symposium [on] Particles, Strings, and Cosmology, Northeastern University, Boston, 25-30 March 1991 / editors, Pran Nath and Stephen Reucroft. Published/Created: Singapore; River Edge, N.J.: World Scientific, c1992. Related Names: Pran Nath, 1939- Reucroft, Stephen. Related Titles: Particles, strings, and cosmology. Description: xvi, 921 p.: ill.; 23 cm. ISBN: 9810209711 Notes: Includes bibliographical references. Subjects: Particles (Nuclear physics)--Congresses. String models--Congresses. Cosmology--Congresses. LC Classification: QC793 .I5617 1991 Dewey Class No.: 539.7/2 20

Proceedings of the Sixth International Symposium Particles, Strings, and Cosmology: PASCOS 98, Northeastern University, Boston, Massachusettes, USA, 22-29 March 1998 / editor, P. Nath. Portion of Title: Particles, strings, and cosmology Published/Created: Singapore; River Edge, N.J.: World Scientific, c1999. Related Names: Pran Nath, 1939- Description: xviii, 856 p.: ill.; 23 cm. ISBN: 9810236123 (alk. paper) Notes: Includes bibliographical references. Subjects: Particles (Nuclear physics)--Congresses. String models--Congresses. Relativistic astrophysics--Congresses. Supersymmetry--Congresses. Cosmology--Congresses. LC Classification: QC793 .I5617 1998 Dewey Class No.: 539.7/2 21

Proceedings of the Summer Workshop on Superstrings: KEK, Tsukuba, Japan: August 29-September 3, 1988 / edited by M. Kobayashi and K. Higashijima. Published/Created: Tsukuba-shi, Ibaraki-ken, Japan: National Laboratory for High Energy Physics, [1989] Related Names: Kobayashi, M. (Makoto) Higashijima, K. (Kiyoshi) Ko-enerugi Butsurigaku Kenkyujo (Japan) Description: iii, 345 p.: ill.; 30 cm. Notes: Cover title. Includes bibliographical references. Subjects: Superstring theories--Congresses. String models--Congresses. Field theory (Physics)--Congresses. Series: KEK report; 88-12 LC Classification: QC794.6.S85 S854 1988 Dewey Class No.: 539.7/2 20

Progress in string theory and M-theory / edited by Laurent Baulieu ... [et al.]. Published/Created: Dordrecht; Boston: Kluwer Academic, 2001. Related Names: Baulieu, Laurent. North Atlantic Treaty Organization. Scientific Affairs Division NATO Advanced Study Institute on Progress in String Theory and M-Theory (1999: Cargese, France) Description: ix, 417 p.: ill.; 25 cm. ISBN: 0792370333 (hardcover: alk. paper) Notes: "Published in cooperation with NATO Scientific Affairs Division." "Proceedings for the NATO Advanced Study Institute on Progress in String Theory and M-Theory, Cargese, France, May 24-June 5, 1999"--T.p. verso. Includes bibliographical references. Subjects: String models--Congresses. Series: NATO science series. Series C, Mathematical and physical sciences; v. 564 LC Classification: QC794.6.S85 P77 2001 Dewey Class No.: 539.7/2 21

Progress in string theory: TASI 2003 lecture notes, Boulder, Colorado, USA, 2-27 June 2003 / editor, Juan M. Maldacena. Portion of Title: TASI 2003 lecture notes Published/Created: Hackensack, N.J.: World Scientific, c2005. Related Names: Maldacena, Juan Martín, 1968- Description: vii, 548 p.: ill.; 24 cm. ISBN: 9812564063 Notes: Includes bibliographical references. Subjects: String models--Congresses. LC Classification: QC794.6.S85 T48 2003 Dewey Class No.: 530.14 22

Progress in string, field, and particle theory / edited by Laurent Baulieu ... [et al.]. Published/Created: Dordrecht; Boston: Kluwer Academic Publishers, c2003.

Related Names: Baulieu, Laurent. North Atlantic Treaty Organization. Scientific Affairs Division. Description: xiii, 507 p.: ill.; 25 cm. ISBN: 1402013604 (acid-free paper) Notes: Includes bibliographical references. Subjects: String models--Congresses. Field theory (Physics)--Congresses. Particles (Nuclear physics)--Congresses. Series: NATO science series. Series II, Mathematics, physics, and chemistry; v. 104 LC Classification: QC794.6.S85 N38 2002 Dewey Class No.: 530.14 21

Quantization, gauge theory, and strings: proceedings of the international conference dedicated to the memory of Professor Efim Fradkin: Moscow, Russia, June 5-10, 2000 / editors A. Semikhatov, M. Vasiliev, V. Zaikin. Published/Created: Moscow: Scientific World, c2001. Related Names: Fradkin, E. S. (Efim Samoĭlovich), 1924- Semikhatov, A. M. VasilÊ¹ev, Mikhail VasilÊ¹evich. Zaikin, V. (Vladimir) Description: 2 v.: ill.; 22 cm. ISBN: 5891761254 Notes: Includes bibliographical references and index. Subjects: Geometric quantization--Congresses. Gauge fields (Physics)--Congresses. String models--Congresses. LC Classification: QC174.17.G46 Q33 2001 Dewey Class No.: 530.14 21

Quantum aspects of gauge theories, supersymmetry, and unification: proceedings of the second international conference held in Corfu, Greece, 20-26 September 1998 / A. Ceresole ... [et al.] (eds.). Published/Created: Berlin; New York: Springer, c1999. Related Names: Ceresole, A. (Anna), 1961- Description: ix, 534 p.; 24 cm. ISBN: 3540660054 (hc: alk. paper) Notes: Includes bibliographical references. Subjects: Gauge fields (Physics)--Congresses. Supersymmetry--Congresses. String models--Congresses. Series: Lecture notes in physics, 0075-8450; 525 LC Classification: QC793.3.F5 Q36 1999 Dewey Class No.: 530.14/3 21

Quantum field theory and string theory / edited by Laurent Baulieu ... [et al.]. Published/Created: New York: Plenum Press, c1995. Related Names: Baulieu, Laurent. North Atlantic Treaty Organization. Scientific Affairs Division. NATO Advanced Research Workshop on New Developments in String Theory, Conformal Models, and Topological Field Theory (1993: Cargèse, France) Description: x, 420 p.: ill.; 26 cm. ISBN: 0306448866 Notes: "Published in cooperation with NATO Scientific Affairs Division". "Proceedings of a NATO Advanced Research Workshop on New Developments in String Theory, Conformal Models, and Topological Field Theory, held May 10-21, 1993, in Cargèse, France"--T.p. verso. Includes bibliographical references and index. Subjects: Quantum field theory--Congresses. String models--Congresses. Quantum gravity--Congresses. Series: NATO ASI series. Series B, Physics; v. 328 LC Classification: QC174.45.A1 Q36265 1995 Dewey Class No.: 530.1/43 20

Quantum field theory of point particles and strings / Brian Hatfield. Published/Created: Redwood City, Calif.: Addison-Wesley, c1992. Related Titles: Point particles and strings. Description: xvii, 734 p.: ill.; 24 cm. ISBN: 020111982X Notes: Includes bibliographical references (p. 721-724) and index. Subjects: Quantum field theory. String models. Series: Frontiers in physics; v. 75 LC Classification: QC174.45 .H29 1992 Dewey Class No.: 530.1/43 19

Quantum field theory, statistical mechanics, quantum groups and topology: proceedings of the NATO advanced research workshop, University of Miami, 7-12 January 1991 / edited by Thomas Curtright, Luca Mezincescu & Rafael Nepomechie.

Published/Created: Singapore; River Edge, NJ: World Scientific, c1992. Related Names: Curtright, Thomas Mezincescu, Luca. Nepomechie, Rafael I., 1958- Description: ix, 347 p.: ill.; 23 cm. ISBN: 9810209592 Cancelled ISBN: 9810219606 (pbk.) Notes: Includes bibliographical references. Subjects: Quantum field theory--Congresses. Statistical mechanics--Congresses. Quantum groups--Congresses. Knot theory--Congresses. String models--Congresses. LC Classification: QC174.45.A1 Q36274 1991 Dewey Class No.: 530.1/4 20

Quantum fields and strings: a course for mathematicians / Pierre Deligne ... [et al.], editors. Published/Created: Providence, R.I.: American Mathematical Society; [Princeton, NJ]: Institute for Advanced Study, c1999. Related Names: Deligne, Pierre. Institute for Advanced Study (Princeton, N.J.) Description: 2 v. (xxii, 1501 p.): ill.; 26 cm. ISBN: 0821811983 (set: acid-free paper) 0821819879 (v. 1: acid-free paper) 0821819887 (v. 2: acid-free paper) Notes: "The written record of the 1996-1997 Special Year in Quantum Field Theory held at the Institute for Advanced Study"--T.p. verso. Includes bibliographical references and index. Subjects: Quantum field theory. String models. Mathematical physics. LC Classification: QC174.45 .Q395 1999 Dewey Class No.: 530.14/3 21

Quantum string theory: proceedings of the Second Yukawa Memorial Symposium, Nishinomiya, Japan, October 23-24, 1987 / editors N. Kawamoto and T. Kugo. Published/Created: Berlin; New York: Springer-Verlag, c1988. Related Names: Kawamoto, N. (Noboru), 1948- Kugo, T. (Taichiro), 1949- Description: ix, 147 p.: ill.; 25 cm. ISBN: 0387503137 (U.S.) Notes: Includes bibliographical references and index. Subjects: String models--Congresses. Quantum field theory--Congresses. Series: Springer proceedings in physics; 31 LC Classification: QC794.6.S85 N57 1988 Dewey Class No.: 530.1/4 19

Random surfaces and quantum gravity / edited by Orlando Alvarez, Enzo Marinari, and Paul Windey. Published/Created: New York: Plenum Press, c1991. Related Names: Alvarez, Orlando, 1953- Marinari, Enzo. Windey, Paul. North Atlantic Treaty Organization. Scientific Affairs Division. Description: viii, 407 p.: ill.; 26 cm. ISBN: 0306439395 Notes: "Proceedings of a NATO Advanced Research Workshop on Random Surfaces and Quantum Gravity, held May 27-June 2, 1990, in Cargèse, France"--T.p. verso. "Published in cooperation with NATO Scientific Affairs Division." Includes bibliographical references and index. Subjects: Quantum gravity--Congresses. String models--Congresses. Surfaces (Physics)--Congresses. Series: NATO ASI series. Series B, Physics; v. 262 The Language of science LC Classification: QC178 .N32 1990 Dewey Class No.: 531/.14 20

Recent directions in particle theory: from superstrings and black holes to the standard model: proceedings of the 1992 Theoretical Advanced Study Institute in Elementary Particle Physics, Boulder, Colorado, 1-26 June 1992 / editors, Jeffrey Harvey, Joseph Polchinski. Published/Created: Singapore; River Edge, N.J.: World Scientific, c1993. Related Names: Harvey, Jeffrey. Polchinski, Joseph Gerard. Description: vii, 827 p.: ill.; 23 cm. ISBN: 981021233X Notes: Includes bibliographical references. Subjects: String models--Congresses. Standard model (Nuclear physics)--Congresses. Field theory (Physics)--Congresses. Astrophysics--Congresses. LC Classification: QC794.6.S85 T48 1992 Dewey Class No.: 539.7/2 20

Relativity, supersymmetry, and strings / edited by Arnold Rosenblum. Published/Created: New York: Plenum Press, c1990. Related Names: Rosenblum, A., 1943- Utah State University. International Institute of Theoretical Physics. Description: vii, 128 p.: ill.; 26 cm. ISBN: 0306426803 Notes: "Proceedings of the International Institute of Theoretical Physics School on Relativity, Supersymmetry, and Strings, held February 24-28, 1986, in Logan, Utah"--T.p. verso. Includes bibliographies and index. Subjects: General relativity (Physics)--Congresses. Grand unified theories (Nuclear physics)--Congresses. Supersymmetry--Congresses. String models--Congresses. LC Classification: QC173.6 .I575 1986 Dewey Class No.: 530.1/1 19

Second Paris Cosmology Colloquium: within the framework of the International School of Astrophysics 'Daniel Chalonge': Observatoire de Paris, 2-4 June 1994 / editors, H.J. de Vega, N. Sánchez. Published/Created: Singapore; New Jersey: World Scientific, c1995. Related Names: Vega, H. J. de (Héctor J.), 1949- Sanchez, N. (Norma), 1952- Description: xi, 546 p.: ill. (some col.); 23 cm. ISBN: 981022172X Notes: Includes bibliographical references (p. 525-527). Subjects: Cosmology--Congresses. Astrophysics--Congresses. String models--Congresses. LC Classification: QB980 .P37 1994 Dewey Class No.: 523.1 20

Spatio-temporal chaos and vacuum fluctuations of quantized fields / Christian Beck. Published/Created: New Jersey: World Scientific, c2002. Description: xviii, 272 p.: ill.; 23 cm. ISBN: 9810247982 Summary: "This book deals with new applications for coupled map lattices in quantum field theories and elementary particle physics"--P. xiii. Notes: Includes bibliographical references (p. 253-266) and index. Subjects: Coupled map lattices. Quantum field theory. Stochastic processes. String models. Chaotic behavior in systems. Statistical mechanics. Particles (Nuclear physics) Series: Advanced series in nonlinear dynamics; 21 LC Classification: QC174.85.L38 B43 2002 Dewey Class No.: 530.14/3 21

String gravity and physics at the Planck energy scale / edited by N. Sánchez and A. Zichichi. Published/Created: Dordrecht; Boston: Kluwer Academic Publishers, c1996. Related Names: Sanchez, N. (Norma), 1952- Zichichi, Antonino. North Atlantic Treaty Organization. Scientific Affairs Division. NATO Advanced Study Institute on String Gravity and Physics at the Planck Energy Scale (1995: Erice, Italy) Description: viii, 544 p.: ill. (some col.); 25 cm. ISBN: 0792339908 (hardcover: alk. paper) Notes: "Published in cooperation with NATO Scientific Affairs Division." "Proceedings of the NATO Advanced Study Institute on String Gravity and Physics at the Planck Energy Scale, Erice, Italy, 8-19,September, 1995"--T.p. verso. Includes bibliographical references and index. Subjects: String models--Congresses. Quantum gravity--Congresses. Nuclear astrophysics--Congresses. Cosmology--Congresses. Series: NATO ASI series. Series C, Mathematical and physical sciences; vol. 476 NATO ASI series. Series C, Mathematical and physical sciences; no. 476. LC Classification: QC794.6.S85 S74 1996 Dewey Class No.: 539.7/54 20

String theory / Joseph Polchinski. Published/Created: Cambridge, UK; New York: Cambridge University Press, 1998. Description: 2 v.: ill.; 26 cm. ISBN: 0521633036 (v. 1: hardback) 0521633044 (v. 2: hardback) Contents: v. 1. An introduction to the bosonic string -- v. 2. Superstring theory and beyond. Notes: Includes bibliographical references and indexes. Subjects: String models.

Superstring theories. Series: Cambridge monographs on mathematical physics LC Classification: QC794.6.S85 P65 1998 Dewey Class No.: 530.14 21 Electronic File Information: Publisher description http://www.loc.gov/catdir/description/cam029/98004545.ht ml Table of contents http://www.loc.gov/catdir/toc/cam022/98004545.html

String theory and cosmology: proceedings of Nobel Symposium 127, Sigtuna, Sweden, August 14-19, 2003 / editors, U. Danielsson, A. Goobar, B. Nilsson. Published/Created: Stockholm: Physica Scripta, Royal Swedish Academy of Sciences; Singapore: World Scientific, c2005. Related Names: Danielsson, Ulf H. Goobar, Ariel. Nilsson, Bengt E. W. Kungl. Svenska vetenskapsakademien. Description: 105 p.: ill.; 31 cm. ISBN: 9812564330 Notes: "The contents of this volume were also published as vol. T117 of Physica scripta"--T.p. verso. Includes bibliographical references. Subjects: String models--Congresses. Cosmology--Congresses. LC Classification: QC794.6.S85 N63 2003 Dewey Class No.: 523.1 22

String theory and M-theory: a modern introduction / Katrin Becker, Melanie Becker, and John H. Schwarz. Published/Created: Cambridge; New York: Cambridge University Press, c2007. Related Names: Becker, Melanie. Schwarz, John H. Description: xvi, 739 p.: ill.; 25 cm. ISBN: 9780521860697 (hbk.) 0521860695 (hbk.) Notes: Includes bibliographical references (p. 700-725) and index. Subjects: String models. String models--Problems, exercises, etc. LC Classification: QC794.6.S85 B435 2007 Dewey Class No.: 530.14 22 National Bibliography No.: GBA664617 bnb National Bibliographic Agency No.: 013517166 Uk

String theory and quantum gravity: proceedings of the Trieste Spring School, 23 April-1 May 1990 / edited by M. Green ... [et al.]. Published/Created: Singapore; Teaneck, N.J.: World Scientific, c1991. Related Names: Green, M. (Michael B.) International Centre for Theoretical Physics. International Atomic Energy Agency. Unesco. Description: vii, 173 p.: ill.; 23 cm. ISBN: 9810203721 Notes: "International Centre for Theoretical Physics, International Atomic Energy Acency, United Nations Educational, Scientific and Cultural Organization." Includes bibliographical references. Subjects: String models--Congresses. Quantum gravity--Congresses. LC Classification: QC794.6.S85 T75 1990 Dewey Class No.: 539.7/2 20

String theory in a nutshell / Elias Kiritsis. Published/Created: Princeton: Princeton University Press, c2007. Description: xviii, 588 p.: ill.; 26 cm. ISBN: 9780691122304 (acid-free paper) 069112230X (acid-free paper) Notes: Includes bibliographical references (p. [553]-573) and index. Subjects: String models. LC Classification: QC794.6.S85 K565 2007 Dewey Class No.: 530.14 22

String theory in curved space times: a collaborative research report / editor, N. Sanchez. Published/Created: Singapore; River Edge, NJ: World Scientific, c1998. Related Names: Sanchez, N. (Norma), 1952- Description: xiv, 483 p.: ill.; 23 cm. ISBN: 9810234392 (alk. paper) Notes: Includes bibliographical references. Subjects: String models. Special relativity (Physics) Space and time. LC Classification: QC794.6.S85 S75 1998 Dewey Class No.: 530.11 21

String theory in four dimensions / editor, Michael Dine. Published/Created: Amsterdam; New York: North Holland; New York, NY, USA: Sole distributors for the USA and Canada, Elsevier Science

Pub. Co., 1988. Related Names: Dine, Michael. Related Titles: String theory in 4 dimensions. Description: x, 476 p.: ill.; 27 cm. ISBN: 0444871004 0444871020 (pbk.) Notes: Includes bibliographies. Subjects: String models. Series: Current physics, sources and comments;1 Current physics; 1. LC Classification: QC794.6.S85 S76 1988 Dewey Class No.: 539.7/54 19

String theory, gauge theory and quantum gravity: proceedings of the Trieste Conference on Duality Symmetries in String Theory--II (1-4 April 1997) and Spring School on String Theory, Gauge Theory, and Quantum Gravity (7-12 April 1997), Trieste, Italy / edited by E. Gava ... [et al.]. Published/Created: Amsterdam: North-Holland, c1998. Related Names: Gava, E. Spring School on String Theory, Gauge Theory, and Quantum Gravity (1997: Trieste, Italy) Description: viii, 251 p.: ill.; 27 cm. Notes: Includes bibliographical references and index. Subjects: String models--Congresses. Quantum gravity--Congresses. Series: Nuclear physics. B (Proc. Suppl.), 0920-5632; 67 Nuclear physics. B, Proceedings, supplements; v. 67. LC Classification: QC770 .N772 vol. 67 QC794.6.S85 Dewey Class No.: 539.7 s 530.1/4 21

String theory, gauge theory and quantum gravity: proceedings of the Trieste Spring School and Workshop, ICTP, Trieste, Italy, 27 March-7 April, 1995 / edited by R. Dijkgraaf ... [et al.]. Published/Created: Amsterdam: North-Holland, c1996. Related Names: Dijkgraaf, R. Description: viii, 265 p.: ill.; 28 cm. Notes: Includes bibliographical references and index. Subjects: String models--Congresses. Quantum gravity--Congresses. Series: Nuclear physics. B, Proceedings, supplements, 0920-5632; 45B-C Nuclear physics. B, Proceedings, supplements; v. 45B-C. LC Classification: QC770 .N772 vol. 45B-C QC794.6.S85 Dewey Class No.: 530.1 21

String theory, gauge theory, and quantum gravity '93: proceedings of the Trieste Spring School & Workshop, ICTP, Trieste, Italy, April 19-29, 1993 / edited by R. Dijkgraaf ... [et al.]. Published/Created: Singapore; River Edge, NJ: World Scientific, c1994. Related Names: Dijkgraaf, R. International Centre for Theoretical Physics. International Atomic Energy Agency. Unesco. Description: viii, 339 p.: ill.; 22 cm. ISBN: 9810218060 Notes: "International Centre for Theoretical Physics, International Atomic Energy Agency, United Nations Educational, Scientific and Cultural Organization." Includes bibliographical references. Subjects: String models--Congresses. Quantum gravity--Congresses. LC Classification: QC794.6.S85 T75 1993 Dewey Class No.: 530.1/4 20

String theory, quantum cosmology and quantum gravity: integrable and conformal invariant theories: proceedings of the Paris-Meudon Colloquim, 22-26 September 1986 / editors, H.J. de Vega and N. Sanchez. Published/Created: Singapore; Philadelphia, PA: World Scientific, c1987. Related Names: Vega, H. J. de (Héctor J.), 1949- Sanchez, N. (Norma), 1952- Description: xi, 511 p.; 23 cm. ISBN: 9971502860 9971502992 (pbk.) Notes: Includes bibliographies. Subjects: String models--Congresses. Cosmology--Congresses. Quantum gravity--Congresses. Nuclear astrophysics--Congresses. LC Classification: QC794.6.S85 P37 1986 Dewey Class No.: 530.1 19

String theory, quantum gravity, and the unification of the fundamental interactions / edited by Massimo Bianchi ... [et al.]. Published/Created: Singapore; River Edge, NJ: World Scientific, c1993. Related Names: Bianchi, Massimo. Description: xix, 524 p.: ill., port.; 23 cm. ISBN:

9810211619 Notes: Proceedings of a meeting held in Rome, Sept. 21-26, 1992. Proceedings dedicated to François Englert on the occasion of his 60th birthday. "Publications of F. Englert"--P. ix-xv. Includes bibliographical references. Subjects: Englert, François. String models--Congresses. Quantum gravity--Congresses. LC Classification: QC794.6.S85 S762 1993 Dewey Class No.: 530.1/4 20

String theory: 10th Tohwa University international symposium on string theory, Fukuoka, Japan, 3-7 July 2001 / editors, Hajime Aoki, Tsukasa Tada. Published/Created: Melville, N.Y.: American Institute of Physics, 2002. Related Names: Aoki, Hajime. Tada, Tsukasa. Description: xiii, 366 p.: ill.; 24 cm. ISBN: 0735400512 (casebound) 0735400547 (pbk.) Notes: Includes bibliographical references and index. Subjects: String models--Congresses. Series: AIP conference proceedings, 0094-243X; v. 607 AIP conference proceedings; no. 607. LC Classification: QC794.6.S85 T68 2001 Dewey Class No.: 530.14 21

String theory: an introduction to the bosonic string / Joseph Polchinski. Edition Information: Pbk. ed. Published/Created: Cambridge; New York: Cambridge University Press, c2005 Description: 2 v.: ill.; 26 cm. ISBN: 0521672279 (v. 1: pbk.) 0521672287 (v. 2: pbk.) Contents: v. 1. An introduction to the bosonic string -- v. 2. Superstring theory and beyond. Subjects: String models. Superstring theories. Series: Cambridge monographs on mathematical physics LC Classification: QC794.6.S85 P65 2005 Dewey Class No.: 539.7/258 22 Electronic File Information: Table of contents only http://www.loc.gov/catdir/toc/fy0604/2005284209.html Publisher description http://www.loc.gov/catdir/enhancements/fy0632/2005284209 -d.html

String theory: from Gauge interactions to cosmology / edited by Laurent Baulieu ... [et al.]. Published/Created: Dordrecht, The Netherlands: Springer, c2006. Related Names: Baulieu, Laurent. NATO Public Diplomacy Division. NATO Advanced Study Institute on String Theory: From Gauge Interactions to Cosmology (2004: Cargèse, France) Description: xi, 404 p.: ill.; 25 cm. ISBN: 1402037317 (hbk.) 1402037333 (e-book) Notes: "Proceedings of the NATO Advanced Study Institute on String Theory: From Gauge Interactions to Cosmology, Cargèse, France, 7-19 June 2004"--T.p. verso. "Published in cooperation with NATO Public Diplomacy Division." Includes bibliographical references. Subjects: String models--Congresses. Gauge fields (Physics)--Congresses. Cosmology--Congresses. Series: NATO science series. Series II, Mathematics, physics, and chemistry; v. 208 NATO science series. Series II, Mathematics, physics, and chemistry; v. 208. LC Classification: QC794.6.S85 S66 2006

Strings 2001: proceedings of the Strings 2001 Conference, Tata Institute of Fundamental Research, Mumbai, India, January 5-10, 2001 / Atish Dabholkar, Sunil Mukhi, Spenta R. Wadia, editors. Published/Created: Providence, RI: American Mathematical Society; Cambridge, MA: Clay Mathematics Institute, c2002. Related Names: Dabholkar, Atish. Mukhi, Sunil. Wadia, S. R. (Spenta R.) Description: xx, 490 p.: ill.; 26 cm. ISBN: 0821829815 (alk. paper) Notes: Includes bibliographical references. Subjects: String models--Congresses. Series: Clay mathematics proceedings, 1534-6455; v. 1 LC Classification: QC794.6.S85 S67 2001 Dewey Class No.: 530.14 21

Strings '93: proceedings of the conference, May 24-29, 1993, Berkeley, USA / editors,

M.B. Halpern & G. Rivlis, A. Sevrin. Published/Created: Singapore; River Edge, N.J.: World Scientific, c1995. Related Names: Halpern, M. B. (Martin B.) Rivlis, G. (Gil) Sevrin, A. (Alexander) Description: xii, 490 p.: ill.; 23 cm. ISBN: 9810221878 Notes: Includes bibliographical references. Subjects: String models--Congresses. Field theory (Physics)--Congresses. Gravity--Congresses. LC Classification: QC794.6.S85 S769 1995

Strings and geometry: proceedings of the Clay Mathematics Institute 2002 Summer School on Strings and Geometry, Isaac Newton Institute, Cambridge, United Kingdom, March 24-April 20, 2002 / Michael Douglas, Jerome Gauntlett, Mark Gross, editors. Portion of Title: Proceedings of the Clay Mathematics Institute 2002 Summer School on Strings and Geometry Published/Created: Providence, RI: American Mathematical Society; Cambridge, MA: Clay Mathematics Institute, c2004. Related Names: Douglas, Michael (Michael R.) Gauntlett, Jerome. Gross, M. W. (Mark W.), 1965- Clay Mathematics Institute. Isaac Newton Institute for Mathematical Sciences. Description: ix, 376 p.: ill.; 26 cm. ISBN: 082183715X (alk. paper) Notes: Includes bibliographical references. Subjects: String models--Mathematics--Congresses. Geometry, Algebraic--Congresses. Branes--Congresses. D-branes--Congresses. Series: Clay mathematics proceedings; v. 3 LC Classification: QC794.6.S85 C56 2002 Dewey Class No.: 539.7/258 22

Strings and gravity: tying the forces together: proceedings of the Fifth Francqui Colloquium, 19-21 October 2001, Brussels / Marc Henneaux, Alexander Sevrin (eds.). Portion of Title: Tying the forces together Published/Created: Bruxelles: De Boeck & Larcier, c2003. Related Names: Henneaux, Marc. Sevrin, A. (Alexander) Description: 306 p.: ill. (some col.); 26 cm. ISBN: 2804140946 Notes: Includes bibliographical references. Subjects: String models--Congresses. Gravitation--Congresses. Series: Bibliothèque scientifique Francqui = Francqui scientific library, 1374-0911;5 Francqui scientific library; 5. LC Classification: QC794.6.S85 F73 2001 Dewey Class No.: 530.14 22

Strings and superstrings / Jerusalem Winter School for Theoretical Physics, Jerusalem, 20 Dec., 85-9 Jan., 86; edited by T. Piran and S. Weinberg. Published/Created: Singapore; Teaneck, NJ, USA: World Scientific, 1988. Related Names: Piran, Tsvi, 1949- Weinberg, Steven, 1933- Description: ix, 222 p.: ill.; 24 cm. ISBN: 9971503743 9971503751 (pbk.) Notes: "School met at the Institute for Advanced Study of the Hebrew University"--Foreword. "Volume 3." Includes bibliographies. Subjects: String models--Congresses. Superstring theories--Congresses. LC Classification: QC794.6.S85 J47 1986 Dewey Class No.: 530.1 19

Strings and symmetries: proceedings of the Gürsey Memorial Conference I held at Istanbul, Turkey, 6-10 June 1994 / Gülen Aktaş, Cihan Saçlıoğlu, Meral Serdaroğlu (eds.). Published/Created: Berlin; New York: Springer, c1995. Related Names: Gürsey, Feza. Aktaş, Gülen, 1951- Saçlıoğlu, Cihan, 1948- Serdaroğlu, M. (Meral) Description: xiv, 389 p.: ill.; 25 cm. ISBN: 354059163X (acid-free paper) Notes: Includes bibliographical references. Subjects: String models--Congresses. Symmetry (Physics)--Congresses. Field theory (Physics)--Congresses. Mathematical physics--Congresses. Series: Lecture notes in physics; 447 LC Classification: QC794.6.S85 G87 1994 Dewey Class No.: 530.1/4 20

Strings clássicos relativísticos / por Carlos A.P. Galvão. Published/Created: Rio de Janeiro: CNPq, Centro Brasileiro de Pesquisas Físicas, 1985. Description: 2 v. (viii, 545 p.): ill.; 30 cm. Notes: "CBPF-MO-006/85." Bibliography: v. 2, p. 540-545. Subjects: String models. Relativity (Physics) Quantum field theory. Series: Monografia, 0101-9236 Monografia (Centro Brasileiro de Pesquisas Físicas) LC Classification: QC794.6.S85 G35 1985

Strings, branes and extra dimensions: TASI 2001, Boulder, Colorado, USA, 4-29 June 2001 / editors, Steven S. Gubser, Joseph D. Lykken. Published/Created: River Edge, N.J.: World Scientific, c2004. Related Names: Gubser, Steven Scott, 1972- Lykken, J. D. (Joe D.) Theoretical Advanced Study Institute in Elementary Particle Physics (2001: Boulder, Colo.) Description: xi, 850 p.: ill.; 23 cm. ISBN: 9812387889 9789812387882 Notes: Includes bibliographical references. Subjects: String models--Congresses. Branes--Congresses. LC Classification: QC794.6.S85 S72 2004

Strings, branes, and dualities / edited by Laurent Baulieu ... [et al.]. Published/Created: Dordrecht; Boston: Kluwer Academic Publishers, c1999. Related Names: Baulieu, Laurent. North Atlantic Treaty Organization. Scientific Affairs Division. NATO Advanced Study Institute on Strings, Branes, and Dualities (1997: Cargèse, France) Description: x, 493 p.: ill.; 25 cm. ISBN: 0792353447 (alk. paper) Notes: "Published in cooperation with NATO Scientific Affairs Division." "Proceedings of the NATO Advanced Study Institute on Strings, Branes, and Dualities, Cargèse, France, May 26-June 14, 1997"--T.p. verso. Includes bibliographical references and index. Subjects: String models--Congresses. Supersymmetry--Congresses. Duality (Nuclear physics)--Congresses. Field theory (Physics)--Congresses. Branes--Congresses. Series: NATO ASI series. Series C, Mathematical and physical sciences; vol. 520 NATO ASI series. Series C, Mathematical and physical sciences; no. 520. LC Classification: QC794.6.S85 S7 1999 Dewey Class No.: 539.7/25 21

Strings, branes, and gravity: TASI 99: Boulder, Colorado, USA, 31 May-25 June 1999 / editors, Jeffrey Harvey, Shamit Kachru, Eva Silverstein. Variant Title: Subtitle on t.p. verso: Lecture notes of TASI 99 Portion of Title: TASI 99 Published/Created: Singapore; River Edge, NJ: World Scientific, c2001. Related Names: Harvey, Jeffrey. Kachru, Shamit, 1970- Silverstein, Eva, 1970- Description: vii, 933 p., [3] leaves of plates: ill.; 23 cm. ISBN: 9810247745 Contents: TASI lectures on branes, black holes and anti-De Sitter space / M.J. Duff -- D-brane primer / Clifford V. Johnson -- TASI lectures on black holes in string theory / Amanda W. Peet -- TASI lectures: cosmology for string theorists / Sean M. Carroll -- TASI lectures on matrix theory / Tom Banks -- TASI lectures M theory phenomenology / Michael Dine -- TASI lectures: introduction to the AdS/CFT correspondence / Igor R. Klebanov -- TASI lectures on compactification and duality / David R. Morrison -- Compactification, Geometry and duality: N=2 / Paul S. Aspinwall -- TASI lectures on non-BPS D-brane systems / John H. Schwarz -- Lectures on warped compactifications and stringy brane constructions / Shamit Kachru -- TASI lectures on the holographic principle / Daniela Bigatti and Leonard Susskind. Notes: Includes bibliographical references. Subjects: String models--Congresses. Gravity--Congresses. Compactifications--Congresses. Duality (Nuclear physics)--Congresses. Black holes (Astronomy)--Congresses. Supersymmetry--Congresses. Branes--

Congresses. LC Classification: QC794.6.S85 T48 1999 Dewey Class No.: 530.14 21

Strings, conformal fields, and M-theory / Michio Kaku. Edition Information: 2nd ed. Published/Created: New York: Springer, c2000. Description: xv, 531 p.: ill.; 25 cm. ISBN: 0387988920 (alk. paper) Notes: Includes bibliographical references and index. Subjects: String models. Conformal invariants. Quantum field theory. Series: Graduate texts in contemporary physics LC Classification: QC794.6.S85 K355 2000 Dewey Class No.: 530.14 21

Strings, conformal fields, and topology: an introduction / Michio Kaku. Published/Created: New York: Springer-Verlag, c1991. Description: xiv, 535 p.: ill.; 25 cm. ISBN: 0387974962 (New York: acid-free paper) 3540974962 (Berlin: acid-free paper) Notes: Includes bibliographical references and index. Subjects: String models. Superstring theories. Quantum field theory. Algebraic topology. Series: Graduate texts in contemporary physics LC Classification: QC794.6.S85 K36 1991 Dewey Class No.: 539.7/2 20

Strings, lattice gauge theory, high energy phenomenology: proceedings of the winter school Panchgani, January 25 to February 5, 1986 / [organized by] Tata Institute of Fundamental Research Bombay, India; editors V. Singh and S.R. Wadia. Published/Created: Singapore; Philadelphia, PA: World Scientific, c1987. Related Names: Virendra Singh, 1938- Wadia, S. R. (Spenta R.) Tata Institute of Fundamental Research. Description: x, 598 p.: ill.; 23 cm. ISBN: 9971501570 Notes: Includes bibliographies. Subjects: String models--Congresses. Gauge fields (Physics)--Congresses. Phenomenological theory (Physics)--Congresses. LC Classification: QC794.6.S85 S77 1987 Dewey Class No.: 530.1/4 19

Superstrings '87: proceedings of the Trieste Spring School, 1-11 April 1987 / edited by L. Alvarez-Gaume ... [et al.]. Published/Created: Singapore; [Teaneck] N.J.: World Scientific, 1988. Related Names: Alvarez-Gaumé, Luis. Trieste Spring School (1987: International Centre for Theoretical Physics) Description: ix, 422 p.; 23 cm. ISBN: 9971502976 9971505185 (pbk.) Contents: Introduction -- Lectures on complex manifolds / P. Candelas -- Conformal field theory / M.T. Grisaru -- String calculus: conformal field theory as a tool in string theory / Emil Martinec -- Symmetry and algebras in string theory / Peter Goddard -- Toroidal and orbifold compactification of heterotic string / K.S. Narain -- String theory from a macroscopic point of view / Nathan Seiberg -- Introduction to string field theory / Gary T. Horowitz -- Lectures on closed string field theory / Andrew Strominger -- New methods in string theory / L. Alvarez-Gaume, C. Gomez, and C. Reina. Notes: "Proceedings of the 1987 Spring School on Superstrings ... held at the International Centre for Theoretical Physics at Trieste, Italy"--Pref. Includes bibliographies. Subjects: String models--Congresses. Superstring theories--Congresses. LC Classification: QC794.6.S85 S868 1988 Dewey Class No.: 530.1 19

Superstrings: the first 15 years of superstring theory / edited by John H. Schwarz. Published/Created: Singapore; Philadelphia: World Scientific, c1985. Related Names: Schwarz, John H. Description: 2 v. (xi, 1141 p.); 26 cm. ISBN: 9971978660: 9971978679 (pbk.): Notes: Includes bibliographies. Subjects: Superstring theories. String models. Duality (Nuclear physics) Particles (Nuclear physics) LC Classification: QC794.6.S85 S87 1985 Dewey Class No.: 539.7/54 19

Supersymmetry and string theory: beyond the standard model / Michael Dine. Published/Created: Cambridge; New York: Cambridge University Press, 2007. Description: xx, 515 p.: ill.; 26 cm. ISBN: 9780521858410 0521858410 Notes: Includes bibliographical references (p. 505-510) and index. Subjects: Supersymmetry. String models. LC Classification: QC174.17.S9 D56 2007 Dewey Class No.: 539.725 22 National Bibliography No.: GBA681935 bnb National Bibliographic Agency No.: 013562831 Uk

Teaching music in the urban classroom / edited by Carol Frierson-Campbell. Published/Created: Lanham, Md.: Rowman & Littlefield Education, c2006. Related Names: Frierson-Campbell, Carol, 1961- MENC, the National Association for Music Education (U.S.) Description: 2 v.: music; 23 cm. ISBN: 1578864607 (hardcover: alk. paper) 1578864615 (pbk.: alk. paper) 9781578864607 9781578864614 9781578864645 (v. 2, hardcover: alk. paper) 157886464X (v. 2, hardcover: alk. paper) 9781578864652 (v. 2, pbk: alk. paper) 1578864658 (v. 2, pbk: alk. paper) Contents: v. 1. A guide to survival, success, and reform. pt. I. Cultural responsivity: Defining ourselves as other: envisioning transformative possibilities / Cathy Benedict. Cultural clashes: the complexity of identifying urban culture / Donna T. Emmanuel. Building confianza: using dialogue journals with English-language learners in urban schools / Regina Carlow. White teacher, students of color: culturally responsive pedagogy for elementary general music in communities of color / Kathy M. Robinson; pt. 2. Music teacher stories. The challenges of urban teaching: young urban music educators at work / Janice Smith. Teaching music in urban landscapes: three perspectives / Carlos R. Abril; pt. 3. Teaching strategies: Motivating urban music students / Elizabeth Ann McAnally. Differentiating instruction in the choral rehearsal: strategies for choral conductors in urban schools / Daniel Abrahams. Building an instrumental music program in an urban school / Kevin Mixon. The string chorale concept / Jeanne Porcino Dolamore. The small, big city in music education: the impacts of instrumental music education for urban students / Karen Iken; pt. 4. Alternative teaching models: A new sound for urban schools: rethinking how we plan / Frank Abrahams and Patrick K. Schmidt. Music of every culture has something in common and can teach us about ourselves: using the Aesthetic Realism Teaching Method / Edward Green and Alan Shapiro. Music educators in the urban school reform conversation / Carol Frierson-Campbell -- v. 2. A guide to leadership, teacher education, and reform. pt 1. Educational leadership; pt. 2. Teacher education; pt. 3. Partnerships -- pt. 4. School reform. Notes: "Published in partnership with National Association for Music Education." Includes bibliographical references. Subjects: Music--Instruction and study. Education, Urban--Social aspects. City children--Education. LC Classification: MT1 .T387 2006 Dewey Class No.: 780.71 22

The history of the American guitar: from 1833 to the present day / by Tony Bacon. Portion of Title: American guitar Published/Created: [New York, N.Y.]: Friedman/Fairfax Publishers: Distributed by Sterling Pub. Co., c2001. Description: 148 p.: col. ill.; 32 cm. ISBN: 1586632973 Notes: Includes bibliographical references (p. 148) and index. Subjects: Guitar--United States--History. Guitar--United States--Pictorial works. LC Classification: ML1015.G9 B216 2001 Dewey Class No.: 787.87/1973 21

The Large N expansion in quantum field theory and statistical physics: from spin systems to 2-dimensional gravity / editors, Edouard Brézin, Spenta R. Wadia. Published/Created: Singapore; River Edge, NJ: World Scientific, c1993. Related Names: Brézin, E. Wadia, S. R. (Spenta R.) Description: xiv, 1130 p.: ill.; 26 cm. ISBN: 9810204558 9810204566 (pbk.) Notes: Includes bibliographical references. Subjects: Quantum field theory. String models. Gauge fields (Physics) LC Classification: QC174.45 .L37 1993 Dewey Class No.: 530.1/2 20

The new cosmology: Conference on Strings and Cosmology, College Station, Texas, 14-17 March 2004 [and] the Mitchell Symposium on Observational Cosmology, College Station, Texas, 12-16 April 2004 / editors, Roland E. Allen, Dimitri V. Nanopoulos, Christopher N. Pope. Portion of Title: Conference on Strings and Cosmology, College Station, Texas, 14-17 March 2004 Strings and cosmology Mitchell Symposium on Observational Cosmology, College Station, Texas, 12-16 April 2004 Observational cosmology Published/Created: Melville, N.Y.: American Institute of Physics, 2004. Related Names: Allen, Roland E. Nanopoulos, D. V. Pope, Christopher N. Mitchell Symposium on Observational Cosmology (2004: College Station, Tex.) Description: ix, 458 p.: ill. (some col.), col. port.; 28 cm. ISBN: 0735402272 Contents: Observational cosmology -- Strings and cosmology. Notes: Includes bibliographical references and index. Additional Formats: Also available via the World Wide Web. Subjects: Cosmology--Congresses. String models--Congresses. Cosmology--Observations--Congresses. Superstring theories--Congresses. Series: AIP conference proceedings, 0094-243X; v. 743 AIP conference proceedings; no. 743. LC Classification: QB980 .C655 2004 Dewey Class No.: 523.1 22

The Santa Fe TASI-87: proceedings of the 1987 Theoretical Advanced Study Institute in Elementary Particle Physics, Santa Fe, New Mexico, July 5-August 1, 1987 / editors, Richard Slansky, Geoffrey West. Published/Created: Singapore; Teaneck, N.J.: World Scientific, c1988. Related Names: Slansky, Richard. West, Geoffrey B. Description: 2 v. (viii, 1030 p.): ill.; 22 cm. ISBN: 9971504383 (set) 9971504391 (pbk.: set) Notes: Includes bibliographical references. Subjects: Particles (Nuclear physics)--Congresses. String models--Congresses. Superstring theories--Congresses. Astrophysics--Congresses. LC Classification: QC793 .T44 1987 Dewey Class No.: 539.7/21 19

The superworld II / edited by Antonino Zichichi. Published/Created: New York: Plenum Press, c1990. Related Names: Zichichi, Antonino. Related Titles: Superworld two. Superworld 2. Description: viii, 564 p.: ill.; 26 cm. ISBN: 0306434938 Notes: "Proceedings of the Twenty-fifth course of the International School of Subnuclear Physics on the superworld II, held August 6-14, 1987, in Erice, Sicily, Italy"--T.p. verso. Includes bibliographical references. Subjects: String models--Congresses. Particles (Nuclear physics)--Congresses. Series: The Subnuclear series; 25 Subnuclear series; v. 25. LC Classification: QC794.6.S85 I57 1987 Dewey Class No.: 539.7/2 20

The trouble with physics: the rise of string theory, the fall of a science, and what comes next / Lee Smolin. Published/Created: Boston: Houghton Mifflin, c2006. Description: xxiii, 392 p.; 24 cm. ISBN: 9780618551057 0618551050 Notes: Includes bibliographical references (p. [359]-371) and index. Subjects: Physics--Methodology--History--20th century. String models. LC Classification: QC6 .S6535 2006 Dewey Class No.: 530.14 22

Electronic File Information: Table of contents only http://www.loc.gov/catdir/toc/ecip069/2006007235.html Publisher description http://www.loc.gov/catdir/enhancements/fy0634/2006007235-d.html Sample text http://www.loc.gov/catdir/enhancements/fy0666/2006007235-s.html Contributor biographical information http://www.loc.gov/catdir/enhancements/fy0737/2006007235-b.html

Theoretical high energy physics: MRST 2001, a tribute to Roger Migneron: London, Ontario, Canada, 15-18 May 2001 / editors, V. Elias, D.G.C. McKeon, V.A. Miransky. Portion of Title: MRST 2001, a tribute to Roger Migneron Published/Created: Melville, N.Y.: American Institute of Physics, 2001. Related Names: Migneron, Roger, 1937-1999. Elias, Vic, 1948- McKeon, Dennis Gerard Creaser, 1949- Miranskiĭ, V. A. (Vladimir AdolÊ¹fovich) Description: ix, 314 p.: ill.; 25 cm. ISBN: 0735400458 Contents: Machine generated contents note: MIGNERON TRIBUTE SESSION I -- Progress in M -Theory3 -- M. J. Duff -- Perturbative SUSYM vs. AdS/CFT: A Brief Review 19 -- G. W. Semenoff -- Production of Heavy Quarks in Deep-Inelastic Lepton-Hadron -- Scattering40 -- W. L. van Neerven -- Spinor and Supersymmetry in Spaces of Various Dimensions and -- Signatures60 -- D. G. C. McKeon and T. N. Sherry -- Dynamically Generating the Quark-Level SU(2) Linear Sigma Model 66 -- M. D. Scadron -- GRAVITY/GEOMETRY -- Oscillating Metrics and the Cosmological Constant 75 -- B. Holdom -- Hiding a Cosmological Constant in a Warped Extra Dimension 82 -- H. Collins -- 5-Dimensional Warped Cosmological Solutions with -- Radius Stabilization by a Bulk Scalar88 -- J. M. Cline and H. Firouzjahi -- B-PHYSICS -- Determining I|Vub from the B--Xu Iv Dilepton Invariant -- Mass Spectrum97 -- C. W. Bauer, Z. Ligeti, and M. Luke -- B-*Xsl+F in the Vectorlike Quark Model 105 -- M. R. Ahmady, M. Nagashima, and A. Sugamoto -- Renormalon Analysis of Heavy-Light Exclusive B Decays115 -- A. R. Williamson -- MIGNERON TRIBUTE SESSION II -- Composites in Color Superconducting Phase of Dense QCD with Two -- Quark Flavors129 -- V. A. Miransky -- Infrared Dynamics in Vector-Like Gauge Theories: QCD and Beyond 140 -- V. Elias -- Topics on Neutrino Physics148 -- G. Karl and V. Novikov -- QCD Sum-Rule Bounds on the Light Quark Masses151 -- T. G. Steele -- QUARKS, GLUONS, AND MESONS -- String Model Building at Low String Scale: Towards the -- Standard M odel161 -- R. G. Leigh -- Lepton Pair Production in a Charged Quark Gluon Plasma 168 -- A. Majumder and C. Gale -- O(a/s) Estimate for the Longitudinal Cross Section in e+e-- Annihilation to Hadrons175 -- F. A. Chishtie -- Exploring Pseudoscalar Meson Scattering in Linear Sigma Models 182 -- D. Black, A. H. Fariborz, S. Moussa, S. Nasri, and J. Schechter -- Probing Scalar Mesons Using a Toy Model in the Linear -- Sigma M odel188 -- D. Black, A. H. Fariborz, S. Moussa, S. Nasri, and J. Schechter -- FIELD THEORY I -- Cutoff Dependence of the Casimir Effect 197 -- C. R. Hagen -- Pair Production of Arbitrary Spin Particles with EDM and AMM and -- Vacuum Polarization 206 -- S. I. Kruglov -- Analyzing the 't Hooft Model on a Light-Cone Lattice212 -- J. S. Rozowsky -- Fuzzy Non-trivial Gauge Configurations 219 -- B. Ydri -- Generalized Coherent State Approach to Star Products and -- Applications to the Fuzzy Sphere 226 -- G. Alexanian, A. Pinzul, and A. Stem -- FIELD THEORY II -- Spontaneous CPT Violation in Confined QED 235 -- E. J. Ferrer, V. de la Incera, and A. Romeo -- Issues on Radiatively Induced Lorentz and CPT Violation in -- Quantum Electrodynamics242 -- W. F. Chen --

Thermal Conductivity of the 2+1-Dimensional NJL Model in an -- External Magnetic Field253 -- E. J. Ferrer, V. A. Gusynin, and V. de la Incera -- Example of an Asymptotically Free Matrix Model 259 -- A. Agarwal and S. G. Rajeev -- Variational Principle for Large N Matrix Models 267 -- L. Akant, G. S. Krishnaswami, and S. G. Rajeev -- BRANES, STRINGS, AND THINGS -- Analytic Semi-classical Quantization of a QCD String with -- Light Quarks277 -- T. J. Allen, C. Goebel, M. G. Olsson, and S. Veseli -- String Webs from Field Theory282 -- P. Argyres and K. Narayan -- Geometry of Large Extra Dimensions versus Graviton Emission 287 -- F. Leblond -- Quantum Myers Effect and its Supergravity Dual for DO/D4 Systems 295 -- P. J. Silva -- AFTERWORD -- Roger Migneron (1937-1999): Reminiscences on Sharing Life with a -- Physicist305 -- I. P. Migneron -- List of Participants311 -- Author Index313. Notes: Includes bibliographical references and index. Subjects: Particles (Nuclear physics)--Congresses. Series: AIP conference proceedings, 0094-243X; v. 601 AIP conference proceedings; no. 601. LC Classification: QC793 .M77 2001 Dewey Class No.: 539.7/2 21

Topology, geometry and quantum field theory: proceedings of the 2002 Oxford symposium in the honour of the 60th birthday of Graeme Segal / edited by U.L. Tillmann. Published/Created: Cambridge, UK; New York: Cambridge Univeristy Press, 2004. Related Names: Segal, Graeme. Tillmann, U. L. (Ulrike Luise), 1962- Description: xii, 577 p.; 23 cm. ISBN: 0521540496 (pbk.) Notes: Includes bibliographical references and index. Subjects: Homology theory--Congresses. String models--Congresses. Quantum gravity--Congresses. Field theory (Physics)--Congresses. Series: London Mathematical Society lecture note series; 308 LC Classification: QC20.7.H65 S96 2002 Dewey Class No.: 530.15/423 22

Uncommon carriers / John McPhee. Edition Information: 1st ed. Published/Created: New York: Farrar, Straus, and Giroux, 2006. Description: 248 p.; 22 cm. ISBN: 0374280398 (hardcover: alk. paper) 9780374280390 (hardcover: alk. paper) Summary: McPhee's books are about real people in real places. Over the past eight years, McPhee has spent considerable time in the company of people who work in freight transportation. This is his sketchbook of them and of his journeys with them. He rides from Atlanta to Tacoma alongside Don Ainsworth, owner and operator of a sixty-five-foot, eighteen-wheel chemical tanker carrying hazmats. He attends ship-handling school on a pond in the foothills of the French Alps, where, for a tuition of $15,000 a week, skippers of the largest ocean ships refine their capabilities in twenty-foot scale models. He goes up the Illinois River on a "towboat" pushing a triple string of barges, the overall vessel being "a good deal longer than the Titanic." And he travels by canoe up the canal-and-lock commercial waterways traveled by Henry David Thoreau and his brother, John, in a homemade skiff in 1839.--From publisher description. Contents: A fleet of one -- The ships of Port Revel -- Tight-assed river -- Five days on the Concord and Merrimack rivers -- Out in the sort -- Coal train -- A fleet of one: II. Subjects: Freight and freightage. Transportation. LC Classification: HE199.A2 M39 2006 Dewey Class No.: 388/.044 22

Unity from duality: gravity, gauge theory and strings = L'unité de la physique fondamentale: gravité, théorie de jauge et cordes: a NATO Advanced Study Institute, Les Houches, session LXXVI, 30 July-31 August 2001 / edited by C. Bachas ... [et al.]. Portion of Title: Houches, session

LXXVI Parallel Title: Unité de la physique fondamentale Published/Created: Les Ulis [France] EDP Sciences; Berlin; New York: Springer, c2002. Related Names: Bachas, C. Université Joseph Fourier. Institut national polytechnique de Grenoble. North Atlantic Treaty Organization. Scientific Affairs Division. NATO Advanced Study Institute. Description: xxxiv, 663 p.: ill.; 23 cm. ISBN: 3540002766 (Springer: cloth) 2868836259 (EDP Sciences: cloth) Notes: At head of title: Ecole de physique des Houches -- UJF & INPG -- Grenoble. "Published in cooperation with the NATO Scientific Affair [sic] Division." "ISSN 0924-8099 print edition"--T.p. verso. Includes bibliographical references. Preface in English and French; Text in English. Subjects: Grand unified theories (Nuclear physics)--Congresses. Gravitational fields--Congresses. Gauge fields (Physics)--Congresses. String models--Congresses. Supersymmetry--Congresses. LC Classification: QC794.6.G7 E26 2002 Dewey Class No.: 530.14/2 22 National Bibliography No.: GBA3-Z2588

Vvedenie v kvantovuĭı̆, uĭ,¡ teoriĭı̆, uĭ,¡ strun i superstrun / S.V. Ketov; otvetstvennyi redaktor V.G. Bagrov. Published/Created: Novosibirsk: "Nauka," Sibirskoe otd-nie, 1990. Related Names: Bagrov, V. G. (Vladislav Gavrilovich) Description: 365, [4] p.: ill.; 23 cm. ISBN: 5020296600. Includes bibliographical references (p. 342-[366]). Subjects: String models. Superstring theories. Quantum field theory. LC Classification: QC794.6.S85 K48 1990

Warped passages: unraveling the mysteries of the Universe's hidden dimensions / Lisa Randall. Edition Information: 1st ed. Published/Created: New York: Ecco, c2005. Description: xii, 499 p.: ill.; 24 cm. ISBN: 0060531088 9780060531089 Summary: Discusses dimensions of space, early twentieth-century advances, the physics of elementary particles, string theory and branes, and proposals for extra-dimension universes. Contents: Introduction -- Entryway passages: demystifying dimensions -- Restricted passages: rolled-up extra dimensions -- Exclusive passages: branes, braneworlds, and the bulk -- Approaches to theoretical physics -- Relativity: the evolution of Einstein's gravity -- Quantum mechanics: principled uncertainty, the principal uncertainties, and the uncertainty principle -- The standard model of particle physics: matter's most basic known structure -- Experimental interlude: verifying the standard model -- Symmetry: the essential organizing principle -- The origin of elementary particle masses: spontaneous symmetry breaking and the Higgs mechanism -- Scaling and grand unification: relating interactions at different lengths and energies -- The hierarchy problem: the only effective trickle-down theory -- Supersymmetry: a leap beyond the standard model -- Allegro (ma non troppo) passage for strings -- Supporting passages: brane development -- Bustling passages: braneworlds -- Sparsely populated passages: multiverses and sequestering -- Leaky passages: fingerprints of extra dimensions -- Voluminous passages: large extra dimensions -- Warped passage: a solution to the hierarchy problem -- The warped annotated "Alice" -- Profound passage: an infinite extra dimension -- A reflective and expansive passage -- Extra dimensions: are you in or out? -- (In)conclusion. Notes: Includes bibliographical references and index. Subjects: Physics--Philosophy. Particles (Nuclear physics) Cosmology. String models. Branes. LC Classification: QC6 .R26 2005 Dewey Class No.: 530/.01 22

Workshop on Unified String Theories: 29 July-16 August 1985, Institute for Theoretical Physics, University of California, Santa

Barbara / edited by M. Green and D. Gross. Published/Created: Singapore; [Philadelphia]: World Scientific, c1986. Related Names: Green, M. (Michael B.) Gross, D. (David) University of California, Santa Barbara. Institute for Theoretical Physics. Description: viii, 745 p.: ill.; 23 cm. ISBN: 9971500310 9971500329 (pbk.) Notes: Includes bibliographies. Subjects: String models--Congresses. Unified field theories--Congresses. Superstring theories--Congresses. LC Classification: QC794.6.S85 W67 1985 Dewey Class No.: 530.1/42 19

INDEX

A

accelerator, 136
achievement, 124
acid, 199, 211, 212, 214, 217, 219
adaptation, 180
Aharonov-Bohm effect, 67, 70
alternative, 3, 40, 46, 47, 52, 55, 59, 89, 104
ambiguity, 4, 89
amplitude, 86, 117, 148
angular velocity, 137
animals, 197
annihilation, 23, 32, 34, 103, 140, 144
applied mathematics, 203
Argentina, 179
argument, 110
ASI, 196, 197, 199, 200, 206, 210, 211, 212, 213, 218
assignment, 88
attention, 184, 191

B

baryon(s), 2, 5, 54, 147, 150
basis set, 84
behavior, 4, 54, 145, 213
Beijing, 201
binding, 196
black hole, viii, 47, 179, 192, 196, 199, 200, 212, 218
blocks, vii, 181
boson(s), 2, 21, 22, 31, 84, 86, 88, 92, 106
bounds, 81
brane tension, 120
branes, vii, viii, 4, 5, 39, 41, 42, 47, 48, 49, 51, 52, 53, 54, 55, 81, 82, 112, 113, 115, 116, 117, 118, 119, 120, 121, 122, 123, 125, 126, 134, 137, 199, 200, 206, 217, 218, 224
Britain, 208
building blocks, vii

C

C++, 154
Calabi-Yau, viii, 81, 82, 83, 111, 112, 113, 114, 115, 116, 125, 208
calculus, 219
California, 199, 202, 209, 224
Canada, 196, 197, 214, 222
candidates, 4, 83, 125
carbon, 208
cation, 138
Cauchy problem, 138
CERN, 149
chaos, 213
chemical reactions, 196
Chicago, 197, 199
children, 220
China, 152, 201
chirality, 120
classes, 111, 112, 113, 123, 168
classical mechanics, 133, 141, 152
classification, 6, 103, 140, 146
classroom, 220
closed string, 29, 31, 32, 33, 35, 36, 37, 38, 39, 120, 122, 134, 138, 145, 219
cognition, 203
College Station, 221
collisions, 140
community, 82
compatibility, 158, 170
complexity, viii, 133, 220
complications, 103

components, 84, 85, 86, 94, 95, 97, 102, 103, 105, 106, 121, 125, 157, 159, 160, 164
composition, 147
computation, 144, 146, 147, 180, 181, 190, 192, 193
computers, 146, 147
computing, 112, 181
concentrates, 136
concentration, 140
concrete, 6, 103, 104, 142
condensation, 99, 100, 111, 113, 120, 121, 122
condensed matter, vii, 1, 5, 55, 67, 68
configuration, 35, 40, 41, 44, 45, 46, 48, 49, 50, 51, 66, 114, 136
confinement, vii, 1, 3, 4, 5, 52, 53, 76, 77, 136, 147, 200
conjecture, 4, 5, 39, 42, 53, 54, 55, 61, 76, 193
conjugation, 64, 183
conservation, 8, 121, 135, 138, 140, 192
constraints, viii, 6, 85, 86, 87, 88, 92, 93, 94, 95, 109, 110, 117, 121, 122, 123, 135, 139, 140, 141, 143, 144, 155, 156, 157, 158, 160, 161, 162, 163, 164, 165, 166, 167, 168, 170, 172, 173, 175, 190, 191, 192
construction, viii, 21, 81, 82, 83, 88, 91, 104, 105, 115, 119, 120, 124, 126, 137, 138, 142, 146, 190, 208
Copenhagen, 148
correlation, 5, 18, 20, 25, 104, 179, 180, 181, 182, 183, 184, 188, 189, 190, 192, 194
correlation function, 18, 25, 179, 180, 181, 182, 183, 184, 188, 189, 190, 192, 194
correspondence principle, 135, 142
Coulomb, viii, 4, 46, 53, 54, 179, 180, 181, 183, 184, 185, 187, 189, 190, 191, 192, 193
couples, 158
coupling, 4, 39, 40, 41, 42, 43, 44, 49, 82, 83, 86, 92, 93, 97, 99, 100, 101, 102, 104, 110, 111, 112, 114, 115, 120, 121, 125, 190, 191
covering, 112
critical value, 145, 146
criticism, 3, 4
culture, 220
cycles, 117, 123, 124

D

D-branes, 39, 41, 55, 112, 116, 117, 118, 119, 122, 200, 217
decay, 83, 110, 112, 115, 125, 141, 147
decomposition, 32, 114
decoupling, 5, 41, 42, 47, 48, 52, 53, 89, 90, 94, 101, 105, 109, 110, 125
definition, 4, 24, 31, 82, 89, 136, 139

degenerate, 19, 20, 85, 181, 183
degrees of freedom, 37, 84, 100, 105, 106, 137, 141, 158
demand, 40, 86
density, 65, 66, 70, 82, 138, 139, 140, 162
derivatives, 113, 139, 156, 173
discrete variable, 158
dispersion, 36
displacement, 7
divergence, 184
dynamical systems, 152

E

education, 220
educators, 220
effective field theory, vii, 81, 83, 105, 107
eigenvalue, 27, 57, 58, 73, 172
Einstein, Albert, 39, 177
electric charge, 2, 90, 134, 136, 137
electric field, 66
electromagnetic, vii, 2, 66, 134, 136
electron(s), vii, 1, 2, 4, 5, 68, 69, 70, 71, 72, 73, 74, 90, 94, 98, 99, 110, 115, 116
elementary particle, 2, 82, 136, 138, 141, 147, 198, 209, 213, 224
energy, vii, 2, 4, 5, 9, 10, 11, 14, 17, 24, 25, 26, 36, 39, 41, 42, 43, 44, 45, 49, 50, 51, 52, 53, 70, 72, 74, 75, 81, 83, 88, 97, 103, 105, 107, 111, 113, 120, 121, 124, 135, 136, 138, 140, 156, 167, 168, 174, 191, 197, 201, 202, 203, 208, 213, 219, 222
energy momentum tensor, 174
England, 154
enlargement, 166
environment, 146
equipment, 146
Euclidean space, 7, 8, 15, 133
Euler, 29, 30, 31, 138, 140
Euler equations, 138
evidence, 55, 67, 102, 103, 145, 191
evolution, 28, 36, 135, 138, 139, 196, 224

F

failure, 206
family, 6, 17, 120, 122, 123
fear, 40
fermions, 83, 84, 85, 88, 89, 92, 104, 105, 117, 123
field theory, vii, viii, 1, 2, 3, 5, 6, 14, 15, 17, 22, 31, 42, 47, 55, 66, 77, 81, 83, 86, 97, 105, 107, 124, 136, 137, 141, 145, 179, 180, 181, 182, 192, 194,

198, 199, 200, 201, 202, 205, 206, 208, 209, 211, 212, 213, 218, 219, 221, 223, 224
financial support, 193
fine tuning, 109
first generation, 112
fish, 199
fixation, 143, 144
flatness, 93, 94, 95, 102, 103, 109, 110, 111, 113, 123
flavor, 89
flexibility, 103
fluctuations, 47, 213
Fock space, 33
Fourier, 32, 143, 158, 168, 169, 190, 224
fractal structure, vii, 1, 74
France, 77, 155, 194, 196, 210, 211, 212, 216, 218, 223
freedom, 1, 30, 31, 37, 66, 68, 84, 94, 100, 105, 106, 137, 138, 141, 145, 158, 191
freezing, 120
fusion, 184, 192

G

gauge fields, 2, 65
gauge group, 65, 83, 85, 86, 88, 89, 92, 93, 100, 102, 104, 107, 108, 109, 111, 114, 115, 116, 117, 118, 119, 122, 123, 126, 146
gauge invariant, 2, 3, 4, 66, 67, 70, 71, 72, 73, 74, 76, 103, 110, 143, 144
gauge theory, vii, 1, 2, 4, 5, 41, 47, 52, 53, 54, 55, 56, 61, 65, 66, 73, 76, 211, 215, 219, 223
Gaussian, 22, 145
generalization, 106, 156, 181
generation, 83, 89, 90, 91, 94, 97, 98, 99, 102, 104, 107, 108, 110, 112, 116, 119, 121, 122, 123, 147
glueballs, 148
gluons, 136, 140
goals, 209
grants, 193
gravitation, 138, 140, 198
gravity, vii, 4, 5, 29, 41, 52, 53, 54, 76, 89, 124, 180, 181, 184, 194, 195, 196, 201, 202, 203, 207, 211, 212, 213, 214, 215, 216, 217, 218, 220, 223, 224
Great Britain, 208
Greece, 211
groups, 93, 103, 107, 108, 115, 116, 122, 123, 124, 134, 135, 154, 212

H

hadrons, 2, 4, 136, 137, 138, 140, 141, 144, 147

Hamiltonian, 2, 3, 23, 24, 31, 33, 34, 35, 44, 68, 69, 70, 71, 72, 74, 134, 135, 138, 139, 140, 145, 160, 162
hands, 51
height, 35
helicity, 157, 160, 161, 171
Hermitian operator, 56, 58, 142
Higgs, vii, 81, 82, 83, 88, 90, 94, 96, 97, 98, 99, 102, 103, 104, 108, 109, 112, 114, 115, 116, 117, 119, 121, 122, 123, 125, 126, 224
Higgs field, 103, 123
Hilbert, 6, 16, 35, 85, 135, 144, 146, 160, 166, 172, 190, 191
Hilbert space, 6, 16, 35, 85, 135, 144, 146, 160, 166, 172, 190, 191
Hofstadter Butterfly diagram, vii, 1, 67
Hubble, 124
human cognition, 203
hypothesis, 141

I

identity, 17, 18, 186
illusion, 196, 198
images, 117
implementation, 135, 141, 143
inclusion, 137
independent variable, 8, 40, 142
India, 216, 219
indication, 2
indices, 157, 165, 180, 181, 184, 186, 191
inequality, 113
infinite, 1, 6, 7, 8, 11, 21, 124, 134, 136, 138, 141, 143, 144, 157, 160, 163, 171, 224
insertion, 184
instruction, 220
instruments, 206
integration, 134, 145, 166, 183, 184, 202
intensity, 134
interaction(s), 2, 3, 41, 55, 77, 101, 112, 134, 136, 137, 138, 146, 150, 156, 175, 192, 196, 203, 205, 215, 216, 224
International Atomic Energy Agency, 206, 214, 215
internet, 146, 151
interpretation, 55, 56, 77, 103, 107, 140, 180, 190, 191, 192
interval, 6, 56
invariants, 141, 143, 144, 198, 199, 219
IR, 54
isospin, 144
Italy, 151, 195, 200, 206, 213, 215, 219, 221

J

Japan, 1, 77, 202, 210, 212, 216

K

kernel, 166
knots, 203
Korea, 200, 207

L

Lagrangian density, 65, 66, 138
land, 199
landscapes, 220
language, 55, 62, 64, 65, 103, 107, 119, 220
lattices, 124, 213
laws, 124, 135, 207
lead, vii, viii, 37, 102, 104, 109, 115, 117, 124, 142, 143, 179
leadership, 220
learners, 220
lepton, 2, 96, 97, 98, 99, 102, 108, 109, 116, 118
Lie algebra, 12, 124, 157, 166, 167, 175, 208
lifetime, 113, 141
linear function, 142
linear systems, 146
literature, 145, 179, 184, 185, 197
location, 86, 114
London, 149, 195, 202, 208, 222, 223
long distance, 54

M

magnet, 136
magnetic field, vii, 1, 2, 66, 68, 69, 70, 71, 74, 117
manifolds, 110, 115, 197, 219
mapping, 139, 142
Maryland, 204
Massachusetts, 153
mathematics, 151, 197, 203, 205, 206, 208, 216, 217
matrix, 19, 20, 29, 83, 84, 85, 86, 92, 96, 97, 98, 99, 108, 109, 125, 138, 146, 147, 197, 218
meanings, 28, 82
memory, 147, 211
men, 189, 204
mesons, 2, 144, 148
Mexico, 221
Miami, 127, 211
mixing, 83, 95, 125, 160
modeling, 146, 152, 203

models, viii, 5, 6, 9, 10, 15, 19, 21, 22, 25, 35, 40, 46, 53, 55, 62, 64, 81, 82, 83, 84, 86, 87, 88, 89, 91, 96, 97, 101, 102, 103, 104, 105, 108, 110, 111, 112, 115, 116, 117, 118, 119, 120, 121, 122, 123, 124, 125, 126, 133, 134, 135, 136, 137, 138, 141, 180, 181, 184, 193, 195, 196, 197, 198, 199, 200, 201, 202, 203, 204, 205, 206, 207, 208, 209, 210, 211, 212, 213, 214, 215, 216, 217, 218, 219, 220, 221, 223, 224, 225
modules, 16, 19
momentum, 15, 35, 36, 37, 39, 60, 62, 63, 68, 69, 71, 72, 73, 74, 75, 135, 136, 137, 138, 140, 143, 145, 157, 158, 164, 165, 167, 168, 174, 182, 191, 192
monodromy, 181
Moscow, 147, 150, 152, 154, 211
motion, 22, 30, 60, 68, 134, 136, 140, 141, 151, 161, 162, 165, 173
motivation, 179
M-theory, vii, 81, 103, 200, 210, 214, 219
multiplication, 55, 62, 67, 85, 136
multiplicity, 116
multiplier, 156, 160, 161
music, 203, 220
mutation, 168

N

natural sciences, 203
Netherlands, 197, 198, 216
neutrinos, 198
New Jersey, 200, 203, 213
New Mexico, 221
New York, 126, 147, 148, 149, 154, 193, 195, 196, 197, 198, 199, 201, 203, 204, 205, 206, 207, 208, 211, 212, 213, 214, 216, 217, 219, 220, 221, 223, 224
Newton, 41, 217
noise, 154
noncommutativity, vii, 1, 54, 55, 59, 61, 62, 64, 67, 69
nonlinear dynamics, 213
North Atlantic Treaty Organization (NATO), 196, 197, 199, 200, 205, 210, 211, 212, 213, 216, 218, 223, 224
North Carolina, 204, 209

O

one dimension, 160
open string, 29, 31, 32, 34, 35, 36, 37, 40, 41, 61, 103, 116, 134, 138, 140, 141, 146, 151, 157, 164, 165

operator(s), 6, 11, 12, 13, 14, 15, 16, 17, 22, 24, 25, 26, 27, 28, 33, 35, 36, 54, 57, 58, 63, 72, 74, 78, 85, 86, 135, 157, 170, 174, 182, 183, 188, 192, 223
orbit, 143
orientation, 123

P

pairing, 84
parameter, 28, 36, 39, 53, 54, 55, 62, 67, 103, 109, 110, 111, 119, 134, 137, 138, 139, 141, 143, 170
Paris, 155, 201, 202, 213, 215
particle mass, 83, 125, 224
particle physics, vii, 209, 213, 224
particles, vii, viii, 2, 55, 82, 83, 136, 138, 140, 141, 147, 155, 156, 157, 177, 198, 205, 209, 211, 224
partition, 28, 84, 191
partnership, 220
path integrals, 145
PBC, 7
pedagogy, 220
perception, 124
periodicity, 31, 140
personal computers, 146
pH, 52
phase transitions, 202
phenomenology, 82, 83, 89, 94, 98, 102, 103, 109, 125, 218, 219
photographs, 197
photons, 136
physical sciences, 200, 210, 213, 218
physics, vii, 1, 2, 4, 5, 22, 40, 55, 62, 66, 67, 68, 76, 81, 82, 104, 124, 125, 136, 137, 146, 150, 151, 195, 196, 197, 198, 199, 200, 201, 202, 203, 204, 205, 206, 207, 208, 209, 210, 211, 212, 213, 214, 215, 216, 217, 218, 219, 220, 221, 222, 224
plane waves, 160
Poincaré, viii, 208
Poincare group, 133, 134, 137
Poland, 200, 202, 205, 208
polarization, 145
polynomial functions, 135
polynomials, 110
potential energy, 52, 113
power, 101, 207
prediction, 144
production, 125, 206
program, 209, 220

Q

QCD, 2, 3, 4, 41, 43, 53, 54, 66, 76, 77, 136, 137, 147, 148, 208, 222
QED, 2, 3, 4, 66, 137, 222
quantization, viii, 4, 55, 60, 73, 120, 133, 135, 137, 138, 141, 142, 143, 144, 145, 150, 151, 152, 156, 157, 166, 175, 208, 211
quantum chromodynamics, vii, 136
quantum cosmology, 215
quantum electrodynamics, vii, 1, 2, 55, 137, 144
quantum field theory, 2, 3, 55, 77, 137, 141, 202, 205, 206, 220, 223
quantum gravity, 4, 124, 201, 212, 214, 215
quantum groups, 211
quantum Hall effect, 67
quantum mechanics, vii, 1, 2, 5, 55, 63, 67, 135, 141, 146, 154
quantum theory, vii, 9, 124, 142, 143, 145, 162
quarks, 2, 3, 4, 43, 45, 53, 76, 83, 94, 102, 104, 108, 109, 110, 112, 114, 116, 121, 122, 136, 137, 145, 147, 148

R

radius, 7, 21, 27, 40, 44, 68, 88, 104, 111, 119, 190
range, 101, 103, 111, 113, 114, 122, 186
reality, 124
recognition, 103
recombination, 119, 140
reduction, 39, 104, 110, 112, 123, 146
reference frame, 133, 135
reflection, 182, 183, 184, 185
relationship, 74
relativity, vii, viii, 56, 133, 138, 141, 213, 214
renormalization, 2, 86
residues, 180
resolution, 55, 111
Rio de Janeiro, 218
Romania, 198
Rome, 216
rotations, 133, 136
Royal Society, 208
rubber, 199
Russia, 133, 151, 211

S

SA, 202, 207
sample, 125
scalar, 22, 222

scalar field, 6, 22, 23, 24, 29, 30, 31, 33, 35, 89, 92, 104, 110, 122, 125
scaling, 15, 28
scatter, 140
scattering, viii, 138, 179, 180, 184, 190, 191, 192
school, 77, 219, 220, 223
science, 81, 124, 196, 199, 201, 206, 210, 211, 212, 216, 221
SCT, 13
search(es), 82, 94, 102, 103, 111, 123, 148, 206
second generation, 94, 102
Seiberg-Witten map, 55
selecting, 143
separation, 56, 136, 140
Serbia, 204
series, 19, 55, 61, 191, 195, 197, 199, 200, 201, 206, 208, 210, 211, 212, 213, 216, 218, 221, 223
severity, 105
shape, 133
shares, viii, 138, 155, 156
sign(s), 89, 90, 95, 108, 110
signalling, 156
signals, 162
Singapore, 149, 150, 152, 195, 200, 202, 203, 204, 207, 208, 209, 210, 212, 213, 214, 215, 217, 218, 219, 221, 224
software, 146
spacetime, vii, 84, 87, 88, 89, 103, 105, 156, 157, 174, 203, 207
space-time, 5, 36, 38, 41, 55, 56, 57, 58, 59, 61, 133, 134, 135, 136, 137, 140, 145, 151, 172, 175
special relativity, 56, 133
special theory of relativity, viii, 133
species, 35
spectrum, vii, viii, 1, 2, 5, 24, 35, 41, 53, 54, 55, 57, 58, 59, 70, 72, 74, 75, 87, 88, 89, 93, 102, 104, 112, 114, 116, 117, 118, 119, 120, 137, 142, 144, 145, 148, 155, 156, 157, 175, 191, 192
speculation, 82
spin, viii, 13, 15, 28, 38, 88, 92, 136, 137, 138, 142, 155, 156, 157, 158, 159, 160, 165, 170, 175, 177, 208, 220
spinor fields, 136
stability, 112, 113, 114, 115
stabilization, 111, 120, 121, 122, 126
standard model, vii, 81, 82, 83, 85, 87, 88, 89, 91, 93, 95, 97, 99, 101, 102, 103, 105, 107, 109, 111, 113, 115, 117, 119, 120, 121, 123, 124, 125, 127, 129, 131, 201, 204, 206, 212, 220, 224
storage, 147
strategies, 220
strength, 65, 66, 93, 117, 123, 137
stress, 9, 10, 11, 14, 17, 24, 25, 26, 160, 182, 183

string theory, vii, viii, 1, 4, 5, 6, 22, 36, 39, 40, 41, 42, 47, 54, 55, 56, 59, 76, 81, 82, 83, 116, 124, 125, 134, 135, 136, 137, 138, 140, 141, 142, 143, 144, 145, 146, 148, 151, 155, 156, 167, 168, 177, 179, 190, 191, 195, 196, 197, 198, 199, 200, 201, 202, 203, 204, 206, 207, 208, 210, 211, 212, 216, 218, 219, 220, 221, 224
strong interaction, 2, 136
students, 220
subtraction, 45
supergravity, vii, 1, 4, 39, 41, 42, 44, 46, 53, 101, 103, 123, 128
supermanifolds, 198
superstrings, 149, 200, 212, 217
supersymmetry, viii, 39, 83, 84, 88, 89, 92, 93, 94, 99, 100, 101, 109, 110, 111, 113, 114, 125, 126, 155, 172, 206, 211, 213
suppression, 91, 97, 98, 99, 102, 105, 110
survival, 220
SUSY, 103, 111, 112, 114, 120, 121, 123, 175
Sweden, 214
Switzerland, 203
symmetry, 2, 3, 4, 5, 6, 18, 21, 23, 39, 40, 53, 56, 57, 76, 77, 88, 89, 93, 101, 103, 105, 107, 108, 109, 113, 114, 116, 122, 125, 136, 139, 141, 143, 175, 181, 183, 190, 205, 206, 224
synthesis, 154
systems, 103, 135, 136, 137, 146, 149, 152, 156, 203, 213, 218, 220

T

teaching, 124, 220
technology, 103
tension, viii, 29, 42, 54, 120, 134, 136, 137, 138, 155, 156, 191
tensor field, 38
Texas, 221
textbooks, 137, 143
theory, vii, viii, 1, 2, 3, 4, 5, 6, 9, 14, 15, 17, 21, 22, 31, 36, 39, 40, 41, 42, 43, 44, 46, 47, 49, 52, 53, 54, 55, 56, 59, 61, 65, 66, 73, 76, 77, 81, 82, 83, 86, 87, 97, 103, 113, 116, 118, 119, 120, 123, 124, 125, 133, 134, 135, 136, 137, 138, 139, 140, 141, 142, 143, 144, 145, 146, 147, 148, 150, 151, 152, 153, 154, 155, 156, 162, 167, 168, 175, 176, 177, 179, 180, 181, 182, 183, 184, 185, 190, 191, 192, 193, 194, 195, 196, 197, 198, 199, 200, 201, 202, 203, 204, 205, 206, 207, 208, 209, 210, 211, 212, 213, 214, 215, 216, 217, 218, 219, 220, 221, 223, 224
thermodynamics, 6
threat, 95

threshold, 115
time, 2, 4, 5, 7, 12, 28, 31, 32, 33, 36, 38, 41, 43, 53, 55, 56, 57, 58, 59, 61, 64, 84, 114, 133, 134, 135, 136, 137, 140, 145, 151, 160, 163, 172, 175, 201, 214, 223
top quark, 97, 99, 102, 108
topology, 113, 135, 138, 211, 219
torus, 88, 105, 106, 107, 117, 121, 123, 124
total energy, 45, 49, 51
trajectory, 137, 140, 145
transactions, 208
transformation(s), 7, 8, 9, 11, 13, 14, 29, 30, 39, 40, 56, 57, 59, 64, 65, 68, 70, 74, 95, 110, 133, 135, 136, 139, 141, 142, 160, 175
transitions, 133, 138, 202
translation, 8, 13, 35, 40, 122
transportation, 223
tuition, 223
Turkey, 217

U

UIR, 157, 160
uncertainty, 57, 58, 59, 140, 224
uniform, 68, 70, 71
United Kingdom (UK), 202, 203, 213, 217, 223
United Nations, 206, 214, 215
United States, 220
universality, 6
universe, 81, 82, 124, 125, 196, 199, 201, 204, 207
UV, 54

V

vacuum, viii, 3, 6, 14, 23, 24, 33, 43, 82, 84, 85, 91, 104, 108, 109, 113, 124, 125, 136, 140, 145, 147, 155, 170, 172, 190, 213
values, 19, 21, 52, 57, 59, 84, 91, 94, 97, 101, 102, 103, 108, 112, 113, 120, 125, 136, 141, 142, 146, 181, 186, 191
variable(s), 8, 14, 40, 51, 56, 135, 139, 142, 143, 144, 146, 152, 158, 160, 161, 164, 165, 166, 167, 182
variation, 8, 11, 13, 22, 31, 104, 112, 119, 124, 142
vector, 29, 34, 37, 66, 68, 83, 84, 85, 86, 88, 90, 94, 100, 103, 105, 107, 108, 109, 111, 112, 113, 114, 115, 116, 134, 135, 136, 139, 143, 158, 165, 166
velocity, 56, 68, 137, 140
visualization, 146, 152, 154

W

war, 199
Warsaw, 208
waterways, 223
weak coupling limit, 41
weak interaction, 2
winter, 219
World Wide Web, 221
writing, 163, 182, 189

Y

Yang-Mills, 4, 41, 42, 43, 44, 46, 49, 52, 53, 54, 64, 138, 208
yield, 85, 97, 99, 107, 108, 166